BLIND SOURCE SEPARATION

BLIND SOURCE SEPARATION

THEORY AND APPLICATIONS

Xianchuan Yu, Dan Hu and Jindong Xu
Beijing Normal University, P.R. China

WILEY

This edition first published 2014

Published by John Wiley & Sons Singapore Pte. Ltd., 1 Fusionopolis Walk, #07-01 Solaris South Tower, Singapore 138628, under exclusive license by Science Press in all media throughout the world excluding Mainland China and excluding Simplified and Traditional Chinese languages.

For details of our global editorial offices, for customer services and for information about how to apply for permission to reuse the copyright material in this book please see our website at www.wiley.com.

Library of Congress Cataloging-in-Publication Data

Yu, Xianchuan.
Blind source separation : theory and applications / Xianchuan Yu, Dan Hu, Jindong Xu.
pages cm
Includes bibliographical references and index.
ISBN 978-1-118-67984-5 (cloth)
1. Blind source separation. I. Hu, Dan, Ph.D. II. Xu, Jindong. III. Title.
TK5102.9.Y8 2014
621.382′2–dc23

2013031750

Set in 11/13pt Times by Laserwords Private Limited, Chennai, India

1 2014

Contents

About the Authors

Xianchuan YU, the director of the school key laboratory of Spatial Multi-Source Information Fusion and Analysis, is currently a Professor at the Computer Science Department, College of Information Science and Technology, Beijing Normal University. He was born in 1967, and received a Ph.D degree in mathematical geology from Jilin University. Now, he is the academic leader of Intelligent Information Processing at Beijing Normal University. He has been an IEEE Senior Member since 2009, a vice director of the China Mathematical Geology and Geological Information Processing professional committee since 2012, a vice chairman of the Branch China National Committee of the International Mathematical Geosciences Society (IAMG) since 2009, an executive member of the Chinese Computer Graphics and GIS Society, a committee standing member of the Chinese Aerospace Optical Society of Technical, and a professional committee member of CCF Collaborative Computing. His current research interests include blind source separation, modeling uncertainty in geosciences, remote image processing, fuzzy sets, and mineral resources appraisement.

Dan HU was born in 1977, received the B.S. degree and M.S. degree in mathematics from Sichuan Normal University, Chengdu, China, in 1999 and 2002, respectively and the Ph.D. degree in in applied mathematics from Beijing Normal University, Beijing, China. She is currently an Associate Professor at college of Information Science and Technology, Beijing Normal University, working in the field of data mining. Her main areas of interest are the theory and application of blind signal processing, rough sets, fuzzy sets, and attribute importance appraisement.

Jindong XU, a Ph. D. candidate at college of Information Science and Technology, Beijing Normal University, was born in 1980, received the B. S. Degree in communication engineering from Shandong University, Jinan, China, in 2003, and M. S. Degree in communication information and system from Beijing Normal University, Beijing, China, in 2008. From 2003 to 2005 and from 2008 to 2011, he is a senior lecturer in Qufu Normal University. He is a student member of China Computer Federation, and International Association for Mathematical Geosciences member. His research interests include blind signal processing, data mining and pattern recognition.

Preface

At a vibrant cocktail party, there are the mixed sounds of friends chatting, cheering voices, and faint background music. If these mixed sounds are picked up by a microphone, how can one identify a single voice of interest from the mixed sound?

In the field of remote sensing image analysis, ground information, mineral information, and other interference information is all mixed together in the received signal. How can one determine the ground and mineral information from this signal?

Modern battlefield environments are generally complex and unfavorable for military communication. How, then, can one correctly and accurately capture, separate, and distinguish communication originating from the enemy and our troops from the mixed radio signals?

These problems and many more are exactly what blind source separation aims to solve and is currently solving.

With the rapid development in network, communication, and computer technology, we have advanced into a digital and information era, with digital signal processing one of the key links. As the main concept in blind signal processing, blind source separation is defined as: with unknown signal source and transmission channel parameters, the observed signals are detected and all components are separated in the input source signal according to their statistical characteristics. This process requires knowledge of many fields, including information theory, statistical signal processing, and artificial neural networks. It is also a very attractive option for use in many fields such as remote sensing image processing, mobile communication, voice processing, biomedical engineering, economics, and sonar and seismic signal processing applications. Blind source separation is an emerging technology which, although it has only been around for the past 20 years, has quickly become a research hot spot that has attracted scholars in every field owing to its significant theoretical value and wide application prospects.

In April 1984, the French scholars Herault and Jutten presented recurrent neural networks and learning algorithms based on the Hebb learning rule and implemented blind separation of two signal sources. The broad application potential thereof immediately attracted the attention of other scholars, thereby starting a new research chapter in the field of signal processing, namely, blind source separation. After more than 20 years of development, blind source separation has made great progress, both in basic theory and actual applications. Currently, there is a theory system with a blind source separation algorithm (based on information entropy or likelihood estimation) as the foundation, independent component analysis as the core, and non-negative matrix decomposition, sparse components analysis, and other emerging algorithms as the frontier. Moreover, blind source separation has been successfully applied in signal, image, and voice processing.

Since 1998, with support from the National 863 Project, the National Natural Science Fund Project, and the Beijing Natural Science Fund Project, Beijing Normal University Multi-source Space Information Fusion and Analysis University-level Key Laboratory (formerly known as the Beijing Normal University Information Science and Technology College Space Information Processing Research Center) has been committed to research on blind source separation optimization algorithms and the application thereof in image feature information analysis, image fusion, image pixel unmixing, image target recognition, remote sensing information processing, functional magnetic resonance imaging (fMRI) medical image processing, geochemistry, geophysics, information processing, mineral resources prediction, and ore information identification. Based on this research, several theoretical achievements have been realized. At the same time, during the research and learning process, we found that there are no reference books either locally or internationally that cover the theoretical basis, core algorithms, advanced algorithms, and related application examples of blind source separation. This is our main reason for writing this book.

Some of the algorithms in this book have been integrated into a software system; if readers are interested in these, please contact Professor Xianchuan YU (yuxianchuan@bnu.edu.cn, chuan.yu@ieee.org). We welcome your feedback on all aspects of this book.

Acknowledgements

The contents of this book have been based on the authors' many years of research and teaching in this area and with reference to a large number of local and international publications. Results from laboratory tests and experiments presented in the book were obtained through the combined effort of the research team consisting of: Xianchuan YU, Dan HU, Jindong XU, Guian WANG, Ligen FANG, Nan ZHANG, Ting ZHANG, Wei ZOU, Jiamian REN, Na JI, Zhongni WANG, Hengzhi CAO, Bixin HAO, Chunping YANG, Zhi HUANG, Di PENG, Siliang LONG, Yiqing BAI, Chen YU, Yan FU, Meng YANG, Junlan ZhANG, Ye LUO, Zengji KANG, Yi FANG, Jianguang LI, Xin ZHOU, Qi ZHOU, and Wenjing PEI. We also wish to thank all the lecturers and students at the laboratory, especially Hui HE, Weijie AN, Sha DAI, Weike DENG, Feng NI, Tingting CAO, and Xiaofeng CHU for their assistance on various chapters. All these individuals have contributed greatly to the completion of this manuscript. In addition, we wish to acknowledge the hard work of the translation and proofreading team, consisting of Jindong XU, Jie ZHANG, Haihua XING, Wenjing PEI, Genxia WANG, Xin ZHOU, LiPing XIONG, Zhonghua LV, Yali REN, Cheng YANG, Yinggang ZHANG, Wei ZHOU, Guangyin GAO, and Wu HE, without whom the English version of this book would not have materialized. We hope that this book can help the reader systematically understand blind source separation algorithms and applications.

During the preliminary acquisition of results, manuscript preparation, writing and publication, we received financial support from the Chinese Academy of Sciences, Scientific Publication Fund, National Natural Science Fund (40372129, 40672195, 41072245, 41272359), National 863 Plan (2007AA12Z156), Ministry of Education for new century excellent talents (NCET-06-0131), Specialized Research Fund for the Doctoral Program of Higher Education of China (20120003110032), Fundamental Research Funds for the Central Universities (2012LZD05), Beijing Natural Science Foundation (4062020, 4102029), and the University-level Key Laboratories of Beijing Normal University Fund. We also acknowledge the proofreading and editing assistance of Academician Xiaowen LI of Beijing Normal University, Academician Yirong WU of the Chinese Academy of Sciences Electronics Institute, Chairmen Qiuming CHENG (Canada York University) of the International Association for Mathematical Geology, Professor Guoxue WU of Jilin University, Professor WangLu PENG,

Professor ChaoJi ZHU, Associate Professor Guocheng DING, and Project Supervisor Jie DAI of Beijing Normal University. To Clarissa Lim at John Wiley & Sons, Yanfen ZHANG at China Science Press, Shaohua LIU at the Chinese Geophysics Journal, researcher Keyan XIAO at the China Geological Academy of Sciences Mineral Resources Institute, Senior Engineer Liangang LIU and Senior Engineer Jinglian WEI of Beijing Institute of Geology, 722 Geological Brigade Captain Shihua LIU and Senior Engineer Hongzhen LI at the Guangdong Geological Prospecting Bureau, and 719 Geological Brigade Captain Liwen LIU at the Guangdong Geological Prospecting Bureau, we are grateful for your assistance.

Xianchuan YU, Dan HU, Jindong XU
April 2013

Glossary

$\boldsymbol{A}^{\mathrm{T}}$	transposition of matrix $\boldsymbol{A}$
$\boldsymbol{A}^{\mathrm{H}}$	conjugate transposition of matrix $\boldsymbol{A}$
$\boldsymbol{A}^{+}$	(Moore–Penrose) pseudo-inverse of matrix $\boldsymbol{A}$
$\boldsymbol{A}^{-1}$	inverse of matrix $\boldsymbol{A}$
$\mathbf{R}$	real field
$\operatorname{diag}\boldsymbol{A}$	vector whose components are the diagonal of matrix $\boldsymbol{A}$
$\operatorname{tr}(\boldsymbol{A})$	trace of matrix $\boldsymbol{A}$
$\det(\boldsymbol{A})$	determinant of matrix $\boldsymbol{A}$
$\operatorname{rank}(\boldsymbol{A})$	rank of matrix $\boldsymbol{A}$
$\operatorname{kurt}(\boldsymbol{x})$	The kurtosis of random vector $\boldsymbol{x}$
$\operatorname{Mom}[\boldsymbol{x}]$	Moment of random vector $\boldsymbol{x}$
$\operatorname{Com}[\boldsymbol{x}]$	Cumulant of random vector $\boldsymbol{x}$
$E\{\boldsymbol{x}\}$	Mean of random vector $\boldsymbol{x}$
$H(\boldsymbol{x})$	Entropy of random vector $\boldsymbol{x}$
Φ	The joint characteristic function
$\widetilde{\Psi}$	The joint second characteristic function
PDF	Probability density function
$\otimes$	Kronecker product between matrices

1

Introduction

1.1 Overview of Blind Source Separation

Blind source separation (BSS) is a powerful signal processing method that was proposed in the late 1980s. As the product of artificial neural networks, statistical signal processing, and information theory, BSS has become an important topic in research and development in many areas, especially in the biomedical sciences, speech signal communication, image processing, earth science, econometrics, and text data mining. At present, the BSS problem is a research hotspot in signal processing, artificial neural networks, and other disciplines internationally, and thus has great practical value.

BSS, which is a traditional and challenging problem in signal processing, involves extracting and recovering the underlying source signals from multivariable statistical data. "Blind" implies that the source signal is unknown (unobserved), and that either the characteristics of the hybrid system are not known in advance or there is only a small amount of a priori knowledge (such as non-Gaussian, cycle-stability, and statistical independence). In scientific and engineering applications, many of the observed signals can be seen as a mixture of a plurality of source signals, that is, the observed mixed signal is a series of sensor outputs, with each sensor receiving different combinations of the source signals. The main task of BSS is to recover the source signal that we are interested in from the observed data.

A typical example is the "cocktail party" problem. Suppose that you are attending a cocktail party with a variety of sounds coming from the surroundings: talking (possibly in a variety of different languages), music, and even a whistle from outside the window. If sufficient microphones are placed at different positions to record these sounds, then each microphone can record signals mixed according to different weights (Figure 1.1). Although there may be a great deal of interference, you would be able to focus on the words of your friend, despite also being able to identify a side conversation or two and listen to the music. Given only the received speech signal from the microphone, how is it possible to separate the desired speaker's voice when the locations of the microphones and the sound source with the information we require

Blind Source Separation: Theory and Applications, First Edition. Xianchuan Yu, Dan Hu and Jindong Xu.

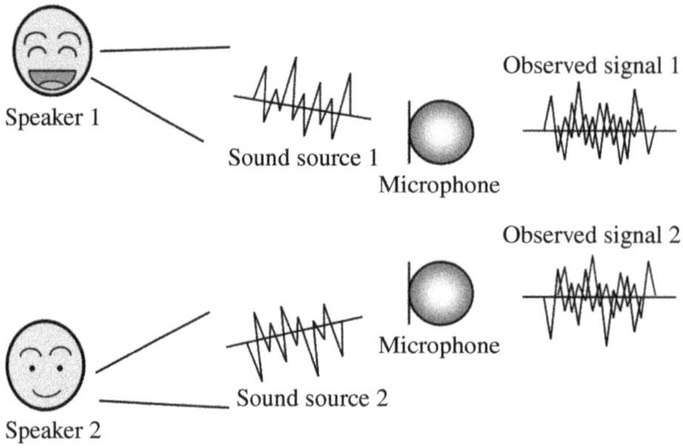

Figure 1.1 Signal mixing diagram

are not known beforehand? BSS was introduced to solve this exact problem (Holland, 1978; Bylund, 2001; Si and Zhang, 2002; Haykin and Chen, 2005).

A more general expression of the BSS problem is given below.

Assume that the measured sensor signal from a multiple input, multiple output (MIMO) nonlinear dynamic system is denoted as $x(k) = [x_1(k), x_2(k), \dots, x_m(k)]^T$ and we need to find a reverse system to reconstruct the original source signal $\mathbf{s}(k) = [s_1(k), s_2(k), \dots, s_n(k)]^T$. Both source signal $\mathbf{s}(k)$ and how the source signals are mixed to produce the observed signal are unknown, thus reflecting the "blindness" characteristic of the problem. The output can be expressed by Equation 1.1, while the processing module for general BSS is illustrated in Figures 1.2 and 1.3.

$$\boldsymbol{y}(k) = \boldsymbol{W}\boldsymbol{x}(k) = \boldsymbol{W}\boldsymbol{A}\boldsymbol{s}(k) = \boldsymbol{C}\boldsymbol{s}(k)\pi \tag{1.1}$$

where $\boldsymbol{C} = \boldsymbol{W}\boldsymbol{A}$ is the mixing-separating matrix with dimensions $r \times n$.

For the overcomplete ($m < n$) BSS problem, $\boldsymbol{W}$ may not exist. In addition, we need to determine mixing matrix $\boldsymbol{A}$ and make use of prior knowledge, such as the independence or sparseness of the unknown source signals, to estimate the source signal.

The simplest situation is when $\boldsymbol{x}(k)$ is linear and simultaneously mixed by $\boldsymbol{s}(k)$, that is, $\boldsymbol{x}(k) = \boldsymbol{H} * \boldsymbol{s}(k)$. Here, $\boldsymbol{H}$ is the mixing matrix with dimensions $m \times n$, and the BSS problem is simplified to calculate the unmixing matrix $\boldsymbol{W}$, thus obtaining the output:

$$\boldsymbol{y}(k) = \boldsymbol{W}\boldsymbol{x}(k) \approx \boldsymbol{s}(k) \tag{1.2}$$

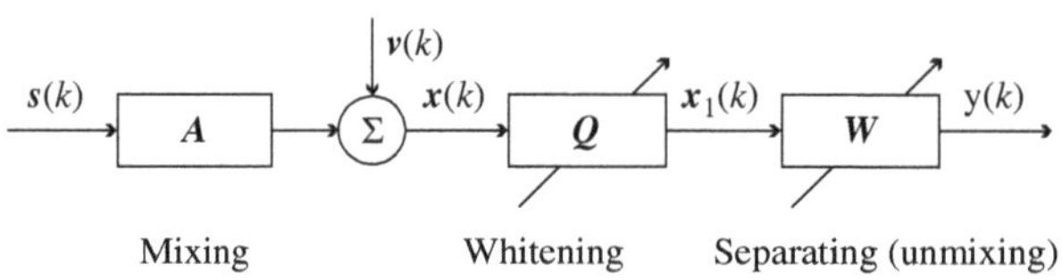

Figure 1.2 Processing module for a general BSS problem

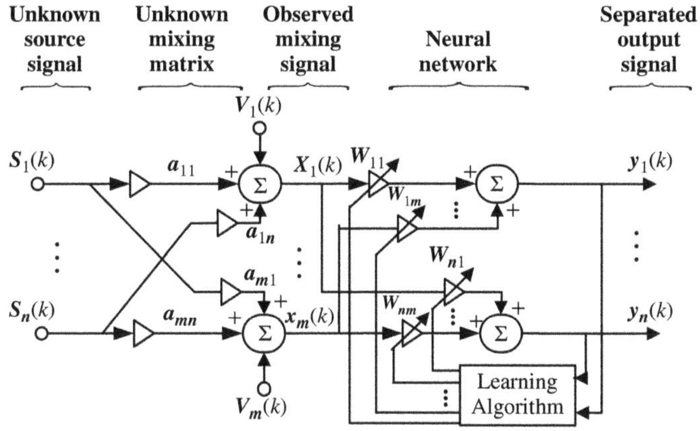

Figure 1.3 Detailed processing diagram for a linear BSS problem

In Equation 1.2, $y(k)$ is an estimation and approximation of the real source signal.

The greatest benefit of BSS is that it takes full consideration of prior knowledge, such as statistical independence, sparseness, space–time independence, and smoothness to estimate the different sources, thereby providing various robust and efficient algorithms (Choi, 2005). The basic processing steps are shown in Figure 1.4. From the diagram we can see that in order to extract reliable, important, and physically meaningful components, the preprocessing and post processing of the data are very important. Therefore, most of the BSS methods are unsupervised learning methods based on a priori information or a theoretically constructed objective function. Independent component analysis (ICA) has become a powerful tool for signal processing and data analysis as a BSS method, while non-negative matrix factorization (NMF) and sparse component analysis (SCA) have also begun to reveal powerful data analysis capabilities in signal separation and related applications. ICA is an analysis method based on higher-order statistical characteristics of the signal for decomposing mutually independent signal components. Using non-negative constraints of the local features, NMF seeks to express the source data and uses a certain optimization algorithm to obtain the source separation matrix. SCA is

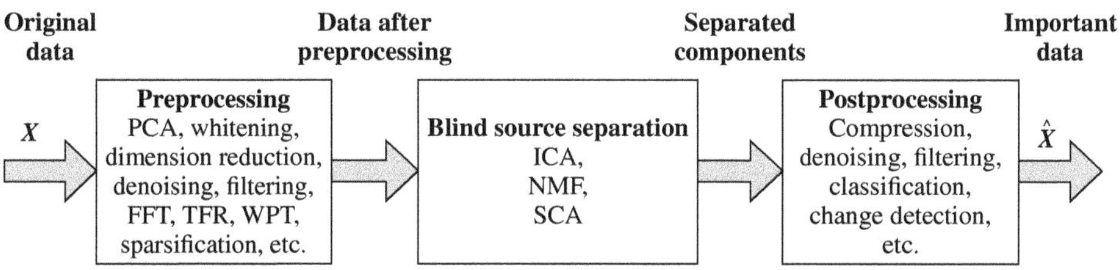

Figure 1.4 Basic steps for efficient decomposition and signal extraction based on the BSS method

usually used in the time–frequency domain to express the source through sparse ingredients of mixed data. Traditional principal component analysis (PCA) achieves data dimensionality reduction by retaining data containing more than 80% of the main features of the information. Generally, we can have different mixing rules, but the objective function basically follows the same principle, albeit with different optimization algorithms (Lee *et al.*, 2000; Chen and Huang, 2008).

1.2 History of BSS

Generally, the earliest substantive study of the BSS problem dates back to 1986, when Herault and Jutten presented the H-J algorithm (i.e., a feedback neural network model with a Hebb-based learning algorithm) to separate two mixed independent source signals at a Neural Network for Computing Conference held in America (Herault and Jutten, 1986). The H–J algorithm can separate two mixed statistically independent source signals using the search method. This work opened a new chapter in the field of signal processing, and, since then, the BSS problem has attracted extensive attention from research scholars.

In fact, solving BSS problems is a very difficult task, because we do not have any information about the source signals. In the algorithm proposed by Herault and Jutten, two assumptions are made: it is assumed that the source signals are statistically independent and that the statistical distribution of the known source signal is known. If the source signals have a Gaussian distribution, it is easy to see that there is no general solution for the BSS problem, as any linear mixing Gaussian distribution is still Gaussian. The Herault–Jutten network model assumes that the source signals are sub-Gaussian signals, which means that the kurtosis of each signal should be less than 0 (the kurtosis of a Gaussian signal is 0) (Cohen and Andreou, 1992).

As the solvability and solvability conditions of the BSS problem were not resolved by the H–J algorithm, Herault and Jutten, as well as many other scholars, have carried out further research to overcome its drawbacks. Linsker proposed the maximum mutual information criterion (Linsker, 1988, 1989), which is most suitable for establishing a self-organizing model and characteristic mapping. Giannakis and Swami introduced a third-order cumulate based on an exhaustive search to solve the BSS identification problem (Giannakis and Swami, 1987). In 1989, the first international workshop on higher-order spectral analysis was convened. In this workshop, early papers on ICA by Cardoso (1989) and Comon (1989) were presented. These works provided a generally clear framework for ICA, whereby if the original signals are statistically independent, mutually statistically independent source signals can be obtained. Since then, ICA theory has gradually been perfected. In 1991, Herault and Jutten published a classic article on the BSS problem in *Signal Processing*, introducing an artificial neural network algorithm for BSS (Jutten and Herault, 1991). This study was the beginning of a new research area. Their learning algorithm was heuristic, and did not clearly point out the need to take advantage of higher order statistics of the observed signals; nevertheless, the iterative formula provided a prototype for a later algorithm.

Over the past 20 years since these introductory studies, the BSS problem has become a hotspot in the field of signal processing. Further in-depth research has greatly improved both the theory and its practical application. In 1994, Comon (1994) proposed a popular minimum mutual information-based ICA method, while in the following year Bell and Sejnowski (1995) proposed the Infomax principle based maximum entropy approach. This algorithm was further refined by the Japanese scholar Amari and his coworkers using the natural gradient (Amari, Cichocki, and Yang, 1996; Amari, 1998), and its fundamental connections to maximum likelihood estimation, as well as to the Cichocki–Unbehauen algorithm, were established. A few years later, the Finnish scholars Hyvärinen, Oja, and Pajunen presented the fixed-point or FastICA algorithm (Hyvärinen and Oja, 1997, 2000; Hyvärinen, 1999; Hyvärinen and Pajunen, 1999), which has contributed to the application of ICA to large-scale problems owing to its computational efficiency. Until now, the standard ICA methods, such as the fast algorithm (FastICA), Infomax algorithm, extended information maximization algorithm (Girolami, 1999; Lee, Girolami, and Sejnowski, 1999), and EASI (equivariant adaptive separation by independence) algorithm (Cardoso and Laheld, 1996), among others, have been more complete. Standard ICA uses an idealized mathematical model, but users now tend to focus on various expansions of ICA, such as noise ICA (Hyvärinen, 2001; Zhong *et al.*, 2004a,b), sparse and overcomplete representations (Zhong *et al.*, 2004a; Lewicki and Sejnowski, 2000; Girolami, 2001), nonlinear ICA (Hyvärinen, 1999; Lappalainen and Honlela, 2000; Lee, Kohler, and Orglmeister, 1997; Taleb and Jutten, 1997; Harmeling *et al.*, 2001, 2003), non-stationary ICA (Pham and Cardoso, 2000; Sanchez, 2002), and convolution ICA (Hyvärinen, 1998, 2001).

With the rapid development of digital signal processing technology and related disciplines, a large number of effective BSS algorithms are continually being proposed, which has gradually resulted in the BSS problem becoming the most popular topic today in the field of information processing.

Since the mid-1990s, there has been an increase in the number of papers, workshops, and special sessions devoted to BSS. The first international workshop on ICA was held in Aussois, France, in January 1999, and the second workshop followed in June 2000 in Helsinki, Finland. Both were attended by more than 100 researchers working on ICA and blind signal separation. Since the publication in 1991 of the first international papers in *Signal Processing* (Jutten and Herault, 1991; Comon, Jutten, and Hérault, 1991), a variety of international journals have contributed to the dissemination of BSS research, including *Traitement du Signal* (in French), *Signal Processing*, *IEEE Transactions on Signal Processing*, *IEEE Transactions on Circuits and Systems*, *Neural Computation*, and *Neural Networks*. In addition, a technical committee devoted to blind techniques was created in July 2001 in the IEEE Circuits and Systems Society, and BSS is a current "EDICS" in the *IEEE Transactions on Signal Processing* and in many conferences. These events have all contributed to the transformation of BSS into an established and mature field of research.

From an algorithm perspective, BSS algorithms can be divided into two groups: adaptive and batch algorithms. On the other hand, from the perspective of assumptions, BSS algorithms can be divided into three groups based on the statistical independence of the source signal, the sparse characteristics of the source signal, and non-negative constraints of the source signal, respectively. From the point of view of the cost function or guidelines, BSS algorithms can be divided into second-order statistics-based independent methods (such as the AMUSE (a minimally-unsatisfiable subformula extractor) algorithm), higher-order statistics-based approaches, neural network-based methods, and methods based on nonlinear functions.

Most of the existing BSS algorithms are used to solve linear instantaneous mixture problems, with some successful algorithms achieving good performance. For example, the ICA-based algorithm has been applied to the field of signal processing, biomedical signal processing (for example, electroencephalography (EEG) and magnetoencephalography (MEG) signals), speech recognition systems, and in areas such as earthquake prediction. In recent years, nonlinear approaches and technologies to address the BSS problem (Zadeh, Jutten, and Nayebi, 2004) have been developed. Jutten and Babaie-Zadeh proposed several algorithms to solve the BSS problem in the post-nonlinear (PNL) model (Valpola and Karhunen, 2002) with these algorithms having achieved practical value in sensor array signal processing, microwave communications, satellite communications, and many biological systems. Valpola incorporated Bayesian ensemble learning into nonlinear BSS and realized improved results (Martinez and Bray, 2003). A general nonlinear Bayesian blind separation method has become an important research focus in recent years. The nuclear-based nonlinear BSS algorithm (Taleb, Jutten, and Olympieff, 1998) and local linear BSS method have also attracted much attention. However, owing to the inherent complexity of nonlinear problems, there is no generally applicable algorithm that can be applied to all kinds of practical problems. Therefore, a series of different model-based nonlinear BSS methods (Woo and Sali, 2002; Tan, Wang, and Zurada, 2001) has been proposed. With the development of neural networks, radial basis function (RBF) networks (Woo and Khor, 2004), multilayer perceptron (MLP) networks (Tan, Wang, and Zurada, 2001), polynomial neural networks (PNN) (Rojas *et al.*, 2001), and genetic algorithms (GA) (Lee, 1998) have arisen amid wide interest because of their flexible nonlinear ability and, therefore, we can take advantage of neural networks to solve nonlinear problems. Under nonlinear conditions, the RBF network-based decomposition method has the fastest convergence speed, but low signal recovery accuracy. An MLP network produces the most accurate source signal, but has high computational complexity. A PNN network with a flexible hidden neuron activation function can prevent an "over optimized" network, and make the solutions more regular. The question thus arises: How do we select the most suitable neural network? We need to estimate the speed, accuracy, and complexity considering the requirements of the actual practical application. As the MLP network is more generalized than the others, it is more suitable if there is no prior knowledge about the hybrid systems. An RBF network has rapid convergence, and is thus a good choice when considering speed preference, whereas a PNN can provide clear separation results but requires prior knowledge of the source signal (Rojas *et al.*, 2001).

Currently, BSS algorithms can be divided into three main groups: ICA, NMF, and SCA. More details about these three algorithm groups are introduced in later chapters.

Famous international research institutions and well-known scholars include the following: T. J. Sejnowski and A. J. Bell from Salk Institutes of America; Professor H. S. Seung from the Massachusetts Institute of Technology (MIT); Professor P. Paatero from Clarkson University; Professors S. Amari and A. Cichowski from the Institute of Brain Science; E. Oja and A. Hyvärinen from Aalto University School of Science and Technology in Finland (originally Helsinki University of Technology); and P. Comon and J. F. Cardoso from the French National Center for Scientific Research (CNRS). There are also many well-known books about the BSS problem such as, Independent Component Analysis, Theory and Applications (Lee, 1998), Independent Component Analysis (Hyvärinen, Karhunen, and Oja, 2001), Adaptive Blind Signal and Image Processing (Cichocki and Amari, 2002), Independent Component Analysis: a Tutorial Introduction (Stone, 2004), and Handbook of Blind Source Separation and Independent Component Analysis and Applications (Comon and Jutten, 2010).

In China, there have also been many research achievements and innovations regarding BSS problems. Several research institutions, such as Tsinghua University, Northwestern Polytechnic University, Southeast University, Shanghai Jiaotong University, Xi'an University of Electronic Science and Technology, South China University of Technology, Beijing Normal University, and the Psychological Research Institute of the Chinese Academy of Sciences are involved in research on blind signal processing as well as its related applications. Many famous Chinese scholars (Ling, 1996; Wang and He, 1997; Zhang and Bao, 2001; Zhang, Zhu, and Bao, 2002; Liu, Nie, and He, 2001; Zhang and Shi, 2001; Yang, Zhuang, and Wu, 2002; Su *et al.*, 2002; Zhang, He, and Xie, 2004; You and Chen, 2004; He, Wang, and Xie, 2005; Li, 2005; Yang and Zheng, 2003) have also contributed several novel innovative methods to this topic and applied BSS methods to solve a variety of problems in their own research fields. There are also many good Chinese books systematically introducing the basic theory of BSS problems, such as Artificial Neural Networks and Blind Signal Processing (Yang and Zheng, 2003), Blind Signal Processing and Applications (Zhang, Zhang, and Zhang, 2006), Blind Signal Processing (Ma, Niu, and Chen, 2006), Independent Component Analysis (Zhou, Dong, and Xu, 2007), and Blind Signal Processing – Theory and Practice (Shi, 2008).

The author of this book has invested a great deal of effort in researching BSS theory and its applications in image feature extraction, remote sensing image fusion, mixed-pixel decomposition of synthetic-aperture radar (SAR) images, image object recognition in medical image processing of functional magnetic resonance imaging (fMRI), geochemical and geophysical data mining, mineral resource prediction, and geo-anomaly information recognition supported by the Natural Science Foundation of Beijing, the National High Technology Research and Development Program of China, the National Natural Science Foundation of China, Specialized Research Fund for the Doctoral Program of Higher Education of China, the Fundamental Research Funds for the Central Universities and the Program for New Century Excellent Talents at the University of China.

This book contains all the studies and achievements of authors' research team and others. We have striven to present a systematic, comprehensive, case-based reference book on BSS.

1.3 Applications of BSS

1.3.1 Speech Signal Separation

One of the original motivations for BSS research was the cocktail-party problem, as reviewed in the beginning of this chapter. The basic idea is that there are several observed signals, which are mixed by many source signals, and the aim of the BSS algorithm is to separate the source signals. In an actual speech recognition system, the observed signal is the convolution of impulse responses produced by the comprehensive interaction of the source speech signal, the sensor, and the surrounding environment. Since in real-life situations the positions of the microphones with respect to the sources can be rather arbitrary, the mixing process is not known, and thus has to be estimated blindly. In this situation, BSS algorithms are important.

In addition, convolutive BSS algorithms tend to be preferably applied in the music field. Douglas (2002) separated two voices from a cappella using convolutive BSS techniques, while Vincent (2005) applied ICA technology to obtain the sounds of different instruments in an audio soundtrack. Although polyphony identification is a very difficult problem, researchers from the Mitsubishi Research Institute and MIT successfully identified every single tone in a recorded polyphony using NMF. Not only is the NMF algorithm simple, but it also does not rely on a knowledge base (Kawamoto *et al.*, 2000).

1.3.2 Data Communication and Array Signal Processing

BSS has wide application in wireless digital communication (Zhang and Hu, 2000; Cances, Mohammadkhani, and Meghdadi, 2006). Various scholars have proposed using an antenna array receiving system and spatial filtering techniques as well as a BSS algorithm to realize separation of the co-channel multi-user signal, so that a plurality of user signals can be transmitted over the same channel at the same time, thereby greatly improving the capacity of the channel. In a code division multiple access (CDMA) system, multiple users share the same channel bandwidth and transmit signals simultaneously. Users can distinguish their own signals by the spread codes. However, owing to multipath fading, the signals at the receiving end are no longer orthogonal, and therefore, using ICA can help improve the signal-to-noise ratio (SNR) at the receiver side. In 2006, the Japanese RIKEN Research Institute and Chaosware Company, which have for a long time been engaged in research on human brain information processing mechanisms and signal separation technology (ICA), developed a new technique to separate a phone hybrid electromagnetic wave precisely, thus greatly improving phone communication transmission speed and clarity (http://www.sina.com.cn 2006, December 16, 2011 10:29 Xinhua net).

In the array sensors, each sensor receives a mixed signal, with the source signals as well as the mixing characteristics completely unknown, which is a typical BSS problem. Many applications such as marine sonar detection, passive radar detection, and even stealth aircraft detection have been explored using BSS techniques to solve the problem (Zhang, Qiu, and Zhang, 2003).

1.3.3 Image Processing and Recognition

Currently, BSS is widely used in the field of image processing, with blind separation technology used in image feature extraction, face recognition, moving object detection, digital image watermarks, image denoising, image separation, and image restoration (Tukr and Pentland, 1991; Bartlett, Lades, and Sejnowski, 1998; Bartlett, Movellan, and Sejnowski, 2002). For example, in an image restoration and reconstruction application, available BSS methods can help recover the original content of the images from contaminated ones and eliminate bad effects, such as camera shake, lens distortion, and noise superimposed during the transmission process (Ma, Niu, and Chen, 2006), introduced while acquiring the images.

1.3.4 Geological Spatial Information Processing

In geoscience spatial information processing, BSS techniques have been applied to many aspects of remote sensing data processing, mainly including remote sensing image fusion, mixed pixel decomposition, and remote sensing image classification and recognition (Si and Zhang, 2002; Wang, Yu, and Zhang, 2007; Huang *et al.*, 2008; Wang, Yu, and Zhang, 2008). In mineral prediction, BSS methods have been applied successfully to geophysical and geochemical data analysis (Yu, Cao, and Yang, 2009; Cao and Yu, 2009; Liu *et al.*, 2007). BSS techniques can also be used in seismic exploration (Liu, Liu, and Zheng, 2003). The use of BSS techniques in soil explosive blasting can help distinguish different lithological characteristics from the received reflected or diffracted signals.

1.3.5 Biomedical Signal Processing

This is a very promising area of application for BSS techniques, not only because it is an area of rapid growth and great importance, but also because certain kinds of brain imaging data seem to be quite well described by the BSS model. This is especially true for electrocardiogram (ECG), EEG, electromyogram (EMG), MEG, as well as fMRI signals, which are recordings of electric and magnetic fields of signals emerging from neural currents within the brain (Choi, 2005; Cichocki and Amari, 2002; Brookings, Ortigueb, and Graftonb, 2009). For example, in fetal ECG signal processing, the ECG signal measured from a pregnant woman actually includes the signals from both the pregnant women and the unborn baby, while the parameters of the transmission medium are unknown, which is a typical application of the BSS technique. Using BSS methods, we can distinguish and separate the ECG signal of a pregnant woman from that of the unborn baby, thereby providing accurate data for further diagnosis. Furthermore, ICA has successfully been applied to fMRI signal processing of the brain to obtain hidden feature information, providing useful data for scientific research about human behavior (Fan, Tang, and Tang, 2003; Long *et al.*, 2003; Yu, Ren, and Zhang, 2007; Yang *et al.*, 2008).

As the decomposition results of NMF are nonnegative, many scholars have applied NMF to genetics and drug discovery. Experimental results have shown that the results of NMF-analyzed genomic DNA molecule sequences are more reliable. Similarly, NMF can also help to select the most effective pharmaceutical ingredients with the least negative effects (Gao and Church, 2005).

1.3.6 Text Document Analysis

The key issue in intelligent text processing is to capture the semantics or related information representing the whole text document. However, traditional methods only consider word statistics, ignoring other information. With the extensive application of the BSS technique, scholars have applied NMF to text analysis and processing with the results appearing to be positive (Wei, Xin, and Yihong, 2003). NMF can help decrease the dimensions of the word-text matrix and filter out noise-characteristic items, thereby possibly providing the definition of an effective similarity index.

Nevertheless, BSS technology is far from being limited to the above-mentioned research fields. It has gone far beyond the field of signal processing, and has been applied to financial data analysis, data encryption, and analysis of chemical reactions.

1.4 Contents of the Book

In recent years, BSS theory and practice have been greatly enriched and updated with new algorithms and applications. This book systematically summarizes the new content, focusing on the theoretical foundations of BSS, the relevant cutting-edge algorithms, as well as relevant application instances. I trust that the publication of this book will play a valuable role in the relevant fields.

The book consists of 16 chapters, divided into three parts.

Part I (Chapters 2–4) gives the *mathematical preliminaries*. It introduces the general mathematical concepts needed throughout the book. Chapter 2 introduces a basic knowledge of mathematics, including matrix operations, probability theory, and information theory. Then, the basic mathematical models and classic algorithms for BSS are presented in Chapter 3. Chapter 4 focuses on the definition of the BSS algorithm evaluation criteria. More confident readers may prefer to skip some or all of the introductory chapters in Part 1 and continue directly to the principles of BSS in Part 2.

Part II (Chapters 5–14) covers the *core algorithm for BSS, namely, ICA*. ICA is a multi-channel signal processing method developed in recent years. This part describes the different ICA algorithms including fast ICA (Chapter 6), maximum likelihood ICA (Chapter 7), overcomplete ICA (Chapter 8), core ICA (Chapter 9), natural gradient flexible ICA (Chapter 10), non-negative ICA (Chapter 11), constrained ICA (Chapter 12), optimized ICA (Chapter 13), and supervized learning (ICA) (Chapter 14) as well as various applications in the different fields.

Part III (Chapters 15 and 16) presents the *current state of BSS algorithms and their applications*. BSS technology has developed rapidly and opened up many new areas of research in recent years. This part focuses on sparse component analysis (Chapter 16) and non-negative matrix factorization algorithms (Chapter 15), as well as applications thereof.

References

Amari, S. (1998) Natural gradient works efficiently in learning. *Neural Computation*, **10**(2), 251–276.

Amari, S., Cichocki, A., Yang, H. (1996) A new learning algorithm for blind source separation, in *Advances in Neural Information Processing System*, MIT Press, Cambridge, MA, pp. 757–763.

Bartlett, M.S., Lades, H.M., and Sejnowski, T.J. (1998) Independent component representations for face recognition. Proceedings of the SPIE: Conference on Human Vision and Electronic Imaging, pp. 528–539.

Bartlett, M.S., Movellan, J.R., and Sejnowski, T.J. (2002) Face recognition by independent component analysis. *IEEE Transactions on Neural Networks*, **13**(6), 1450–1464.

Bell, J., and Sejnowski, T.J. (1995) An information maximization approach to blind separation and deconvolution. *Neural Computation*, **7**(6), 1129–1159.

Brookings, T., Ortigueb, S., and Graftonb, S. (2009) Using ICA and realistic BOLD models to Obtain joint EEG/fMRI solutions to the problem of source localization. *Neuroimage*, **44**(2), 411–420.

Bylund, J. (2001) A humble attempt to solve the "cocktail-party" problem using blind source separation. http://www.rcc.uq.edu.au/education/uni-courses/phys3301/2001/johan-report.pdf

Cances, J., Mohammadkhani, G., and Meghdadi, V. (2006) Turbo interference mitigation in layered space-time MIMO DS-CDMA uplink with TTCM(Turbo Trellis Coded Modulation). *Wireless Personal Communications*, **37**(l), 105–121.

Cao, H.Z., and Yu, X.C. (2009) Mixed pixel decomposition of SAR images based on SL-ICA. *Journal of Remote Sensing*, **13**(2), 217–224 (In Chinese).

Cardoso, J.F. (1989) Blind identification of independent components with higher-order statistics. Workshop on Higher-order Spectral Analysis, Vail, CO, pp. 174–179.

Cardoso, J.F., and Laheld, B. (1996) Equivariant adaptive source separation. *IEEE Transactions on Signal Processing*, **45**(2), 434–444.

Chen, X.M., and Huang, S.Y. (2008) Blind source separation review -issues, principles and methods. *Electronic Information Against Technology*, **23**(2), 1–5 (in Chinese).

Choi, S. (2005) Blind source separation and independent component analysis:a review. *Neural Information Processing-Letters and Reviews*, **6**(1), 1–57.

Cichocki, A., and Amari, S. (2002) *Adaptive Blind Signal and Image Processing: Learning Algorithms and Applications*, John Wiley & Sons, Ltd, Chichester.

Cohen, M.H., and Andreou, A.G. (1992) Current-mode subthreshold MOS implementation of the auto adaptive network. *IEEE Journal of Solid-State Circuits*, **27**(5), 714–727.

Comon, P. (1989) Separation of stochastic processes. WorkShop on Higher-Order Spectral Analysis, Vail, CO, pp. 157–160.

Comon, P. (1994) Independent component analysis, a new concept? *Signal Processing*, **36**, 287–314.

Comon, P., and Jutten, C. (2010) *Handbook of Blind Source Separation, Independent Component Analysis and Applications*, Academic Press, Waltham, MA.

Comon, P., Jutten, C., and Hérault, J. (1991) Blind separation of sources, part II: problem statement. *Signal Processing*, **24**, 11–20.

Douglas, S.C. (2002) *Blind Signal Separation and Blind Deconvolution*, CRC Press, New York.

Fan, L.W., Tang, H.W., and Tang, Y.Y. (2003) Applying independent component analysis to fMRI data analysis. *Journal of Dalian University of Technology*, **1**(4), 399–402.

Gao, Y., and Church, G. (2005) Improving molecular cancer class discovery through sparse non-negative matrix factorization. *Bioinformatics*, **21**(21), 3970–3975.

Giannakis, B., and Swami, A. (1987) New results on state-space and input-output identification of non-Gaussian processing using cumulants. *Proceedings of the SPIE*, **826**, 199–205.

Girolami, M. (1999) *Self-Organizing Neural Networks-Independent Component Analysis and Blind Source Separation*, Springer-Verlag, London.

Girolami, M. (2001) A variational method for learning sparse and overcomplete representations. *Neural Computation*, **13**, 2517–2532.

Harmeling, S., Ziehe, A., Kawanabe, M. *et al.* (2001) Nonlinear blind source separation using kernel feature spaces, in *Advances in Independent Component Analysis*, Springer, pp. 102–107.

Harmeling, S., Ziehe, A., Kawanabe, M. *et al.* (2003) Kernel-based nonlinear blind source separation. *Neural Computation*, **15**, 1089–1124.

Haykin, S., and Chen, Z. (2005) The cocktail party problem. *Neural Computation*, **17**, 1875–1902.

He, W.X., Wang, L., and Xie, J.Y. (2005) An adaptive blind source separation algorithm of non-stationary convolutive mixtures. *Journal of System Simulation*, **17**(l), 196–202(In Chinese).

Herault, J., and Jutten, J. (1986) Space or time adaptive signal processing by neural network models, in *AIP Conference Proceedings 151 on Neural Networks for Computing*, American Institute of Physics Inc., pp. 206–211.

Holland, J.G. (1978) Behaviorism part of the problem or part of the solution. *Journal of Applied Behavior Analysis*, **11**(1), 163–174.

Huang, Z., Yu, X.C., Wang, G.A., and Wang, Z.N. (2008) Non-negative matrix factorizations applied to image fusion in remote sensing. *Journal of Beijing Normal University (Natural Science)*, **44**(6), 599–601 (in Chinese).

Hyvärinen, A. (1998) Independent component analysis for time-dependent stochastic process. Proceedings of the International Conference on Artificial Neural Networks (ICANN'98), Skovde, Sweden, pp. 135–140.

Hyvärinen, A. (1999) Fast and robust fixed-point algorithm for independent component analysis. *IEEE Transactions on Neural Networks*, **10**(3), 626–634.

Hyvärinen, A. (2001) Complexity pursuit: separating interesting components from time-series. *Neural Computation*, **13**(4), 883–898.

Hyvärinen, A., Karhunen, J., and Oja, E. (2001) *Independent Component Analysis*, John Wiley & Sons, Ltd, Chichester.

Hyvärinen, A., and Oja, E. (1997) A fast fixed-point algorithm for independent component analysis. *Neural Computation*, **9**(7), 1483–1492.

Hyvärinen, A., and Oja, E. (2000) Independent component analysis: algorithms and applications. *Neural Networks*, **13**(4–5), 411–430.

Hyvärinen, A., and Pajunen, P. (1999) Nonlinear independent component analysis: existence and uniqueness. *Neural Networks*, **12**(3), 429–439.

Jutten, J., and Herault, J. (1991) Blind separation of sources, part I: an adaptive algorithm based on neuromimetic. *Signal Processing*, **24**(1), 1–10.

Kawamoto, T., Hotta, K., Mishima, T. *et al.* (2000) Estimation of single tones from chord sounds using non-negative matrix factorization. *Neural Network World*, **3**, 429–436.

Lappalainen, H., and Honlela, A. (2000) Bayesian nonlinear independent component analysis by multi-layer perceptrons, in *Advances in Independent Component Analysis*, Springer, pp. 93–121.

Lee, T.W. (1998) *Independent Component Analysis: Theory and Applications*, Kluwer Academic Publishers, Boston, MA.

Lee, T.W., Girolami, M., Bell, A.J., and Sejnowski, T.J. (2000) A unifying information-theoretic framework for independent component analysis. *Computers and Mathematics with Applications*, **39**(11), 1–21.

Lee, T.W., Girolami, M., and Sejnowski, T.J. (1999) Independent component analysis using an extended infomax algorithm for mixed sub-Gaussian and super-Gaussian sources. *Neural Computation*, **11**(2), 417–441.

Lee, T., Kohler, B., and Orglmeister, R. (1997) Blind source separation of nonlinear mixing models. IEEE XIII Workshop on Neural Networks for Signal Processing, pp. 406–415.

Lewicki, M.S., and Sejnowski, T.J. (2000) Learning overcomplete representations. *Neural Computation*, **12**, 337–365.

Li, L.M. (2005) New blind source separation method based on genetic algorithm. *Journal of Xi'an Jiaotong University*, **39**(7), 740–743 (in Chinese).

Ling, X.T. (1996) Blind separation of wide-band signals for the near-field situation. *Acta Electronica Sinica*, **24**(7), 87–92 (in Chinese)..

Linsker, R. (1988) Self-organization in a perceptual network. *Computer*, **21**(3), 105–117.

Linsker, R. (1989) An application of the principle of maximum information preservation to linear systems, in *Advances in Neural Information Processing Systems*, Morgan Kaufmann Publishers Inc, San Francisco, CA, pp. 485–494.

Liu, S.H., Huang, Z., Yu, X.C. *et al.* (2007) An improved mineral prediction method based on ICA. *Journal of Minerals*, **27**, 450–451 (in Chinese).

Liu, X.W., Liu, H., and Zheng, T.Y. (2003) Separation of converted-wave from seismic multiples by independent component analysis. *Journal of Seismology*, **23**(1), 11–19 (in Chinese).

Liu, J., Nie, K.B., and He, Z.Y. (2001) Blind extraction of independent signals from their linear mixtures. *Journal of Applied Sciences*, **19**(3), 24–29 (in Chinese).

Long, Z., Yao, L., Zhao, X. *et al.* (2003) Spatial independent component analysis of multitask-related activation in fMRI data. International Conference on Artificial Neural Networks, pp. 515–522.

Ma, J.C., Niu, Y.L., and Chen, H.Y. (2006) *Blind Signal Processing*, National Defence Industry Press, Beijing (in Chinese).

Martinez, D., and Bray, A. (2003) Nonlinear blind source separation using kernels. *IEEE Transactions on Neural Network*, **14**(1), 228–235.

Oja, E. (1997) The nonlinear PCA learning rule in independent component analysis. *Neurocomputing*, **17**, 25–45.

Pham, T., and Cardoso, J.F. (2000) Blind separation of instantaneous mixtures of non stationary sources. *IEEE Transactions on Signal Processing*, **49**(9), 1837–1848.

Rojas, F., Rojas, I., Clemente, R. *et al.* (2001) Nonlinear bind source separation using genetic algorithm. Proceedings of Independent Component Analysis and Signal Separation, San Diego, CA, pp. 400–405.

Sanchez, V.D. (2002) Frontiers of research in BSS/ICA. *Neurocomputing*, **49**, 7–23.

Shi, X.Z. (2008) *Blind Signal Processing - Theories and Practice*, Shanghai Jiaotong University Press, Shanghai (in Chinese).

Si, H.L., and Zhang, L.M. (2002) *Intelligent Visual Image Processing*, Shanghai Science and Technology Education Press, Shanghai (in Chinese).

Stone, J.V. (2004) *Independent Component Analysis: A Tutorial Introduction*, MIT Press, Cambridge.

Su, Y.P., He, L., Yang, R.Z., and Zhu, X.G. (2002) An improved higher order cumulants based blind speech separation method. *Acta Electronica Sinica*, **7**, 956–958 (in Chinese).

Taleb, A., and Jutten, C. (1997) Nonlinear source separation: the post-nonlinear mixtures. Proceedings of ESANN'97, pp. 279–284.

Taleb, A., Jutten, C., and Olympieff, S. (1998) Source separation in post nonlinear mixtures: an entropy-based algorithm. International Conference on Acoustics Speech, and Signal Processing, Vol. 4, pp. 2089–2092.

Tan, Y., Wang, J., and Zurada, J. (2001) Nonlinear blind source separation using a radial basis function network. *IEEE Transactions on Neural Network*, **12**(1), 124–134.

Tukr, M., and Pentland, A. (1991) Face recognition using eigenfaces. Proceedings of IEEE Conferences on Computer Vision and Pattern Recognition, Maui, HI, pp. 586–591.

Valpola, H., and Karhunen, J. (2002) An unsupervised ensemble learning method for nonlinear dynamic state-space models. *Neural Computation*, **14**(11), 2647–2692.

Vincent, E. (2005) Musical source separation using time-frequency source priors. *IEEE Transactions on Speech and Audio Processing*, **14**(1), 91–98.

Wang, J., and He, Z.Y. (1997) Blind source separation of instantaneous mixtures. *Acta Electronica Sinica*, **25**(4), 1–5 (in Chinese).

Wang, Z.N., Yu, X.C., and Zhang, L.B. (2007) A remote sensing image fusion algorithm based on ordinal fast independent component analysis. Workshop on Knowledge Discovery and Data Mining, pp. 142–145.

Wang, Z.N., Yu, X.C., and Zhang, L.B. (2008) A remote sensing image fusion algorithm based on non-negative matrix factorization. *Journal of Beijing Normal University (Natural Science)*, **44**(4), 387–390 (in Chinese).

Wei, X., Xin, L., and Yihong, G. (2003) Document clustering based on non-negative matrix factorization. Proceedings of the 26th Annual International ACM SIGIR Conference on Research and Development in Information Retrieval, Toronto, Canada, pp. 267–273.

Woo, L., and Khor, L.C. (2004) Blind restoration of nonlinearly mixed signals using multilayer polynomial neural network. *IEE Proceedings-Vision, Image and Signal Processing*, **151**(1), 51–61.

Woo, L., and Sali, S. (2002) General multilayer perceptron demixer scheme for nonlinear blind signal separation. *IEE Proceedings on Vision, Image and Signal Processing*, **149**(5), 253–262.

Yang, Z., LaConte, S., Weng, X.C. *et al.* (2008) Ranking and averaging independent component analysis by reproducibility. *Human Brain Mapping*, **29**, 711–725.

Yang, X.J., and Zheng, J.L. (2003) *Artificial Neural Network and Blind Signal Processing*, Tsinghua University Press, Beijing (in Chinese).

Yang, J.A., Zhuang, Z.Q., and Wu, B. (2002) An improved fast ICA algorithm based on negentropy maximization. *Journal of Circuits and Systems*, **7**(4), 37–40 (in Chinese).

You, R.Y., and Chen, Z. (2004) A fast algorithm of blind signal separation based on ICA. *Acta Electronica Sinica*, **32**(4), 669–672 (in Chinese).

Yu, X.C., Cao, T.T., and Yang, C.P. (2009) Remote sensing image classification based on sparse component analysis. *Progress in Geophysics*, **24**(6), 2274–2279 (in Chinese).

Yu, X.C., Ren, J.M., and Zhang, N. (2007) An optimal ICA algorithm applied to fMRI data. Proceedings of 2007 International Conference on Machine Learning and Cybernetics, pp. 3565–3570.

Zadeh, B., Jutten, C., and Nayebi, K. (2004) Differential of mutual information. *IEEE Signal Processing Letters*, **11**(1), 48–51.

Zhang, X.D., and Bao, Z. (2001) Blind source separation. *Acta Electronica Sinica*, **29**(12A), 1766–1771 (in Chinese).

Zhang, J.L., He, Z.S., and Xie, S.L. (2004) Sequential blind signal extraction in order based on genetic algorithm. *Acta Electronica Sinica*, **32**(4), 616–619 (in Chinese).

Zhang, X., and Hu, B. (2000) An application of blind signal separation in wireless digital communication. *Journal of China Institute of Communications*, **21**(2), 73–77 (in Chinese).

Zhang, A.Q., Qiu, T.S., and Zhang, X.H. (2003) Blind separation technique for multipath underwater acoustic signals based on extended joint diagonalization method. *Systems Engineering and Electronics*, **25**(9), 1058–1083 (in Chinese).

Zhang, H.Y., and Shi, X.Z. (2001) An effective algorithm for blind separation of arbitrary source signal. *Acta Electronica Sinica*, **29**(10), 1392–1396 (in Chinese).

Zhang, F.Q., Zhang, B., and Zhang, X.B. (2006) *Blind Signal Processing and Applications*, Xian University of Electronic Science and Technology Press, Xi'an (in Chinese).

Zhang, X.D., Zhu, X.L., and Bao, Z. (2002) Blind source separation based on the stages of learning. *Science of China (Series E)*, **32**(5), 693–702 (in Chinese).

Zhong, M., Tang, H., Chen, H. *et al.* (2004a) An EM algorithm for learning sparse and over complete representations. *Neural Computing*, **57**, 469–476.

Zhong, M., Tang, H., Wang, H. *et al.* (2004b) An EM algorithm for independent component analysis in the presence of Gaussian noise. *Neural Information Processing: Letters and Reviews*, **2**(1), 11–17.

Zhou, Z.T., Dong, G.H., and Xu, X. (2007) *Independent Component Analysis*, Publishing House of Electronics Industry, Beijing (in Chinese).

Part I

Basic Theory of BSS

2

Mathematical Foundation of Blind Source Separation

Some readers may be interested in the mathematical background behind the various blind source separation techniques. In this chapter, we review key concepts and formulas that are essential for this purpose, including matrix analysis, higher-order statistics, and information theory. Readers who have a good knowledge of these concepts can skip most of this chapter.

2.1 Matrix Analysis and Computing

2.1.1 Determinant and Its Properties

1. Determinant

   Assume $\boldsymbol{A}$ is an $n \times n$ square matrix, which can be written as

$$\boldsymbol{A} = \begin{pmatrix} a_{11} & \cdots & a_{1n} \\ \vdots & \ddots & \vdots \\ a_{n1} & \cdots & a_{nn} \end{pmatrix} \tag{2.1}$$

   An nth order determinant of $\boldsymbol{A}$ is defined as

$$D = \sum (-1)^t a_{p_1 1} a_{p_2 2} \dots a_{p_n n} \tag{2.2}$$

   where t is the reverse number of the row mark.

2. Properties of the determinant

   Properties of the determinant are described as follows. Suppose $\boldsymbol{A}$, $\boldsymbol{B}$, $\boldsymbol{C}$, and $\boldsymbol{D}$ are matrices. Then

$$\det \begin{pmatrix} \boldsymbol{A} & \boldsymbol{B} \\ \boldsymbol{D} & \boldsymbol{C} \end{pmatrix} = \det(\boldsymbol{C})\det(\boldsymbol{A} - \boldsymbol{B}\boldsymbol{C}^{-1}\boldsymbol{D}) = \det(\boldsymbol{C} - \boldsymbol{D}\boldsymbol{A}^{-1}\boldsymbol{B}) \tag{2.3}$$

Blind Source Separation: Theory and Applications, First Edition. Xianchuan Yu, Dan Hu and Jindong Xu.

where det(.) represents the determinant. $\boldsymbol{A}$ and $\boldsymbol{C}$ are nonsingular matrices. $\boldsymbol{A}$, $\boldsymbol{B}$, $\boldsymbol{C}$, and $\boldsymbol{D}$ are denoted as $\boldsymbol{A} \in \mathbf{R}^{s\times s}$, $\boldsymbol{B} \in \mathbf{R}^{s\times r}$, $\boldsymbol{C} \in \mathbf{R}^{r\times r}$, and $\boldsymbol{D} \in \mathbf{R}^{r\times s}$, respectively. In particular,

$$\det(\boldsymbol{I}_r + \boldsymbol{AB}) = \det(\boldsymbol{I}_s + \boldsymbol{BA}) \tag{2.4}$$

where $\boldsymbol{I}_r$ is the $r \times r$ identity matrix $\boldsymbol{I}_r \in \mathbf{R}^{r\times r}$ and $\boldsymbol{I}_s$ is the $s \times s$ identity matrix $\boldsymbol{I}_s \in \mathbf{R}^{s\times s}$. $\boldsymbol{A}$ and $\boldsymbol{B}$ are $r \times s$ and $s \times r$ matrices, respectively, denoted as $\boldsymbol{A} \in \mathbf{R}^{r\times s}$ and $\boldsymbol{B} \in \mathbf{R}^{s\times r}$.

$$\det(\boldsymbol{I} + \boldsymbol{XX}^{\mathrm{T}}) = 1 + \|\boldsymbol{X}\|^2 \tag{2.5}$$

where $\boldsymbol{X}$ is an $n \times 1$ -matrix $\boldsymbol{X} \in \mathrm{R}^{n\times 1}$, denoted as $\|\boldsymbol{X}\|^2 = \sum_{i=1}^{n} x_i^2$, with x_i being the ith element of $\boldsymbol{X}$.

2.1.2 Concepts of Matrix

Definition 2.1 Rank of a matrix The largest order in the whole non-zero minor of matrix $\boldsymbol{A}$ is called the rank of $\boldsymbol{A}$, denoted by $E\{\mathrm{tr}\ \mathbf{A}\} = \mathrm{tr}(E\{\mathbf{A}\})$ or Rank $\boldsymbol{A}$.

Definition 2.2 Transpose of a matrix Let $\boldsymbol{A}$ be an $m \times n$ matrix. Then $\boldsymbol{A}^{\mathrm{T}}$, the transpose of $\boldsymbol{A}$, is the matrix obtained by interchanging the rows and columns of $\boldsymbol{A}$. In other words, if $\boldsymbol{A} = [a_{ij}]$, then $(\boldsymbol{A}^{\mathrm{T}})_{ji} = a_{ij}$. Consequently, $\boldsymbol{A}^{\mathrm{T}}$ is an $n \times m$ matrix.

Definition 2.3 Characteristic value of a matrix Given an $n \times n$ matrix $\boldsymbol{A}$, there exist a number λ and a non-zero vector $\boldsymbol{x}$. If λ satisfies the equation $\boldsymbol{Ax} = \lambda\boldsymbol{x}$, then it is called a characteristic value or eigenvalue of $\boldsymbol{A}$, and $\boldsymbol{x}$ is called a characteristic vector or eigenvector of $\boldsymbol{A}$.

Definition 2.4 Trace of a matrix If $\boldsymbol{A}$ is an $n \times n$ matrix, the trace of $\boldsymbol{A}$, denoted by $tr(\boldsymbol{A})$, is the number defined by $tr(\boldsymbol{A}) = a_{11} + a_{22} + \cdots + a_{nn}$.

Definition 2.5 Moore–Penrose of a matrix Let $\boldsymbol{A} \in \mathbf{R}^{r\times s}$ and matrix $\boldsymbol{X}$ be a generalized inverse matrix satisfying one or more of the following properties:

$$\boldsymbol{AXA} = \boldsymbol{A} \tag{2.6}$$

$$\boldsymbol{XAX} = \boldsymbol{X} \tag{2.7}$$

$$(\boldsymbol{AX})^{\mathrm{H}} = \boldsymbol{AX} \tag{2.8}$$

$$(\boldsymbol{XA})^{\mathrm{H}} = \boldsymbol{XA} \tag{2.9}$$

where H represents the conjugate transpose transformation of the matrix. If $\boldsymbol{X}$ satisfies all four properties, it is referred to as the Moore–Penrose and denoted as $\boldsymbol{A}^{+}$.

2.1.3 Matrix Computation Formulas

2.1.3.1 Matrix Inversion Lemma

If the absolute value of a matrix eigenvalue is less than 1, the Neumann expansion of the inverse matrix is

$$(\boldsymbol{I}+\boldsymbol{A})^{-1}=\boldsymbol{I}-\boldsymbol{A}+\boldsymbol{A}^2-\cdots+(-1)^{n-1}\boldsymbol{I}\boldsymbol{A}^{n-1} \tag{2.10}$$

Lemma 2.1 (Matrix inversion lemma)

$$(\boldsymbol{A}+\boldsymbol{BCD})^{-1}=\boldsymbol{A}^{-1}-\boldsymbol{A}^{-1}\boldsymbol{B}(\boldsymbol{C}^{-1}+\boldsymbol{DA}^{-1}\boldsymbol{B})^{-1}\boldsymbol{DA}^{-1} \tag{2.11}$$

where $\boldsymbol{A}$ is an $n\times n$ matrix, $\boldsymbol{B}$ is an $n\times m$ matrix, $\boldsymbol{C}$ is an $m\times m$ matrix, and $\boldsymbol{D}$ is an $m\times n$ matrix.

If formula $\boldsymbol{D}=\boldsymbol{B}^{\mathrm{T}}$ is satisfied,

$$(\boldsymbol{A}+\boldsymbol{BCD}^{\mathrm{T}})^{-1}=\boldsymbol{A}^{-1}-\boldsymbol{A}^{-1}\boldsymbol{B}(\boldsymbol{C}^{-1}+\boldsymbol{B}^{\mathrm{T}}\boldsymbol{A}^{-1}\boldsymbol{B})\boldsymbol{B}^{\mathrm{T}}\boldsymbol{A}^{-1} \tag{2.12}$$

In the special case where $\boldsymbol{B}=\boldsymbol{u}\in\mathbf{R}^n$, $\boldsymbol{D}=\boldsymbol{v}^{\mathrm{T}}\in\mathbf{R}^{n\times n}$, and $\boldsymbol{C}=\boldsymbol{I}$, the Woodbury equality is always true.

$$(\boldsymbol{A}+\boldsymbol{uv}^{\mathrm{T}})^{-1}=\boldsymbol{A}^{-1}-\frac{\boldsymbol{A}^{-1}\boldsymbol{uv}^{\mathrm{T}}\boldsymbol{A}^{-1}}{1+\boldsymbol{v}^{\mathrm{T}}\boldsymbol{A}^{-1}\boldsymbol{u}} \tag{2.13}$$

$$\boldsymbol{A}^{-1}=\begin{pmatrix}\boldsymbol{B}^{-1}\boldsymbol{DF}^{-1}\boldsymbol{CB}^{-1} & -\boldsymbol{B}^{-1}\boldsymbol{DF}^{-1}\\ -\boldsymbol{F}^{-1}\boldsymbol{CB}^{-1} & \boldsymbol{F}^{-1}\end{pmatrix}, \tag{2.14}$$

where $\boldsymbol{A}=\begin{pmatrix}\mathbf{B} & \mathbf{D}\\ \mathbf{C} & \mathbf{Q}\end{pmatrix}$, $\mathbf{F}=\mathbf{Q}-\mathbf{CB}^{-1}\mathbf{D}$, and both are nonsingular matrices.

2.1.3.2 Properties of the Moore–Penrose

1. Matrix $\boldsymbol{A}$ is a non-singular matrix, while $\boldsymbol{A}^+$ is the Moore–Penrose "pseudo-inverse" matrix, satisfying the following properties:

$$\boldsymbol{A}^+=\boldsymbol{A}^{-1} \tag{2.15}$$

$$(\boldsymbol{A}^+)^+=\boldsymbol{A} \tag{2.16}$$

$$(\boldsymbol{A}^{\mathrm{T}})^+=(\boldsymbol{A}^+)^{\mathrm{T}} \tag{2.17}$$

2. Let $\boldsymbol{A}$ be a symmetric idempotent matrix. The matrix satisfies the following formula:

$$\boldsymbol{A}^+=\boldsymbol{A} \tag{2.18}$$

3. If $\boldsymbol{AA}^+$ and $\boldsymbol{A}^+\boldsymbol{A}$ are idempotent matrices, the ranks of the four matrices $\boldsymbol{A}$, $\boldsymbol{A}^+$, $\boldsymbol{AA}^+$, and $\boldsymbol{A}^+\boldsymbol{A}$ are equal. The matrix satisfies the following formulas:

$$\boldsymbol{A}^{\mathrm{T}}\boldsymbol{AA}^+ = \boldsymbol{A}^{\mathrm{T}} = \boldsymbol{A}^+\boldsymbol{AA}^{\mathrm{T}} \tag{2.19}$$

$$\boldsymbol{A}^{\mathrm{T}}(\boldsymbol{A}^+)^{\mathrm{T}}\boldsymbol{A}^+ = \boldsymbol{A}^+ = \boldsymbol{A}^+(\boldsymbol{A}^+)^{\mathrm{T}}\boldsymbol{A}^{\mathrm{T}} \tag{2.20}$$

$$(\boldsymbol{A}^{\mathrm{T}}\boldsymbol{A})^+ = \boldsymbol{A}^+(\boldsymbol{A}^+)^{\mathrm{T}} \tag{2.21}$$

$$(\boldsymbol{AA}^{\mathrm{T}})^+ = (\boldsymbol{A}^+)^{\mathrm{T}}\boldsymbol{A}^+\boldsymbol{A}(\boldsymbol{A}^{\mathrm{T}}\boldsymbol{A})^+\boldsymbol{A}^{\mathrm{T}}\boldsymbol{A} = \boldsymbol{A} = \boldsymbol{AA}^{\mathrm{T}}(\boldsymbol{AA}^{\mathrm{T}})^{\mathrm{T}}\boldsymbol{A} \tag{2.22}$$

$$\boldsymbol{A}^+ = (\boldsymbol{A}^{\mathrm{T}}\boldsymbol{A})^+\boldsymbol{A}^{\mathrm{T}} = \boldsymbol{A}^{\mathrm{T}}(\boldsymbol{AA}^{\mathrm{T}})^+ \tag{2.23}$$

4. Let $\boldsymbol{A}$ be a full column rank matrix. The matrix satisfies the formula $\boldsymbol{A}^+ = (\boldsymbol{A}^{\mathrm{T}}\boldsymbol{A})^{-1}\boldsymbol{A}^{\mathrm{T}}$.
5. Let $\boldsymbol{A}$ be a full row rank matrix. Then

$$\boldsymbol{A}^+ = \boldsymbol{A}^{\mathrm{T}}(\boldsymbol{A}^{\mathrm{T}}\boldsymbol{A})^{-1} \tag{2.24}$$

$$\boldsymbol{A} = 0 \leftrightarrow \boldsymbol{A}^+ = 0 \tag{2.25}$$

$$\boldsymbol{AB} = 0 \leftrightarrow \boldsymbol{B}^+\boldsymbol{A}^+ = 0 \tag{2.26}$$

$$\boldsymbol{A}^+\boldsymbol{B} = 0 \leftrightarrow \boldsymbol{A}^{\mathrm{T}}\boldsymbol{B} = 0 \tag{2.27}$$

$$(\boldsymbol{A} \otimes \boldsymbol{B})^+ = \boldsymbol{A}^+ \otimes \boldsymbol{B}^+ \tag{2.28}$$

where operator $\otimes$ denotes the Kronecker product.

2.1.3.3 Mathematical Expectations of a Vector and a Matrix

$$E\{\boldsymbol{x}\} = \boldsymbol{m}_x \tag{2.29}$$

$$E\{(\boldsymbol{x} - \boldsymbol{m}_x)(\boldsymbol{x} - \boldsymbol{m}_x)^{\mathrm{T}}\} = \boldsymbol{R}_{xx} \tag{2.30}$$

$$E\{tr\,\boldsymbol{A}\} = tr(E\{\boldsymbol{A}\}) \tag{2.31}$$

$$E\{\boldsymbol{Ax} + \boldsymbol{b}\} = \boldsymbol{Am}_x + \boldsymbol{b} \tag{2.32}$$

$$E\{(\boldsymbol{Ax} + \boldsymbol{a})(\boldsymbol{Bx} + \boldsymbol{b})^{\mathrm{T}}\} = \boldsymbol{AR}_{xx}\boldsymbol{B}^{\mathrm{T}} + (\boldsymbol{Am}_x + \boldsymbol{a})(\boldsymbol{Bm}_x + \boldsymbol{b})^{\mathrm{T}} \tag{2.33}$$

$$E\{\boldsymbol{xx}^{\mathrm{T}}\} = \boldsymbol{R}_{xx} + \boldsymbol{m}_x\boldsymbol{m}_x{}^{\mathrm{T}} \tag{2.34}$$

$$E\{\boldsymbol{xa}^{\mathrm{T}}\boldsymbol{x}\} = (\boldsymbol{R}_{xx} + \boldsymbol{m}_x\boldsymbol{m}_x{}^{\mathrm{T}})\boldsymbol{a} \tag{2.35}$$

$$E\{(\boldsymbol{Ax})(\boldsymbol{Ax})^{\mathrm{T}}\} = \boldsymbol{A}(\boldsymbol{R}_{xx} + \boldsymbol{m}_x\boldsymbol{m}_x{}^{\mathrm{T}})\boldsymbol{A}^{\mathrm{T}} \tag{2.36}$$

$$E\{(\boldsymbol{x}+\boldsymbol{a})(\boldsymbol{x}+\boldsymbol{a})^{\mathrm{T}}\} = \boldsymbol{R}_{xx} + (\boldsymbol{m}_x+\boldsymbol{a})(\boldsymbol{m}_x+\boldsymbol{a})^{\mathrm{T}} \tag{2.37}$$

$$E\{(\boldsymbol{Ax}+\boldsymbol{a})^{\mathrm{T}}(\boldsymbol{Bx}+\boldsymbol{b})\} = tr(\boldsymbol{AR}_{xx}\boldsymbol{B}^{\mathrm{T}}) + (\boldsymbol{Am}_x+\boldsymbol{a})^{\mathrm{T}}(\boldsymbol{Bm}_x+\boldsymbol{b}) \tag{2.38}$$

$$\begin{aligned} &E\{(\boldsymbol{Ax}+\boldsymbol{a})(\boldsymbol{Bx}+\boldsymbol{b})^{\mathrm{T}}(\boldsymbol{Cx}+\boldsymbol{c})\} \\ &= \boldsymbol{AR}_{xx}\boldsymbol{B}^{\mathrm{T}}(\boldsymbol{Cm}_x+\boldsymbol{c}) + \boldsymbol{AR}_{xx}\boldsymbol{C}^{\mathrm{T}}(\boldsymbol{Bm}_x+\boldsymbol{b}) + tr(\boldsymbol{BR}_{xx}\boldsymbol{C}^{\mathrm{T}})(\boldsymbol{Am}_x+\boldsymbol{a}) + \\ &(\boldsymbol{Am}_x+\boldsymbol{a})(\boldsymbol{Bm}_x+\boldsymbol{b})^{\mathrm{T}}(\boldsymbol{Cm}_x+\boldsymbol{c}) \end{aligned} \tag{2.39}$$

$$E\{\boldsymbol{xx}^{\mathrm{T}}\} = tr(\boldsymbol{R}_{xx}) + \boldsymbol{m}_x^{\mathrm{T}}\boldsymbol{m}_x = tr(\boldsymbol{R}_{xx}+\boldsymbol{m}_x\boldsymbol{m}_x^{\mathrm{T}}) \tag{2.40}$$

$$E\{\boldsymbol{x}^{\mathrm{T}}\boldsymbol{ax}^{\mathrm{T}}\} = \boldsymbol{a}^{\mathrm{T}}(\boldsymbol{R}_{xx}+\boldsymbol{m}_x\boldsymbol{m}_x^{\mathrm{T}}) \tag{2.41}$$

$$E\{\boldsymbol{x}^{\mathrm{T}}\boldsymbol{Ax}\} = tr(\boldsymbol{AR}_{xx}) + \boldsymbol{m}_x^{\mathrm{T}}\boldsymbol{Am}_x = tr(\boldsymbol{R}_{xx}+\boldsymbol{m}_x\boldsymbol{m}_x^{\mathrm{T}}) \tag{2.42}$$

$$E\{(\boldsymbol{Ax})^{\mathrm{T}}(\boldsymbol{Ax})\} = tr(\boldsymbol{AR}_{xx}\boldsymbol{A}^{\mathrm{T}}) + \boldsymbol{m}_x^{\mathrm{T}}\boldsymbol{A}^{\mathrm{T}}\boldsymbol{Am}_x = tr(\boldsymbol{R}_{xx}+\boldsymbol{m}_x\boldsymbol{m}_x^{\mathrm{T}})\boldsymbol{A}^{\mathrm{T}} \tag{2.43}$$

$$E\{(\boldsymbol{x}+\boldsymbol{a})^{\mathrm{T}}(\boldsymbol{x}+\boldsymbol{a})\} = tr(\boldsymbol{R}_{xx}) + (\boldsymbol{m}_x+\boldsymbol{a})^{\mathrm{T}}(\boldsymbol{m}_x+\boldsymbol{a}) = tr(\boldsymbol{R}_{xx}) + \|\boldsymbol{m}_x+\boldsymbol{a}\|^2 \tag{2.44}$$

$$E\{\boldsymbol{xx}^{\mathrm{T}}\boldsymbol{x}\} = 2\boldsymbol{R}_{xx}\boldsymbol{m}_x + (tr(\boldsymbol{R}_{xx}) + \boldsymbol{m}_x\boldsymbol{m}_x^{\mathrm{T}})\boldsymbol{m}_x \tag{2.45}$$

2.1.3.4 Kronecker Product of a Matrix

1. Definition of the Kronecker product
   If $\boldsymbol{A}$ is an $m \times n$ matrix and $\boldsymbol{B}$ is a matrix $k \times l$, then the Kronecker product $\boldsymbol{A} \otimes \boldsymbol{B}$ is the $mk \times nl$ block matrix:

$$\boldsymbol{A} \otimes \boldsymbol{B} = \begin{pmatrix} \mathrm{a}_{11}\boldsymbol{B} & \dots & \mathrm{a}_{1\mathrm{n}}\boldsymbol{B} \\ \vdots & \ddots & \vdots \\ \mathrm{a}_{\mathrm{m}1}\boldsymbol{B} & \cdots & \mathrm{a}_{mn}\boldsymbol{B} \end{pmatrix} \tag{2.46}$$

2. Basic properties of the Kronecker product
   a. Let k be a constant. Then,

$$k(\boldsymbol{A} \otimes \boldsymbol{B}) = (k\boldsymbol{A}) \otimes \boldsymbol{B} = \boldsymbol{A} \otimes (k\boldsymbol{B}) \tag{2.47}$$

   b. Let $\boldsymbol{A}_1$ and $\boldsymbol{A}_2$ be same order matrices. Then,

$$(\boldsymbol{A}_1 + \boldsymbol{A}_2) \otimes \boldsymbol{B} = \boldsymbol{A}_1 \otimes \boldsymbol{B} + \boldsymbol{A}_2 \otimes \boldsymbol{B} \tag{2.48}$$

$$\boldsymbol{B} \otimes (\boldsymbol{A}_1 + \boldsymbol{A}_2) = \boldsymbol{B} \otimes \boldsymbol{A}_1 + \boldsymbol{B} \otimes \boldsymbol{A}_2 \tag{2.49}$$

   c. The Kronecker product of matrices satisfies the associative law.

$$(\boldsymbol{A} \otimes \boldsymbol{C}) \otimes \boldsymbol{B} = \boldsymbol{A} \otimes (\boldsymbol{C} \otimes \boldsymbol{B}) \tag{2.50}$$

d. The transpose of the Kronecker product satisfies the following formula:

$$(\boldsymbol{A} \otimes \boldsymbol{B})^{\mathrm{T}} = \boldsymbol{A}^{\mathrm{T}} \otimes \boldsymbol{B}^{\mathrm{T}} \tag{2.51}$$

e. The product of the Kronecker product satisfies the following formula:

$$(\boldsymbol{A} \otimes \boldsymbol{B})(\boldsymbol{C} \otimes \boldsymbol{D}) = (\boldsymbol{AC}) \otimes (\boldsymbol{BD}) \tag{2.52}$$

f. If $\boldsymbol{A}$ and $\boldsymbol{B}$ are invertible, then

$$(\boldsymbol{A} \otimes \boldsymbol{B})^{-1} = \boldsymbol{A}^{-1} \otimes \boldsymbol{B}^{-1} \tag{2.53}$$

3. Hadamard product
The Hadamard product produces another matrix where each element ij is the product of elements ij in the two original matrices with the same dimensions. Let $\boldsymbol{A}$ and $\boldsymbol{B}$ be matrices with the same dimensions. Then, the Hadamard product of the matrices is defined as

$$\begin{pmatrix} a_{11}b_{11} & \dots & a_{1n}b_{1n} \\ \vdots & \ddots & \vdots \\ a_{m1}b_{m1} & \cdots & a_{mn}b_{mn} \end{pmatrix} \tag{2.54}$$

4. Matrix trace
a. The trace of matrix $\boldsymbol{A}$ is defined as the sum of the elements on the main diagonal of $\boldsymbol{A}$, denoted as tr($\boldsymbol{A}$). The trace is invariant under cyclic permutations:

$$\mathrm{tr}(\boldsymbol{ABC}) = \mathrm{tr}(\boldsymbol{BCA}) = \mathrm{tr}(\boldsymbol{CAB}) \tag{2.55}$$

b. The trace is a linear map. That is,

$$tr(\boldsymbol{A} + \boldsymbol{B}) = \mathrm{tr}(\boldsymbol{A}) + \mathrm{tr}(\boldsymbol{A}) \tag{2.56}$$

$$\mathrm{tr}(\alpha \boldsymbol{A}) = \alpha \mathrm{tr}(\boldsymbol{A}) \tag{2.57}$$

c. If a random vector $\boldsymbol{x} \in \mathbf{R}^{n \times l}$ follows a normal distribution, the following formulas are satisfied for the expectations of the Hermite quadratic form.

$$E\{\boldsymbol{x}^{\mathrm{T}}\boldsymbol{A}\boldsymbol{x}\} = E\{\mathrm{tr}(\boldsymbol{x}^{\mathrm{T}}\boldsymbol{A}\boldsymbol{x})\} = E\{\mathrm{tr}(\boldsymbol{A}\boldsymbol{x}\boldsymbol{x}^{\mathrm{T}})\} = \mathrm{tr}(\boldsymbol{A}E\{\boldsymbol{x}\boldsymbol{x}^{\mathrm{T}}\}) = \mathrm{tr}\boldsymbol{A}\boldsymbol{R}_{xx} \tag{2.58}$$

$$var\{\boldsymbol{x}^{\mathrm{T}}\boldsymbol{A}\boldsymbol{x}\} = E\{(\boldsymbol{x}^{\mathrm{T}}\boldsymbol{A}\boldsymbol{x})^2\} = 2\mathrm{tr}(\boldsymbol{A}\boldsymbol{R}_{xx})^2 + 4\boldsymbol{m}_x\boldsymbol{A}\boldsymbol{R}_{xx}\boldsymbol{A}\boldsymbol{m}_x \tag{2.59}$$

5. Vector, matrix, and the differential of their function
If $y(\boldsymbol{x})$ is the function of vector $\boldsymbol{x}$, then

$$y(\boldsymbol{x} + \mathrm{d}\boldsymbol{x}) = y(\boldsymbol{x}) + \left(\frac{\partial y}{\partial \boldsymbol{x}}\right) \mathrm{d}\boldsymbol{x} + \frac{1}{2}\mathrm{d}\boldsymbol{x}^{\mathrm{T}} \left(\frac{\partial^2 y}{\partial \boldsymbol{x}^{\mathrm{T}} \partial \boldsymbol{x}}\right) \mathrm{d}\boldsymbol{x} + o^2 \mathrm{d}\boldsymbol{x} \tag{2.60}$$

a. Differential of the trace
If $\boldsymbol{A}$ and $\boldsymbol{B}$ are square matrices, the differential of the trace can take several forms as defined below.

$$\frac{\partial \mathrm{tr}(\boldsymbol{X})}{\partial \boldsymbol{X}} = \boldsymbol{I} \tag{2.61}$$

$$\frac{\partial \mathrm{tr}(\boldsymbol{XA})}{\partial \boldsymbol{X}} = \frac{\partial \mathrm{tr}(\boldsymbol{AX})}{\partial \boldsymbol{X}} = \boldsymbol{A}^{\mathrm{T}} \tag{2.62}$$

$$\frac{\partial \mathrm{tr}(\boldsymbol{BX}^{\mathrm{T}}\boldsymbol{A})}{\partial \boldsymbol{X}} = \frac{\partial \mathrm{tr}(\boldsymbol{A}^{\mathrm{T}}\boldsymbol{XB})}{\partial \boldsymbol{X}} = \boldsymbol{AB} \tag{2.63}$$

$$\frac{\partial \mathrm{tr}(\boldsymbol{AXBX}^{\mathrm{T}})}{\partial \boldsymbol{X}} = \boldsymbol{AXB} + \boldsymbol{A}^{\mathrm{T}}\boldsymbol{XB}^{\mathrm{T}} \tag{2.64}$$

$$\frac{\partial \mathrm{tr}(\boldsymbol{AXBX})}{\partial \boldsymbol{X}} = \boldsymbol{A}^{\mathrm{T}}\boldsymbol{X}^{\mathrm{T}}\boldsymbol{B}^{\mathrm{T}} + \boldsymbol{B}^{\mathrm{T}}\boldsymbol{X}^{\mathrm{T}}\boldsymbol{A}^{\mathrm{T}} \tag{2.65}$$

$$\frac{\partial \mathrm{tr}(\boldsymbol{BX}^{-1}\boldsymbol{A})}{\partial \boldsymbol{X}} = -(\boldsymbol{X}^{-1}\boldsymbol{ABX}^{-1})^{\mathrm{T}} \tag{2.66}$$

$$\frac{\partial \mathrm{tr}\left(\left(\boldsymbol{X}^{\mathrm{T}}\boldsymbol{CX}\right)^{-1}\boldsymbol{A}\right)}{\partial \boldsymbol{X}} = \frac{\partial \mathrm{tr}\left(\boldsymbol{A}\left(\boldsymbol{X}^{\mathrm{T}}\boldsymbol{CX}\right)^{-1}\right)}{\partial \boldsymbol{X}} = -\boldsymbol{CX}(\boldsymbol{X}^{\mathrm{T}}\boldsymbol{CX})^{-1}(\boldsymbol{A} + \boldsymbol{A}^{\mathrm{T}})(\boldsymbol{X}^{\mathrm{T}}\boldsymbol{CX})^{-1} \tag{2.67}$$

b. The differential of the determinant is given by:

$$\frac{\partial |\boldsymbol{AXB}|}{\partial \boldsymbol{X}} = |\boldsymbol{AXB}|\boldsymbol{X}^{-\mathrm{T}}|\boldsymbol{X}| \tag{2.68}$$

$$\frac{\partial \ln|\boldsymbol{X}|}{\partial \boldsymbol{X}} = \boldsymbol{X}^{-\mathrm{T}} \tag{2.69}$$

$$\frac{\partial \ln|\boldsymbol{X}(z)|}{\partial z} = \mathrm{tr}\left(\boldsymbol{X}^{-T}\frac{\partial \boldsymbol{X}}{\partial z}\right) \tag{2.70}$$

Let $\boldsymbol{A}$ be a square matrix. Then,

$$\frac{\partial \ln|\boldsymbol{X}^{\mathrm{T}}\boldsymbol{AX}|}{\partial \boldsymbol{X}} = (\boldsymbol{A} + \boldsymbol{A}^{\mathrm{T}})\boldsymbol{X}(\boldsymbol{X}^{\mathrm{T}}\boldsymbol{AX})^{-1} \tag{2.71}$$

In particular, if $\boldsymbol{A}$ is an identity matrix and $\boldsymbol{X}$ is a full column rank matrix,

$$\frac{\partial \ln|\boldsymbol{X}^{\mathrm{T}}\boldsymbol{X}|}{\partial \boldsymbol{X}} = \boldsymbol{X}(\boldsymbol{X}^{\mathrm{T}}\boldsymbol{X})^{-1} \tag{2.72}$$

c. The differential of a vector and that of a matrix are defined as:

$$\frac{\partial(\boldsymbol{X}^{-1})}{\partial z} = -\boldsymbol{X}^{-1}\frac{\partial \boldsymbol{X}}{\partial z}\boldsymbol{X}^{-1} \tag{2.73}$$

$$\frac{\partial(\boldsymbol{AXB})}{\partial z} = \boldsymbol{A}\frac{\partial \boldsymbol{X}}{\partial z}\boldsymbol{B} \tag{2.74}$$

$$\frac{\partial(\boldsymbol{x}^{\mathrm{T}}\boldsymbol{A})}{\partial \boldsymbol{x}} = \boldsymbol{A}, \ \frac{\partial \boldsymbol{x}^{\mathrm{T}}}{\partial \boldsymbol{x}} = \boldsymbol{I} \tag{2.75}$$

$$\frac{\partial(\boldsymbol{x}^{\mathrm{T}}\boldsymbol{A}\boldsymbol{x}\boldsymbol{x}^{\mathrm{T}})}{\partial \boldsymbol{x}} = (\boldsymbol{A}+\boldsymbol{A}^{\mathrm{T}})\boldsymbol{x}\boldsymbol{x}^{\mathrm{T}} + \boldsymbol{x}^{\mathrm{T}}\boldsymbol{A}\boldsymbol{x}\boldsymbol{I} \tag{2.76}$$

d. The differential of a vector's function and that of a matrix's function are given by:

$$\frac{\partial(\boldsymbol{a}^{\mathrm{T}}\boldsymbol{x})}{\partial \boldsymbol{x}} = \frac{\partial(\boldsymbol{x}^{\mathrm{T}}\boldsymbol{a})}{\partial \boldsymbol{x}} = \boldsymbol{a} \tag{2.77}$$

$$\frac{\partial(\boldsymbol{x}^{\mathrm{T}}\boldsymbol{A}\boldsymbol{x})}{\partial \boldsymbol{x}} = (\boldsymbol{A}+\boldsymbol{A}^{\mathrm{T}})\boldsymbol{x} \tag{2.78}$$

$$\frac{\partial(\boldsymbol{Ax}+\boldsymbol{b})^{\mathrm{T}}\boldsymbol{C}(\boldsymbol{Dx}+\boldsymbol{e})}{\partial \boldsymbol{x}} = \boldsymbol{A}^{\mathrm{T}}\boldsymbol{C}(\boldsymbol{Dx}+\boldsymbol{e}) + \boldsymbol{D}^{\mathrm{T}}\boldsymbol{C}^{\mathrm{T}}(\boldsymbol{Ax}+\boldsymbol{b}) \tag{2.79}$$

$$\frac{\partial(\boldsymbol{a}^{\mathrm{T}}\boldsymbol{X}\boldsymbol{b})}{\partial \boldsymbol{X}} = \boldsymbol{a}\boldsymbol{b}^{\mathrm{T}}, \ \frac{\partial(\boldsymbol{a}^{\mathrm{T}}\boldsymbol{X}^{\mathrm{T}}\boldsymbol{b})}{\partial \boldsymbol{X}} = \boldsymbol{b}\boldsymbol{a}^{\mathrm{T}} \tag{2.80}$$

$$\frac{\partial\left((\boldsymbol{Xa}+\boldsymbol{b})^{\mathrm{T}}\boldsymbol{C}(\boldsymbol{Xa}+\boldsymbol{b})\right)}{\partial \boldsymbol{X}} = (\boldsymbol{C}+\boldsymbol{C}^{\mathrm{T}})(\boldsymbol{Xa}+\boldsymbol{b})\boldsymbol{a}^{\mathrm{T}} \tag{2.81}$$

2.1.3.5 The Symmetric Hessian Matrix

Let $\boldsymbol{x}$ be an $m \times 1$ column vector and $\boldsymbol{y}$ be an $n \times 1$ column vector. Then, the Jacobian matrix $J_{m\times n} = \partial y^{\mathrm{T}}/\partial x$ is a matrix of the first-order partial derivatives of $\boldsymbol{y}$ with respect to $\boldsymbol{x}$. $\partial \boldsymbol{y}/\partial \boldsymbol{x}$ is an $mn \times 1$ vector and $\partial y^{\mathrm{T}}/\partial x^{\mathrm{T}} = (\partial y/\partial x)^{\mathrm{T}}$ is a $1 \times mn$ vector. $\partial y/\partial x^{\mathrm{T}}$ is an $n \times m$ matrix representing the differential of $\boldsymbol{y}$ with respect to $\boldsymbol{x}$, and is denoted as $J^{\mathrm{T}}_{m\times n}$. If f is the scalar-valued function of $\boldsymbol{x}$, the Hessian symmetric matrix is defined as:

$$H = \frac{\partial^2 f(\boldsymbol{x})}{\partial \boldsymbol{x}^2} = \frac{\partial}{\partial \boldsymbol{x}^{\mathrm{T}}}\left(\frac{\partial f(\boldsymbol{x})}{\partial \boldsymbol{x}}\right) \tag{2.82}$$

If the Hessian matrix is a positive definite matrix, a negative definite matrix, or a non-definite matrix, the values of $\partial f(\boldsymbol{x})/\partial \boldsymbol{x} = 0$ correspond to the lowest point, the highest point, and the saddle point, respectively.

$$\frac{\partial^2(\boldsymbol{Ax}+\boldsymbol{b})^T\boldsymbol{C}(\boldsymbol{Dx}+\boldsymbol{e})}{\partial \boldsymbol{x}^2} = \boldsymbol{A}^T\boldsymbol{C}\boldsymbol{D} + \boldsymbol{D}^{\mathrm{T}}\boldsymbol{C}^{\mathrm{T}}\boldsymbol{A} \tag{2.83}$$

2.1.3.6 Some Important Inequalities

1. Cauchy–Schwartz inequality

$$(\boldsymbol{x}^T\boldsymbol{y})^2 \le (\boldsymbol{x}^T\boldsymbol{x})(\boldsymbol{y}^T\boldsymbol{y}) \tag{2.84}$$

This equality holds if and only if $\boldsymbol{x}$ and $\boldsymbol{y}$ have a linear relationship.

$$\left(\sum_i x_i y_i\right)^2 \le \left(\sum_i x_i^2\right)\left(\sum_i y_i^2\right) \tag{2.85}$$

$$(\boldsymbol{x}^T\boldsymbol{A}\boldsymbol{y})^2 \le (\boldsymbol{x}^T\boldsymbol{A}\boldsymbol{x})(\boldsymbol{y}^T\boldsymbol{A}\boldsymbol{y}) \tag{2.86}$$

If $\boldsymbol{A}$ is a positive semi-definite matrix, and $\boldsymbol{Ax}$ and $\boldsymbol{Ay}$ have a linear relationship, the equality in Equation 2.86 holds.

$$(\boldsymbol{x}^{\mathrm{T}}\boldsymbol{A}\boldsymbol{y})^2 \le (\boldsymbol{x}^{\mathrm{T}}\boldsymbol{A}\boldsymbol{x})(\boldsymbol{y}^{\mathrm{T}}\boldsymbol{A}^{-1}\boldsymbol{y}) \tag{2.87}$$

If $\boldsymbol{A}$ is a positive definite matrix, and $\boldsymbol{x}$ and $\boldsymbol{A}^{-1}\boldsymbol{y}$ have a linear relationship, the equality in Equation 2.87 holds.

$$\left(\int_\Omega \boldsymbol{f}\boldsymbol{g}\mathrm{d}v\right)^2 \le \int_\Omega \boldsymbol{f}^2\mathrm{d}v\int_\Omega \boldsymbol{g}^2\mathrm{d}v \tag{2.88}$$

The equality in Equation 2.88 is satisfied, if and only if f and g have a linear relationship.

2. Lagrange's inequality

$$\left(\sum_i x_i^2\right)\left(\sum_i y_i^2\right) - \left(\sum_i x_i y_i\right)^2 \le \sum_{i\le j}(x_i y_i - x_j y_j)^2 \tag{2.89}$$

3. Minkowski's inequality

$$\left[\sum_i (x_i + y_i)^k\right]^{\frac{1}{k}} \le \left[\sum_i (x_i)^k\right]^{\frac{1}{k}} + \left[\sum_i (y_i)^k\right]^{\frac{1}{k}} \tag{2.90}$$

4. Hölder's inequality If $x_i \ge 0, y_i \ge 0, (i = 1, 2, \ldots, n)$, and $\left(\frac{1}{p}\right) + \left(\frac{1}{q}\right) = 1$, $(p < 0)$, then

$$\sum_i x_i y_i \le \left(\sum_i x_i^p\right)^{\frac{1}{p}}\left(\sum_i y_i^q\right)^{\frac{1}{q}} \tag{2.91}$$

If $\frac{1}{p} + \frac{1}{q} + \frac{1}{r} + \cdots \le 1$, then

$$\sum_i x_i y_i z_i \le \left(\sum_i x_i^p\right)^{\frac{1}{p}}\left(\sum_i y_i^q\right)^{\frac{1}{q}}\left(\sum_i z_i^r\right)^{\frac{1}{r}} \tag{2.92}$$

5. Hadamard's inequality

$$\det(\boldsymbol{B})^2 \leq \prod_{i=1}^{n}\left(\sum_{k=1}^{n} b_{ik}^2\right) \tag{2.93}$$

If $\sum_{k=1}^{n} b_{ik}^2 = 1, i = 1, 2, \cdots, n$, then

$$\det(\boldsymbol{B}) \leq 1 \tag{2.94}$$

If $|b_{ij}| \leq M$, then

$$\det(\boldsymbol{B}) \leq M^n n^{\frac{n}{2}} \tag{2.95}$$

If $\boldsymbol{B}$ is a positive semi-definite matrix, then

$$\det(\boldsymbol{B}) \leq \prod_{i=1}^{n} b_{ii} \tag{2.96}$$

6. Jensen's inequalities of a convex function

$$E\{f(\boldsymbol{x})\} \geq f[E\{\boldsymbol{x}\}] \tag{2.97}$$

2.2 Foundation of Probability Theory for Higher-Order Statistics

In spectrum estimation, the signal, which is seen as a statistically irrelevant superimposed multi-harmonic component, is used to estimate the distribution of power in the multi-frequency component, while ignoring the phase relationship between the multi-frequency components. The information contained in the power spectrum is mainly present in the autocorrelation sequence. If the mean is known together with a statistical description of a Gaussian process, the estimation is complete. If the statistical description is nonlinear and deviates from the Gaussian, we need to obtain the information from higher-order statistics.

Higher-order statistics generally include higher-order moments, higher-order cumulants, and their spectra, namely, the higher-order moment spectrum, higher-order cumulant spectrum (also called the polyspectrum) and inverted higher-order cumulant spectrum (also called the inverted multi-spectrum). The third-order spectrum is called the bispectrum, which is the Fourier transform of the third-order cumulant sequence in a stationary random process. The trispectrum is the Fourier transform of the fourth-order cumulants. The higher-order moment spectrum is applied to periodic and aperiodic deterministic signals, while the cumulant spectrum is applied to smooth and non-stationary random signals. Higher-order statistics are derived from the characteristic function of the random variable.

2.2.1 Moment

Let $p(x)$ be the probability density function of a random variable $\boldsymbol{x}$, and $E\{e^{sx}\}$ be the characteristic function of $\boldsymbol{x}$. Then

$$\Phi(s) = \int_{-\infty}^{\infty} p(x)e^{sx}dx = E\{e^{sx}\} \tag{2.98}$$

where $E\{\cdot\}$ represents the expectation.

The kth-order moment m_k of random variable $\boldsymbol{x}$ is defined as

$$m_k = E\{\boldsymbol{x}^k\} \tag{2.99}$$

We can derive the k-order derivative of $\Phi(s)$ at the origin using Equation 2.99.

$$\Phi^k(s)|_{s=0} = E\{x^k e^{sx}\}_{s=0} = E\{x^k\} = m_k \tag{2.100}$$

Given a set of real random variables $\{x_1, x_2, \cdots, x_n\}$, the joint moment of $y = k_1 + k_2 + \cdots + k_n$ is defined as

$$\begin{aligned}\text{Mom}[x_1^{k_1}, x_2^{k_2}, \cdots, x_n^{k_n}] &\equiv E\{x_1^{k_1}, x_2^{k_2}, \cdots, x_n^{k_n}\} \\ &= (-j)^r \frac{\partial^r \Phi(w_1, w_2, \cdots w_n)}{\partial w_1^{k_1} \partial w_2^{k_2} \cdots \partial w_n^{k_n}}|_{w_1=w_2=\cdots=w_n}\end{aligned} \tag{2.101}$$

where $\Phi(w_1, w_2, \cdots, w_n) \equiv E\{\exp(j(w_1x_1 + w_2x_2 + \cdots + w_nx_n))\}$ is the joint characteristic function.

2.2.2 Cumulant

The joint cumulant (also known as the semi-invariant) is defined as the Taylor expansion coefficients of the second characteristic function at the zero point.

$$\text{Cum}[x_1^{k_1}, x_2^{k_2}, \cdots, x_n^{k_n}] = (-j)^r \frac{\partial^r \widetilde{\Psi}(w_1, w_2, \cdots w_n)}{\partial w_1^{k_1} \partial w_2^{k_2} \cdots \partial w_n^{k_n}}|_{w_1=w_2=\cdots=w_n=0} \tag{2.102}$$

where $\Psi(w_1, w_2, \cdots w_n) \equiv \ln[\Phi(w_1, w_2, \cdots w_n)]$.

The relationships of the moment and cumulant are expressed by the following formulas. First we give the expressions for the moment of random variable x_1:

$$\begin{aligned} m_1 &= \text{Mom}[x_1] = E\{x_1\} \\ m_2 &= \text{Mom}[x_1, x_1] = E\{x_1{}^2\} \\ m_3 &= \text{Mom}[x_1, x_1, x_1] = E\{x_1{}^3\} \\ m_4 &= \text{Mom}[x_1, x_1, x_1, x_1] = E\{x_1{}^4\} \end{aligned} \tag{2.103}$$

Listed below are the expressions for the cumulants, assuming that the mean $E\{x_1\}$ of x_1 is non-zero.

$$
\begin{aligned}
c_1 &= \mathrm{Cum}[x_1] = m_1 \\
c_2 &= \mathrm{Cum}[x_1, x_1] = m_2 - {m_1}^2 \\
c_3 &= \mathrm{Cum}[x_1, x_1, x_1] = m_3 - 3m_2m_1 + 2{m_1}^3 \\
c_4 &= \mathrm{Cum}[x_1, x_1, x_1, x_1] = m_4 - 4m_3m_1 - 3{m_2}^2 + 12m_2{m_1}^2 - 6{m_1}^4
\end{aligned}
\tag{2.104}
$$

The relationship between the moment and cumulant is denoted as:

$$
\Phi(w_1) = 1 + jw_1m_1 - \frac{{w_1}^2}{2!}m_2 \cdots + \frac{(jw)^k}{k!}m_k + \cdots \tag{2.105}
$$

The relationship between the nth-order moment of the random variable $\{x_1, x_2, \ldots x_n\}$ and the joint cumulant $Cum[x_1, x_2, \ldots x_n]$ is denoted as

$$
\begin{aligned}
&\mathrm{Cum}[x_1, x_2, \cdots, x_n] \\
&= \sum (-1)^{p-1}(p-1)! \cdot E\left\{\prod_{i\in s_1} x_i\right\} \cdot E\left\{\prod_{i\in s_2} x_i\right\} \cdot \cdots \cdot E\left\{\prod_{i\in s_p} x_i\right\}
\end{aligned}
\tag{2.106}
$$

We add all divisions $(s_1, s_2, \ldots . s_p)$ of datasets in Equation 2.106, where $p = 1, 2, 3, \ldots n$. For example, the dataset $(1, 2, 3)$ can be divided as follows:

$$
p = 1, s_1 = \{1, 2, 3\} \tag{2.107}
$$

$$
\begin{aligned}
p = 2, s_1 &= \{1\},\ s_2 = \{2, 3\} \\
s_1 &= \{2\},\ s_2 = \{1, 3\} \\
s_1 &= \{3\},\ s_2 = \{1, 2\} \\
p = 3, s_1 &= \{1\},\ s_2 = \{2\}, s_3 = \{3\}.
\end{aligned}
\tag{2.108}
$$

Hence, we can derive the joint cumulant of the dataset using Equation 2.106.

$$
\begin{aligned}
\mathrm{Cum}[x_1, x_2, x_3] = {} & E\{x_1x_2x_3\} - E\{x_1\} \cdot E\{x_2x_3\} - E\{x_2\} \cdot E\{x_1x_3\} \\
& - E\{x_3\} \cdot E\{x_1x_2\} - 2E\{x_1\} \cdot E\{x_2\} \cdot E\{x_3\}
\end{aligned}
\tag{2.109}
$$

The moment m_n^x of the stationary stochastic process $\{X(k)\}$ is denoted as

$$
m_n^x(\tau_1, \tau_2, \cdots, \tau_{n-1}) \equiv E\{X(k), X(k+\tau_1), \cdots X(k+\tau_{n-1})\} \tag{2.110}
$$

Similarly, the nth-order cumulant of $\{X(k)\}$ is given by

$$
c_n^x(\tau_1, \tau_2, \cdots, \tau_{n-1}) \equiv \mathrm{Cum}[X(k), X(k+\tau_1), \cdots X(k+\tau_{n-1})] \ \ldots \tag{2.111}
$$

Using Equations 2.109–2.111, we can obtain the relationship between the moment sequence and cumulant sequence.

The first-order cumulant is

$$c_1^x = m_1^x = E\{X(k)\}(\text{Mean}) \tag{2.112}$$

The second-order cumulant is

$$c_2^x(\tau_1) = m_2^x(\tau_1) - \left(m_1^x\right)^2 = m_2^x(-\tau_1) - \left(m_1^x\right)^2 = c_2^x(-\tau_1)\,(\text{Covariance sequence}) \tag{2.113}$$

where $m_2^x(\tau_1)$ is the autocorrelation sequence.

The third-order cumulant is

$$c_3^x(\tau_1,\tau_2) = m_3^x(\tau_1,\tau_2) - m_1^x\left(m_2^x(\tau_1) + m_2^x(\tau_2) + m_2^x(\tau_2-\tau_1)\right) + 2\left(m_1^x\right)^3 \tag{2.114}$$

where $m_3^x(\tau_1,\tau_2)$ represents the third-order cumulant and $\tau_1, \tau_1, \ldots, \tau_{n-1}$ represents the time interval.

The nth-order (for $n = 3, 4$ only) cumulant of a non-Gaussian stationary random process $X(k)$ can be written as

$$c_n^x(\tau_1,\tau_2,\cdots,\tau_{n-1}) = m_n^x(\tau_1,\tau_2,\cdots,\tau_{n-1}) - m_n^G(\tau_1,\tau_2,\cdots,\tau_{n-1}) \tag{2.115}$$

where $m_n^x(\tau_1,\tau_2,\cdots,\tau_{n-1})$ represents the nth moment function of $X(k)$, and $m_n^G(\tau_1,\tau_2,\ldots,\tau_{n-1})$ represents the nth moment function of a Gauss process with the same mean and autocorrelation sequence. If $X(k)$ is Gaussian, then $m_n^x(\tau_1,\tau_2,\ldots,\tau_{n-1}) = m_n^G(\tau_1,\tau_2,\ldots,\tau_{n-1})$. Thus,

$$c_n^x(\tau_1,\tau_2,\ldots,\tau_{n-1}) = 0 \tag{2.116}$$

Let $\tau_1 = \tau_2 = \tau_3 = 0$ and $m_1^x = 0$; we then obtain the following formulas.

$$r_2^x = E\{X^2(k)\} = c_2^x(0)(\text{Variance}) \tag{2.117}$$

$$r_3^x = E\{X^3(k)\} = c_3^x(0,0)(\text{Skewness}) \tag{2.118}$$

$$r_4^x = E\{X^4(k) - 3[r_2^x]\} = c_4^x(0,0,0)(\text{Kurtosis}) \tag{2.119}$$

The normalized kurtosis is defined as $\frac{r_4^x}{[r_2^x]^2}$. In some literature, kurtosis is denoted as $k(x)$.

Kurtosis is often used as a quantitative measure of the non-Gaussianity of a random variable or signal. The signal is divided into three categories according to the kurtosis of the random variable x in blind signal separation. If the kurtosis of random variable x is zero ($k(x) = 0$), x has a Gaussian distribution. If the kurtosis is positive ($k(x) > 0$), the associated distribution is called super-Gaussian, whereas if the kurtosis is negative ($k(x) < 0$), the associated distribution is called sub-Gaussian. Compared with the probability density function, a sub-Gaussian probability density tends to be flatter than a Gaussian one. A typical super-Gaussian probability density has a sharper peak and long tail.

2.2.3 Properties of Moments and Cumulants

The properties of moments and cumulants are summarized as follows:

$$\text{Mom}[a_1x_1, a_2x_2, \cdots, a_nx_n] = a_1a_2\cdots a_n\text{Mom}[x_1, x_2, \cdots, x_n] \tag{2.120}$$

$$\text{Cum}[a_1x_1, a_2x_2, \cdots, a_nx_n] = a_1a_2\cdots a_n\text{Cum}[x_1, x_2, \cdots, x_n] \tag{2.121}$$

where $(a_1, a_2, \ldots, a_n)$ are constants.

If moments and cumulants are symmetric function of the independent variable, then

$$\text{Mom}[x_1, x_2, x_3] = \text{Mom}[x_2, x_1, x_3] = \text{Mom}[x_3, x_2, x_1] \tag{2.122}$$

If variables $\{x_1, x_2, \cdots, x_n\}$ can be divided into statistically independent groups, the nth-order cumulant is zero.

$$\text{Cum}[x_1, x_2, \cdots, x_n] = 0 \tag{2.123}$$

In the general case, $\text{Cum}[x_1, x_2, \cdots, x_n] \neq 0$.

If variables $\{x_1, x_2, \cdots, x_m\}$ and $\{x_{m+1}, x_{m+2}, \cdots, x_n\}$ are independent groups, the joint characteristic function is denoted as

$$\Phi(w_1, w_2, \cdots w_n) = \Phi(w_1, w_2, \cdots w_m)\Phi(w_{m+1}, w_{m+2}, \cdots w_n) \tag{2.124}$$

The joint second characteristic function is denoted as

$$\widetilde{\Psi}(w_1, w_2, \cdots w_n) = \widetilde{\Psi}(w_1, w_2, \cdots w_m) + \widetilde{\Psi}(w_{m+1}, w_{m+2}, \cdots w_n) \tag{2.125}$$

Let x $(\{x_1, x_2, \cdots, x_n\})$ and y $(\{y_1, y_2, \cdots, y_n\})$ be statistically independent random vectors with the same dimensions. Then,

$$\text{Cum}[y_1 + x_1, y_2 + x_2, \cdots, y_n + x_n] = \text{Cum}[x_1, x_2, \cdots, x_n] + \text{Cum}[y_1, y_2, \cdots, y_n] \tag{2.126}$$

Under normal circumstances, this is denoted by the following formula.

$$\begin{aligned}\text{Mom}[y_1 + x_1, y_2 + x_2, \cdots, y_n + x_n] &= E\{(y_1 + x_1)(y_2 + x_2)\cdots(y_n + x_n)\} \\ &\neq \text{Mom}[x_1, x_2, \cdots, x_n] + \text{Mom}[y_1, y_2, \cdots, y_n]\end{aligned} \tag{2.127}$$

The formulas for random variables $\{y_1, x_1, x_2, \cdots, x_n\}$ are given below.

$$\text{Cum}[y_1 + x_1, x_2, \cdots, x_n] = \text{Cum}[x_1, x_2, \cdots, x_n] + \text{Cum}[y_1, x_2, \cdots, x_n] \tag{2.128}$$

$$\text{Mom}[y_1 + x_1, x_2, \cdots, x_n] = \text{Mom}[x_1, x_2, \cdots, x_n] + \text{Mom}[y_1, x_2, \cdots, x_n] \tag{2.129}$$

If the distribution of the random vector $\boldsymbol{x}$ $(\{x_1, x_2, \cdots, x_n\})$ is multivariate Gaussian, all distribution information is expressed in its nth-order cumulants $(n \leq 2)$, while the

nth-order ($n > 2$) cumulants are zero. Therefore, cumulants higher than the second order are used to measure non-Gaussianity and non-normality of a time series.

2.3 Basic Concepts of Information Theory

2.3.1 Entropy

Entropy is the basic concept of information theory. Entropy H is defined for a discrete-valued random variable x as

$$H(x) = -\sum_i p(\boldsymbol{x} = a_i) \lg p(\boldsymbol{x} = a_i) \tag{2.130}$$

where the a_i are the possible values of x. Depending on the base of the logarithm, different units of entropy are obtained. Usually, base 2 is used, in which case the unit is called a bit. When the natural logarithm is used, the unit of entropy is called a nat.

Let us define the function f as

$$f(p) = -p \lg p, \; 0 \leq p \leq 1 \tag{2.131}$$

This is a nonnegative function that is zero for $p = 0$ and for $p = 1$, and positive for values in between. The entropy of a random variable can be interpreted as the degree of information that the observation of the variable gives. The more "random", that is, unpredictable and unstructured the variable is, the larger its entropy. We illustrate it with one example: tossing a coin, in which case the random variable x has only two values, a_1 and a_2. The entropy is largest when the value is obtained with a probability of 50% (that is, $p(x = a_1) = p(x = a_2)$). The connection between entropy and randomness can be made more rigorous by considering coding length. Entropy gives roughly the average minimum code length of the random variable.

2.3.2 Differential Entropy, Maximum Entropy, and Negentropy

1. Differential entropy
   The definition of entropy for a discrete-valued random variable can be extended to continuous-valued random variables and random vectors, in which case it is called the differential entropy. The differential entropy could be interpreted as a measure of randomness. If the random variable is concentrated on certain small intervals, its differential entropy is small. The differential entropy h of a random variable x with density $p(x)$ is defined as

$$h(x) = -\int_{-\infty}^{\infty} p(x) \lg p(x) \mathrm{d}x \ldots \tag{2.132}$$

Where the quantization interval of x converges to zero, the definition of entropy is expressed as

$$\begin{aligned} H(x) &= -\lim_{\delta x \to 0} \sum_{i=-\infty}^{\infty} p(x_i)\delta x \lg(p(x_i)\delta x) \\ &= -\lim_{\delta x \to 0} \left[\sum_{i=-\infty}^{\infty} p\,(x_i)\,\delta x \lg p(x_i) + \lg \delta x \sum_{i=-\infty}^{\infty} p(x_i)\delta x \right] \\ &= -\int_{-\infty}^{\infty} p(x) \lg p(x) dx - \lim_{\delta x \to 0} \lg \delta x \int_{-\infty}^{\infty} p(x) dx \end{aligned} \tag{2.133}$$

The first term on the right-hand side of Equation 2.133 is the differential entropy, and the integral of $p(x)$ is equal to 1. The above equation can be written as

$$h(x) = H(x) + \lim_{\delta x \to 0} \lg \delta x \tag{2.134}$$

Because the second term on the right-hand side of Equation 2.134 goes to $-\infty$, the differential entropy of the continuous random variable goes to $-\infty$. We can use $\lg \delta x$ as the reference to compare two differential entropies, thereby keeping the relative entropy positive. Relative entropy will be discussed in the next section.

2. Maximum entropy
Since entropy can be interpreted as a measure of randomness, the maximum entropy density is the most random of all probability density functions satisfying the constraints. Under some regularity conditions, density $p(\zeta)$, which satisfies the constraints and has the maximum entropy of all such densities, takes the form

$$p(\zeta) = A \exp\left(\sum_i a_i F^i\,(\zeta) \right) \tag{2.135}$$

Where A and a_i are constants determined by c_i in the following formula:

$$\int p(\zeta) F^i(\zeta) \mathrm{d}\zeta = c_i, i = 1, 2, \cdots, m \tag{2.136}$$

Now, consider the set of random variables with zero mean and fixed variance, which can take all the values on the real line. The maximum entropy distribution for such variables, with two constraints, is the Gaussian distribution. This distribution has the form

$$p(\zeta) = A \exp(a_1 \zeta^2 + a_2 \zeta) \tag{2.137}$$

A Gaussian variable has the largest entropy of all random variables with unit variance. This means that entropy could be used as a measure of non-Gaussianity. Entropy is small for distributions that are clearly concentrated on certain values, that is, when the variable is clearly clustered, or has a probability density function that is very “spiky”.

3. Negentropy

A measure that is zero for a Gaussian variable and always nonnegative can be simply obtained from differential entropy and is called negentropy. Negentropy J of a random vector $\boldsymbol{x}$ is defined as

$$J(\boldsymbol{x}) = H(\boldsymbol{x}_{\text{gauss}}) - H(\boldsymbol{x}) \tag{2.138}$$

The Gaussian random vector $\boldsymbol{x}_{\text{gauss}}$ and $\boldsymbol{x}$ have the same covariance matrix $\sum$. The entropy thereof can be calculated as

$$H(\boldsymbol{x}_{\text{gauss}}) = \frac{1}{2}\lg\left|\det\sum\right| + \frac{n}{2}[1 + \lg 2\pi] \tag{2.139}$$

where n is the dimension of $\boldsymbol{x}$.

Negentropy is zero if and only if $\boldsymbol{x}$ has a Gaussian distribution, since the maximum entropy distribution is unique. Negentropy is always nonnegative, due to the maximality property of the Gaussian distribution. Negentropy has the additional property that it is invariant for invertible linear transformations. This is because for $\boldsymbol{y} = M\boldsymbol{x}$ we have $E\{\boldsymbol{y}\boldsymbol{y}^{\text{T}}\} = M\sum M^{\text{T}}$, and, using proceeding results, the negentropy can be computed as

$$\begin{aligned}
J(Mx) &= \frac{1}{2}\lg\left|\det\left(M\sum M^{\text{T}}\right)\right| + \frac{n}{2}[1 + \lg 2\pi] - (H(x) + \lg|\det M|) \\
&= \frac{1}{2}\lg\left|\det\sum\right| + 2\times\frac{1}{2}\lg|\det M| + \frac{n}{2}[1 + \lg 2\pi] - (H(\boldsymbol{x}) + \lg|\det M|) \\
&= \frac{1}{2}\lg\left|\det\sum\right| + \frac{n}{2}[1 + \lg 2\pi] - H(\boldsymbol{x}) \\
&= H(\boldsymbol{x}_{\text{gauss}}) - H(\boldsymbol{x}) = J(\boldsymbol{x})
\end{aligned} \tag{2.140}$$

2.3.3 *Mutual Information*

Mutual information is a measure of the information of the two random variables. The joint entropy of two random variables x and y is defined as

$$H(x, y) = -\sum_i p(x = a_i, y = b_i)\lg p(x = a_i, y = b_i) \tag{2.141}$$

If x and y are independent, then

$$\begin{aligned}
p(x, y) &= p(x)p(y) \\
H(x, y) &= H(x) + H(y)
\end{aligned} \tag{2.142}$$

The conditional entropy of x is the entropy of conditional probability $p(x\,|y = b_k)$ denoted as

$$H(x\,|y = b_k) = -\sum_i p(x\,|y = b_k)\lg p(x\,|y = b_k) \tag{2.143}$$

If $H(x\,|y = b_k)$ is the entropy of the variable x conditional on the variable y taking a certain value b_k, then $H(x|y)$ is the result of averaging $H(x\,|y = b_k)$ over all possible values b_k that y may take. The conditional entropy of x, $H(x\,|y)$, is a measure of average uncertainty.

$$H(x\,|y) = -\sum_i p(x, y) \lg p(x\,|y = b_k) \tag{2.144}$$

Mutual information of two random variables x and y can be expressed as

$$\begin{aligned} I(x, y) &= H(x) + H(y) - H(x, y) \\ &= H(x) - H(x|y) \end{aligned} \tag{2.145}$$

where $H(x)$ and $H(y)$ are marginal entropies, $H(x|y)$ and $H(y|x)$ are conditional entropies, and $H(x, y)$ is the joint entropy of x and y.

2.3.4 *Relative Entropy (Kullback–Leibler Divergence)*

Mutual information can be expressed not only as entropy, but also as a Kullback–Leibler (KL) divergence (also called KL-divergence or relative entropy). KL-divergence is defined between two n-dimensional probability density functions $p(x)$ and $q(x)$ as follows:

For distributions p and q of a continuous random variable, KL-divergence is defined to be the integral

$$\begin{aligned} KL(p, q) &= -\int p(x) \lg q(x)\mathrm{d}x - \left(-\int p\,(x) \lg p(x)\mathrm{d}x\right) \\ &= -\int p(x) \lg \left\{ \frac{q\,(x)}{p(x)} \right\} \mathrm{d}x \end{aligned} \tag{2.146}$$

The KL-divergence of probability distributions p and q of a discrete random variable is defined as

$$KL(p, q) = -\sum_i p(x) \lg \frac{q(x)}{p(x)} \tag{2.147}$$

KL-divergence can be considered as a type of distance between the two probability densities, because it is always nonnegative and zero if and only if the two distributions are equal. It should be noted that KL-divergence is not symmetric and is denoted as $KL(p, q) = KL(q, p)$.

For the definition of mutual information, the following relationship between the mutual information of random variables x and y and the KL-divergence of their probability density functions is denoted as

$$I(x, y) = KL\left(p\,(x, y)\,, p(x)p(y)\right) \tag{2.148}$$

If x and y are statistically independent, the mutual information satisfies the following inequality,

$$I(x, y) \geq 0 \tag{2.149}$$

In applications of signal blind source separation, it is often necessary to use differential entropy, mutual information, and KL-divergence and its properties to measure the independence of a signal.

2.3.5 *Important Inequality*

Let $\sum a_i$ and $\sum b_i$ be positive convergent sequences, and $\sum a_i \geq \sum b_i$. Then

$$\sum a_i \log \frac{b_i}{a_i} \leq 0 \tag{2.150}$$

Let $a_i \leq 1$ and $b_i \leq 1$. The inequality is measured for all value of i:

$$2\sum a_i \log \frac{b_i}{a_i} \geq \sum a_i(a_i - b_i)^2 \tag{2.151}$$

2.4 Distance Measure

2.4.1 *Geometric Distance Measure*

The distance measures between two n-dimensional vectors $\boldsymbol{x} = [x_1, x_2, \ldots, x_n]^{\mathrm{T}}$ and $\boldsymbol{y} = [y_1, y_2, \ldots, y_n]^{\mathrm{T}}$ are given below.

1. Minkowski distance
   The Minkowski distance of order p between $\boldsymbol{x}$ and $\boldsymbol{y}$ is defined as

$$d(\boldsymbol{x}, \boldsymbol{y}) = \left(\sum_{i=1}^{n} |x_i - y_i|^p \right)^{\frac{1}{p}}, p \geq 1 \tag{2.152}$$

2. Euclidean distance
   The Euclidean distance between $\boldsymbol{x}$ and $\boldsymbol{y}$ is defined as

$$d(\boldsymbol{x}, \boldsymbol{y}) = \sqrt{\sum_{i=1}^{n} (x_i - y_i)^2} \tag{2.153}$$

3. Manhattan distance
   The Manhattan distance L_1 is given by

$$d(\boldsymbol{x}, \boldsymbol{y}) = \sum_{i=1}^{n} |x_i - y_i| \tag{2.154}$$

4. Canberra distance
   The Canberra distance is defined as

$$d(\boldsymbol{x},\boldsymbol{y}) = \sum_{i=1}^{n} \frac{|x_i - y_i|}{x_i + y_i} \tag{2.155}$$

5. Chebyshev distance
   The Chebyshev distance is given by

$$d(\boldsymbol{x},\boldsymbol{y}) = \max_{i=1,2,\cdots,n} |x_i - y_i| \tag{2.156}$$

6. City Block distance
   Given the weight value $a_i \geq 0$, the City Block distance is defined as

$$d(\boldsymbol{x},\boldsymbol{y}) = \sum_{i=1}^{n} a_i |x_i - y_i| \tag{2.157}$$

7. Quadratic distance
   Given the symmetric positive definite weighted matrix $\boldsymbol{Q}$, the quadratic distance is defined as

$$d(\boldsymbol{x},\boldsymbol{y}) = (\boldsymbol{x} - \boldsymbol{y})^{\mathrm{T}} Q(\boldsymbol{x} - \boldsymbol{y}) \tag{2.158}$$

2.4.2 *The Distance between Datasets*

Let datasets $\boldsymbol{X}$ and $\boldsymbol{Y}$ belong to $\mathbf{R}^n$, and $d(x, y)$ denote the distance between the two datasets $\boldsymbol{X}$ and $\boldsymbol{Y}$ in $\mathbf{R}^n$. Some of the important distances between $\boldsymbol{X}$ and $\boldsymbol{Y}$ are given below.

Nearest neighbor distance

$$D(X, Y) = \min_{x\in X, y\in Y} d(x, y) \tag{2.159}$$

Farthest neighbor distance

$$D(X, Y) = \max_{x\in X, y\in Y} d(x, y) \tag{2.160}$$

Hausdorff distance

$$D(X, Y) = \max(h(X, Y), h(Y, X)) \tag{2.161}$$

where $h(X, Y) = \max_{x\in X} \min_{y\in Y} \|x - y\|$, $h(Y, X) = \max_{y\in Y} \min_{x\in X} \|y - x\|$, and the distance of $\boldsymbol{X}$ and $\boldsymbol{Y}$ is denoted as the norm $\|\cdot\|$.

2.4.3 Distance Measures

The distance between the sets of classes $\boldsymbol{X}$ and $\boldsymbol{Y}$ at point z in space is denoted as $S(\Omega/z)$, where $\Omega = \{X, Y\}$ is the sets of the classes and z is a point in space. The distance measure depends on the following:

1. The priori probabilities of the class sets $\boldsymbol{X}$ and $\boldsymbol{Y}$, are denoted as $P_r(X)$ and $P_r(Y)$, respectively.
2. The conditional probability density functions of the class sets, $p(z/X)$ and $p(z/Y)$.
3. Under condition z, the posterior probabilities of the class sets, $p(X/z)$ and $p(Y/z)$.
4. The joint probability density, $p(z) = P_r(X)p(z/X) + P_r(Y)p(z/Y)$.

Below we list some of the important distance measures.

1. Shannon entropy

$$S(\Omega/z) = E\{-P_r(X/z)\log P_r(X/z) - P_r(Y/z)\log P_r(Y/z)\} \tag{2.162}$$

2. Average quadratic entropy

$$S(\Omega/z) = E\{P_r(X/z)[1 - P_r(X/z)] + P_r(Y/z)[1 - P_r(Y/z)]\} \tag{2.163}$$

3. Bayesian distance

$$S(\Omega/z) = E\{P_r(X/z)^2 + P_r(Y/z)^2\} \tag{2.164}$$

4. Bhattacharyya coefficient

$$S(\Omega/z) = -\log(\rho) \tag{2.165}$$

where $\rho = \int [p(X/z)p(Y/z)]^{1/2}\mathrm{d}x$ and $0 < \rho < 1$.

5. Kullback–Leibler divergence

$$S(\Omega/z) = E\left\{\left[\frac{P_r(X/z)}{P_r(X)} - \frac{P_r(Y/z)}{P_r(Y)}\right]\log\left[\frac{P_r(Y)P_r(X/z)}{P_r(X)P_r(Y/z)}\right]\right\} \tag{2.166}$$

If $0 < s < 1$, the bound of Chernoff is given by

$$S(\Omega/z) = E\{P_r(X/z)^s P_r(Y/z)^{1-s}\} \tag{2.167}$$

If $s > 0$, the Kolmogorov variational distance is defined as

$$S(\Omega/z) = \frac{1}{2}E\{|P_r(X/z) - P_r(Y/z)|^s\} \tag{2.168}$$

6. Generalized Traubert distance measure

$$S(\Omega/z) = E\left\{\frac{P_r(X/z)^n + P_r(Y/z)^n}{P_r(X/z)^{n-1} + P_r(Y/z)^{n-1}}\right\} \tag{2.169}$$

2.5 Solvability of the Signal Blind Source Separation Problem

2.5.1 *Partitioned Matrix*

Let $N(i = 1, 2, \cdots, n)$ be the set of n natural numbers and $N_i(i = 1, 2, \cdots, r)$ be a subset of N, where $N_i \cap N_j = \emptyset\ (i \neq j)$ and $N_1 \cup N_2 \cup \cdots \cup N_r = N$. In other words, the set N is divided into $\boldsymbol{r}$ $(r \leq n)$ disjoint subsets. The $r \times n$ matrix $\boldsymbol{C} = (c_{ij})$ satisfying the following formula is called a partitioned matrix.

$$C_{ij} = \begin{cases} 1, j \in N_i \\ 0, j \notin N_i \end{cases} \tag{2.170}$$

Given r and n, the number of different partitioned matrices of matrix $\boldsymbol{C}$ is given by

$$\sum_{n_1+n_2+\ldots+n_r=n} \frac{n!}{n_1!n_3!\cdots n_r!} \tag{2.171}$$

The partitioned matrix $\boldsymbol{C}$ is row full rank if and only if all subsets of N $(N_i(i = 1, 2, \ldots, r))$ are not empty. A row full rank partitioned matrix consists of elements 0 or 1, with each column of the matrix having only one non-zero element, and each row having at least one non-zero element. If a subset N_k of N is an empty set, the elements in the kth row of the partitioned matrix $\boldsymbol{C}$ are zero. To make the discussion meaningful, we assume that the partitioned matrix $\boldsymbol{C}$ is row full rank.

Let $\boldsymbol{D}$ be a $n \times n$ full rank diagonal matrix given by $\boldsymbol{D} = diag\{d_{11}, d_{22}, \cdots, d_{nn}\}$ with $d_{ii} \neq 0, i = 1, 2, \ldots n$. Then, the generalized partitioned matrix $\boldsymbol{G}$, defined as $\boldsymbol{G} = \boldsymbol{CD}$, is an $r \times n$ matrix. The zero elements of the generalized partitioned matrix $\boldsymbol{G}$ are the same as those in the partitioned matrix $\boldsymbol{C}$, and the non-zero element of $\boldsymbol{G}$ can take any non-zero values(depending on matrix $\boldsymbol{D}$).

2.5.2 *Decomposability of Mixing Matrix*

If the mixed-separation matrix $\boldsymbol{C}$ in the formula $y(k) = \boldsymbol{C}s(k)$ is a partitioned matrix, then the formula can be written as

$$y_i = \sum_{j \in N_i} s_j, i = 1, 2, \cdots, r \tag{2.172}$$

In this way, the source signal is divided into r groups, and the sum of the source signals in each group is one of the outputs of the separation matrix $\boldsymbol{W}$.

If $r = n$, each subset of N contains one element only and each element of the subset is different. The partitioned matrix satisfying the above conditions is an elementary matrix in linear algebra (also called a permutation matrix). If and only if the partitioned matrix has only one non-zero element equal to 1 in each row and each column, and the other elements are zero, Equation 2.172 can be written as

$$y_i = s_j, i, j = 1, 2, \cdots, n \tag{2.173}$$

In this way, we can separate all the source signals using separation matrix $\boldsymbol{W}$, but we cannot obtain the original sequence of the source signal.

If the mixed-separation matrix $\boldsymbol{C}$ in the formula $y(k) = \boldsymbol{C}s(k)$ is a generalized partitioned matrix, Equation 2.172 can be written as

$$y_i = \sum_{j \in N_i} d_{jj} s_j,\ i = 1, 2, \cdots, r \tag{2.174}$$

In this way, the source signal is divided into r groups, and the linear combination of source signals in each group is one of the outputs of separation matrix $\boldsymbol{W}$.

If $r = n$, a generalized partitioned matrix is equal to the product of a permutation matrix and a full rank diagonal matrix, and each row and column in the generalized partition matrix has only one non-zero element. Equation 2.173 or 2.174 can be written as

$$y_i = d_{jj} s_j,\ \ i, j = 1, 2, \cdots, n \tag{2.175}$$

In this way, if the mixed-separation matrix is a generalized partitioned matrix, we can separate all the source signals using separation matrix $\boldsymbol{W}$, however, we cannot obtain the original sequence or the true amplitude of the source signal.

Further Reading

Faqi, Z., Bin, Z. and Xibin, Z. (2006) *Blind Signal Processing and Its Applications*, Xi'an Electronic Science and Technology University Press, Xi'an (in Chinese).

Ma, J., Niu, Y. and Chen, H. (2006) *Blind Signal Processing*, National Defence Industry Press, Beijing (in Chinese).

Shi, X. (2008) *Blind Signal Processing – Theory and Practice*, Shanghai Jiaotong University Press, Shanghai (in Chinese).

3

General Model and Classical Algorithm for BSS

3.1 Mathematical Model

Given that both the source signals and parameters of the transmission channel are unknown, blind source separation BSS is the process of recovering or separating the source signals from the observed signals through statistical properties of the source signals. The term "blind" has two meanings: first, the source signals cannot be observed, and second, how the source signals have been mixed is unknown. Obviously, if it is difficult to build the transmission model between the source signals and sensors, or we are unable to obtain prior knowledge about the transmission, BSS is the natural choice. To address the different mixing styles of the source signals, the BSS problem can be categorized as an instantaneous linear mixing model, linear convolution model, or nonlinear mixing model.

3.1.1 Linear Mixing Models

See Figure 3.1 and the references Comon, 1994; Hyvärinen, Karhunen, and Oja, 2001.

3.1.1.1 Instantaneous Linear Mixing

Assume that there are N statistically independent sources passing through an instantaneous linear mixing system and received by M sensors . Then, each observed signal is a linear combination of these N sources. Thus, the following equation governs the linear time-invariant instantaneous mixing function.

$$x_j(t) = \sum_{i=1}^{N} a_{ji} s_i(t) \tag{3.1}$$

Blind Source Separation: Theory and Applications, First Edition. Xianchuan Yu, Dan Hu and Jindong Xu.

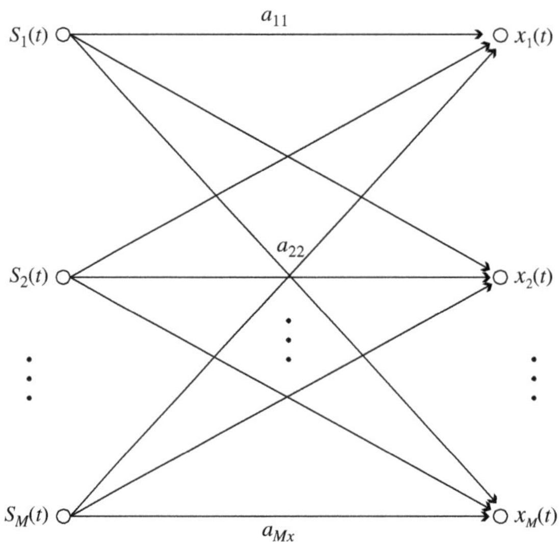

Figure 3.1 Linear mixing model

where $a_{ji}(i \in \{1,2,\ldots,N\}, j \in \{1,2,\ldots,M\})$ are mixing parameters, $s_i(t)$ $(i \in \{1,2,\ldots,N\})$ are the source signals, and $x_j(t)$ $(j \in \{1,2,\ldots,M\})$ are the observed signals.

It is easier to use the vector form to express the mixing process: $\boldsymbol{s}(t) = [s_1(t), \ldots, s_N(t)]^{\mathrm{T}}$, $\boldsymbol{x}(t) = [x_1(t), \ldots, x_N(t)]^{\mathrm{T}}$. Then, Equation 3.1 can be written as

$$\boldsymbol{x}(t) = \boldsymbol{A}\boldsymbol{s}(t) \tag{3.2}$$

where $\boldsymbol{A} \in \mathbf{R}^{m \times n}$ is the mixing matrix and $\boldsymbol{s}(t) \in \mathbf{R}^{m \times n}$ is the source vector, whose elements are the foundation coefficients of $\boldsymbol{A}$.

Because the transmission channel and sensor array could include additive noise that should be taken into account, Equation 3.2 becomes:

$$\boldsymbol{x}(t) = \boldsymbol{A}\boldsymbol{s}(t) + \boldsymbol{n}(t) \tag{3.3}$$

where $\boldsymbol{n}(t) = [n_1(t), n_2(t), \ldots, n_M(t)]^{\mathrm{T}}$ are additive noise vectors.

3.1.1.2 Linear Convolution Mixing Model

The convolution mixing model is closer to an actual mixing model. Assume that there are N statistically independent sources $\boldsymbol{s}_i(t), i = 1,2,\ldots,N$ received by M sensors after the convolution mixing process (Amari et al., 1997). The mixed signals are denoted by $\boldsymbol{x}_i(t), j = 1,2,\ldots,M$, and the convolution mixing model can be expressed as

$$\boldsymbol{x}_j(t) = \sum_{i=1}^{N} a_{ji}(t) * \boldsymbol{s}_i(t) = \sum_{i=1}^{N}\sum_{\tau=0}^{L-1} a_{ji}(\boldsymbol{\tau})\boldsymbol{s}_i(t-\tau) \tag{3.4}$$

where * denotes the convolution operation and $a_{ji}(\tau)$ is the impulse response of the jth sensor about the ith source.

Since each channel can be denoted by an Lth-order finite impulse response (FIR) filter, the convolution mixing system can be represented in the FIR matrix form presented by Lambert as

$$\boldsymbol{x} = \boldsymbol{A}\boldsymbol{s} \tag{3.5}$$

where $\boldsymbol{A}$ is an FIR matrix of the form

$$\boldsymbol{A} = \begin{pmatrix} a_{11}{}^{\mathrm{T}} & \dots & a_{1n}{}^{\mathrm{T}} \\ \vdots & \ddots & \vdots \\ a_{n1}{}^{\mathrm{T}} & \dots & a_{nn}{}^{\mathrm{T}} \end{pmatrix} \tag{3.6}$$

where a_{ji} is a column with dimension L, which denotes the Lth-order FIR filter.

The vector form of Equation 3.5 is

$$\boldsymbol{x}(t) = \sum_{\tau=0}^{L-1} \boldsymbol{A}(\tau)\boldsymbol{s}(t-\tau) \tag{3.7}$$

where $\mathbf{s}(t) = [\boldsymbol{s}_1(t), \boldsymbol{s}_2(t), \dots, \boldsymbol{s}_N(t)]^{\mathrm{T}}, \boldsymbol{x}(t) = [\boldsymbol{x}_1(t), \boldsymbol{x}_2(t), \dots, \boldsymbol{x}_M(t)]^{\mathrm{T}}$, and $\boldsymbol{A}$ is the mixing matrix. For $L = 1$, the convolution model degenerates to an instantaneous mixing model.

3.1.2 Nonlinear Mixing Model

In reality, the observed signals are mostly mixed in a nonlinear way. Solving the BSS problem for nonlinear mixed signals is usually difficult, requiring additional prior information or appropriate constraints. The nonlinear mixing model is an extension of the linear model (Hyvärinen, Karhunen, and Oja, 2001; Tan and Wang, 2001; Woo and Sali, 2002).

$$\boldsymbol{x}(t) = f(\boldsymbol{s}(t)) + \boldsymbol{n}(t) \tag{3.8}$$

where $\boldsymbol{x}(t)$ are M-dimensional observed signal vectors, $\boldsymbol{s}(t)$ are N-dimensional unknown source signal vectors, and $\boldsymbol{n}(t)$ are M-dimensional additive noise vectors that are independent of the source signals. $f : \mathbf{R}^N \to \mathbf{R}^M$ is the unknown reversible nonlinear mixing function.

Based on the observed signals, nonlinear BSS attempts to find a group mapping: $g : \mathbf{R}^N \to \mathbf{R}^M$, which is used to recover the source signals $\boldsymbol{s}(t)$. The mathematical model is given as

$$\boldsymbol{y}(t) = g(\boldsymbol{x}(t)) \tag{3.9}$$

where $\boldsymbol{y}(t)$ is the N-dimensional output.

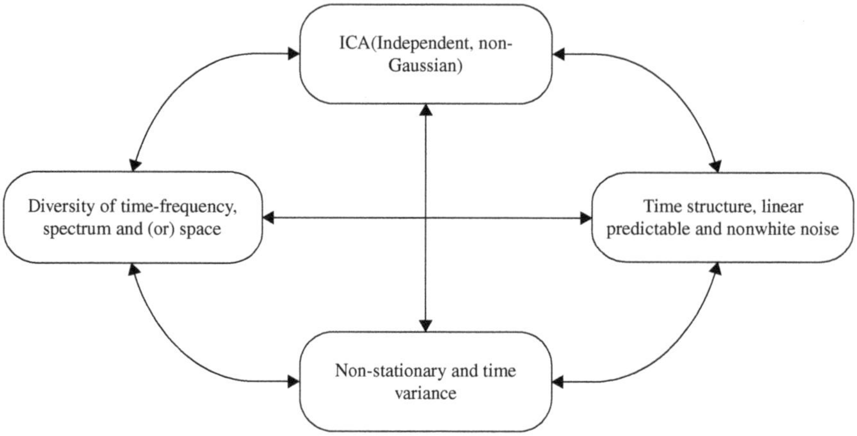

Figure 3.2 Typical BSS methods using some prior knowledge

3.2 BSS Algorithm

There are many BSS algorithms which separate mixtures based on second-order statistics (SOS), higher-order statistics, non-stationarity, or diversity of signals (Cichocki and Amari, 2002; Choi et al., 2005). However, all types of BSS algorithms can be categorized according to the methods illustrated in Figure 3.2.

Next, we introduce some of the BSS algorithms from a mathematical modeling perspective.

3.2.1 BSS Algorithm for Instantaneous Linear Mixing

The instantaneous linear mixing style is the simplest and forms the foundation for the other two mixing styles. The theoretical basis of this algorithm is summarized by the following four methods (Chen and Huang, 2008).

1. Cost function
   The most universal method involves selecting a cost function to measure the independence and non-Gaussianity (or sparseness) of the signals. By assuming that the source signals are independent without a time structure, the fundamental method for solving BSS problems is to use higher-order statistics. However, this method is not suitable for more than one Gaussian source.
2. Second-order statistics
   Using SOS one does not need to consider the Gaussian property of the source signals; that is, probability density function (PDF) styles of the signals and their corresponding kernel functions are not considered. The method utilizes SOS and the time sequence structure blindly to separate mixtures, with the result that the required computation is minimal and it is suitable for engineering data analysis.

Experimental results show that, even in the case of a noisy environment, SOS can separate signals well. If the source signals have time sequence structures and non-zero time sequence related data, which reduce the independence constraints, SOS is able to estimate the mixing matrix and source signals. However, SOS cannot extract signals that are not independent or have the same power spectrum (Belouchrani *et al.*, 1997).

3. Non-stationary (NS) with SOS
   For source signals that change with time, the NS method can be used. Compared with other methods, NS can separate colored Gaussian sources with the same power spectrum, but cannot separate signals with the same NS properties.
4. Diversity of signals
   This method exploits the diversity of signals, such as time, frequency, or time–frequency variations and, more generally, ordinarily unites the spatial–time –frequency variations.

Regarding the methods discussed above, independent component analysis (ICA) belongs to the first class and is an important method for solving BSS problems. A detailed discussion of the theory and application thereof is presented in Chapter 5. To separate or extract source signals with different statistical properties and decrease the influence of noise and other interference, we can integrate several methods to obtain a more complex or more advanced method to solve BSS problems.

3.2.2 Nonlinear Principle Component Analysis (PCA)

3.2.2.1 Introduction

By importing higher-order statistics into principle component analysis (PCA), we can realize signal separation; this is called nonlinear PCA (Oja, 1997). Before separation, we need to whiten the data:

$$\ddot{\boldsymbol{x}}(t) = \boldsymbol{Q}\boldsymbol{x}(t) \tag{3.10}$$

where $\boldsymbol{Q}$ is the whitening matrix making $R_{\ddot{x}} = E\{\ddot{\boldsymbol{x}}\ddot{\boldsymbol{x}}^{\mathrm{T}}\} = \boldsymbol{I}$.

A typical objective function for nonlinear PCA is defined as

$$J(\boldsymbol{W}) = E\{\|\boldsymbol{x} - \boldsymbol{W}f(\boldsymbol{W}^{\mathrm{T}}\boldsymbol{x})\|^2\} \tag{3.11}$$

where $f(\cdot)$ is a nonlinear function. Using the stochastic gradient algorithm, we obtain the updated formula:

$$\boldsymbol{W}(t+1) = \boldsymbol{W}(t) + \boldsymbol{\mu}(t)[\boldsymbol{x}(t)e^{\mathrm{T}}(t)\boldsymbol{W}(t)F(\boldsymbol{x}(t)^{\mathrm{T}}\boldsymbol{W}(t)) + \boldsymbol{e}(t)f(\boldsymbol{x}(t)^{\mathrm{T}}\boldsymbol{W}(t))] \tag{3.12}$$

where

$$\boldsymbol{e}(t) = \boldsymbol{x}(t) - \boldsymbol{W}(t)f(\boldsymbol{W}(t)^{\mathrm{T}}\boldsymbol{x}(t)) \tag{3.13}$$

$$F(\boldsymbol{x}(t)^{\mathrm{T}}\boldsymbol{W}(t)) = diag\left[f\left(\boldsymbol{x}(t)^{\mathrm{T}}\boldsymbol{w}_1(t)\right), \cdots, f\left(\boldsymbol{x}(t)^{\mathrm{T}}\boldsymbol{w}_M(t)\right)\right] \tag{3.14}$$

Under the premise of preserving more information of the original variables, the purpose of PCA is to obtain a linear combination of a group of random variables with zero mean. PCA only uses SOS and the output signals are uncorrelated with each other, with the result that it does not separate signals very well. On the other hand, it is easy for nonlinear PCA to separate signals in an engineering environment, since higher-order statistics are incorporated in the calculation in a cryptic way.

3.2.2.2 Kernel PCA

Kernel PCA is presented as a nonlinear PCA method based on a support vector machine (SVM). The basic idea involves projecting the observed data, which are nonlinear varying, in a linear higher dimensional space and finding a calculable solution. Thus, kernel PCA realizes nonlinear PCA and skillfully applies kernel techniques to avoid solving the nonlinear thorny problem (Shi, 2008).

3.2.3 Separation Algorithm for Nonlinear Mixing

3.2.3.1 Separation of Post Nonlinear (PNL) Mixtures

Separation of post nonlinear (PNL) mixtures was first presented by Taleb, Jutten, and Olympieff (1998) and Taleb and Jutten (1999). The separation approaches usually include two steps (Ma, Niu, and Chen, 2006):

1. Nonlinear stage, in which a nonlinear rectification function $g_i(x_i, \theta_i)$ is used to eliminate the influence of the nonlinear function $f_i(i, \cdots, n)$ and obtain approximate linear mixtures z.
2. Linear stage, in which the approximate linear mixtures z are separated. By using a trained separation matrix $\boldsymbol{W}$ and $\boldsymbol{y} = \boldsymbol{W}z$ to make every component independent, $\boldsymbol{y}$ is obtained as the estimation of the source signal $\boldsymbol{s}$.

Taleb and Jutten (1999) used mutual information $I(\boldsymbol{y})$ of $\boldsymbol{y}(y_1, \ldots, y_n)$ as the cost function and independence criterion; that is, in the linear stage, according to the minimization of mutual information, the following holds:

$$\frac{\partial I(\boldsymbol{y})}{\partial \boldsymbol{W}} = -E\{\boldsymbol{\psi}\boldsymbol{x}^{\mathrm{T}}\} - (\boldsymbol{W}^{\mathrm{T}})^{-1} \tag{3.15}$$

where ψ_i in $\boldsymbol{\psi}(\psi_1, \ldots, \psi_n)$ is the score function of y_i in $\boldsymbol{y}(y_1, \ldots, y_n)$. That is,

$$\boldsymbol{\psi}_i(y_i) = \frac{d}{\mathrm{d}y_i} \log p_i(y_i) = \frac{p_i'(y_i)}{p_i(y_i)} \tag{3.16}$$

where $p_i(y_i)$ is the PDF of y_i and $p_i'(y_i)$ is the derivative of y_i.

In the nonlinear stage, the gradient learning rule is given as

$$\frac{\partial I(\mathbf{y})}{\partial \theta_k} = -E\left\{\frac{\partial \log \left|g_k'\right| (x_k, \theta_k)}{\partial \theta_k}\right\} - E\left\{\sum_{i=1}^{n} \psi_i (y_i) w_{ik} \frac{\partial g_k(x_k, \theta_k)}{\partial \theta_k}\right\} \tag{3.17}$$

where x_k is the kth component of the observed vector $\boldsymbol{x}$, w_{ik} is the element of the separation matrix $\boldsymbol{W}$ in the ith row and kth column, and g_k' is the derivative of the nonlinear function g_k.

Similar to linear BSS, the separation performance of nonlinear BSS mainly relies on the estimation accuracy of the score function, which should be self-adapting with the estimation based on the output vector $\boldsymbol{y}$.

3.2.3.2 General Separation Methods for Nonlinear Mixtures

1. Variational Bayesian inference
   Bayesian inference methods are widely used in the fields of neural networks and statistical information processing. If the assumed model is correct, using this method yields good performance. Since Bayesian inference has the ability to obtain the optimal solution of the model, it is more robust if prior knowledge is used. The major drawback of this method is the computational complexity and it is sometimes necessary to use approximate computing, which increases the number of unknown estimated parameters in the learning process. This hinders the actually application of Bayesian inference in unsupervised or blind learning (Ma *et al.*, 2006).
   Initially, the variational Bayesian method was applied in linear BSS/ICA to process the blind property of various linear models. Later, however, Valpola and Karhunen (2002) presented a new method, based on Bayesian integrated learning, to process nonlinear BSS/ICA problems.
2. Self-organizing map (SOM)
   A SOM was first applied in a nonlinear problem to extract nonlinear features from the mixed data (Pajunen, Hyvärinen, and Karhunen, 1996). Although this method has high flexibility, the complexity of its network structure grows exponentially.
3. Genetic algorithms (GAs)
   BSS/ICA requires a globally optimal solution; however, to a lesser or greater extent, most algorithms fall into local minima, which restricts the actual application of nonlinear BSS/ICA. To improve the performance of parameter estimation of nonlinear functions and to avoid the large number of local minima problems in the learning process, Rojas *et al.* (2001) incorporated GAs into BSS. Compared with the conventional method, the GA starts from a global point and, by successfully avoiding local minima, finds a globally optimal solution.

3.2.3.3 Nonlinear ICA Neural Networks

1. Multilayer perceptron (MLP) model
   Initially, the two level perceptron network had a constant gradient, but in the subsequent forward multilayer perceptron (FMLP) network, the gradient is variable. The objective of the FMLP is to create an inverse system with an optimized cost function, and then to obtain a group of mutually independent signals. Inverse and mixing systems have a similar structure, in which the observed and estimated signals are seen as input and output, respectively.
   The basic idea of a nonlinear ICA neural network is to use an MLP to build a model of the nonlinear projection $f(\cdot)$. As long as there are enough hidden neurons, the network can realize any precise nonlinear mapping (Shi, 2008).
2. Radial basis function (RBF) network
   Based on the PNL network model, Tan, Wang, and Zurada (2001) presented the RBF algorithm. The novelty of this algorithm is that it uses an RBF architecture and new contrast function, which includes mutual information and the cumulant, and can decrease the uncertainty resulting from nonlinearity. The RBF network parameters are obtained by stochastic gradient descent and unsupervised learning approaches. For the local response characteristics of the RBF neural network, this model offers a fast learning convergence rate, neural unsupervised learning, and modularized network structure features. As such, this method is suitable for fast, time varying and multi-channel signal separation.
3. Polynomial neural network (PNN)
   Aimed at a single linear mixing model, Woo and Khor (2004) presented a kind of PNN unmixing machine, which uses a group of nonlinear polynomials in implicit strata. The self-adjusting nonlinear activation function of hidden neurons in the PNN is different from those in previous MLP and RBF networks. Because the fixed nonlinearity leads to an overly optimized network, communication complexity is increased and independent outputs are arbitrarily produced. Thus, a flexible nonlinear activation function must be used.

3.2.4 Fashionable BSS

3.2.4.1 Sparse Component Analysis (SCA)

If the signals are sparse, we can use a small or sparse base subset to represent the big data of interest. Li *et al.* (2003) and Li, Cichocki, and Amari (2004) presented a BSS method based on sparse component analysis (SCA) and sparse signal representation (SSR). This method is mainly used to solve problems involving EG/MEG (magnetoencephalography) signal processing, feature extraction, wavelet denoising, speech coding, spectrum estimation, and disease diagnosis, amongst others (Araki *et al.*, 2003; Choi *et al.*, 2005). Our investigation of the theory and application of SCA is presented in Chapter 16.

3.2.4.2 Nonnegative Matrix Factorization (NMF)

Paatero and Tapper (1994) first introduced nonnegative matrix factorization (NMF), which is sometimes called PMF (positive matrix factorization). When applied to BSS problems, NMF does not consider sparsity and independence, but obtains sparse factorized results. NMF is widely used in processing matrices when the signals have a clear physical significance or have been normalized, such as in the environmental sciences and for mixed pixel factorization (Lee and Seung, 1999, 2001). Chapter 15 focuses on NMF.

References

Amari, S., Douglas, S.C., Cichocki, A. *et al.* (1997) Multichannel blind deconvolution and equalization using the natural gradient. The 1st Signal Processing Workshop on Signal Processing Advances in Wireless Communications, Paris, France, pp. 101–104.

Araki, S., Makino, S., Mukai, R. *et al.* (2003) Equivalence between frequency domain blind source separation and frequency domain adaptive beamforming for convolutive mixtures. *EURASIP Journal on Applied Signal Processing*, **11**(1), 1157–1166.

Belouchrani, A., Meraim, K.A., Cardoso, J. and Moulines, F.E. (1997) A blind source separation technique based on second order statistics. *IEEE Transactions on Signal Processing*, **45**(2), 434–444.

Chen, X.M. and Huang, S.Y. (2008) Blind source separation: problem, principle and method. *Electronic Information Warfare Technology*, **23**(2), 1–5.

Choi, S., Cichocki, A. *et al.* (2005) Blind source separation and independent component analysis: a review. *Neural Information Processing – Letters and Reviews*, **6**(1), 1–57.

Cichocki, A. and Amari, S. (2002) *Adaptive Blind Signal and Image Processing: Learning Algorithms and Applications*, John Wiley & Sons, Inc., New York.

Comon, P. (1994) Independent component analysis, a new concept. *Signal Processing*, **36**(3), 287–314.

Hyvärinen, A., Karhunen, J. and Oja, E. (2001) *Independent Component Analysis*, John Wiley & Sons, Inc., New York.

Lee, D. and Seung, H.S. (1999) Learning the parts of objects by non-negative matrix factorization. *Nature*, **401**(10), 788–791.

Lee, D. and Seung, H.S. (2001) Algorithms for non-negative matrix factorization. *Advances in Neural Information Processing Systems*, **13**(1), 556–562.

Li, Y., Cichocki, A. and Amari, S. (2004) Analysis of sparse representation and blind source separation. *Neural Computation*, **16**(6), 1193–1204.

Li, Y., Cichocki, A., Amari, S. *et al.* (2003) Sparse representation and its applications in blind source separation. 17th Annual Conference on Neural Information Processing Systems, pp. 241–248.

Ma, J.C., Niu, Y.L. and Chen, H.Y. (2006) *Blind Signals Processing*, National Defense Industry Press, Beijing (in Chinese).

Oja, E. (1997) The nonlinear PCA learning rule and signal separation-mathematical analysis. *Neurocomputing*, **17**(3), 25–45.

Paatero, P. and Tapper, U. (1994) Positive matrix factorization: a non-negative factor model with optimal utilization of error estimates of data values. *Environmetrics*, **5**(2), 111–126.

Pajunen, P., Hyvärinen, A. and Karhunen, J. (1996) Nonlinear blind source separation by self-organizing maps, in *Proceedings of the International Conference on Neural Information Processing*, Hong Kong, China, Springer-Verlag, New York, pp. 1207–1210.

Rojas, F., Rojas, I., Clemente, R. *et al.* (2001) Nonlinear blind source separation using genetic algorithm. Proceedings of Independent Component Analysis and Signal Separation, pp. 400–405.

Shi, X.Z. (2008) *Blind Signals Processing – Theory and Practice*, Shanghai Jiaotong University Press, Shanghai (in Chinese).

Taleb, A. and Jutten, C. (1999) Batch algorithm for source separation in post nonlinear mixtures. Proceedings of ICA, pp. 155–160.

Taleb, A., Jutten, C., and Olympieff, S. (1998) Source separation in post nonlinear mixtures: an entropy-based algorithm. Proceedings of the European Symposium on Artificial Neural Networks, pp. 2089–2092.

Tan, Y. and Wang, J. (2001) Nonlinear blind source separation using higher order statistics and a genetic algorithm. *IEEE Transactions on Evolutionary Computation*, **5**(6), 600–612.

Tan, Y., Wang, J. and Zurada, J.M. (2001) Nonlinear blind source separation using a radial basis function network. *IEEE Transactions on Neural Networks*, **12**(1), 134–144.

Valpola, H. and Karhunen, J. (2002) An unsupervised ensemble learning method for nonlinear dynamic state-space models. *Neural Computation*, **14**(11), 2647–2692.

Woo, W.L. and Khor, L.C. (2004) Blind restoration of nonlinearly mixed signals using multilayer polynomial neural network. *IEEE Proceedings-Vision Image and Signal Processing*, **151**(1), 51–61.

Woo, W.L. and Sali, S. (2002) General multilayer perceptron demixer scheme for nonlinear blind signal separation. *IEEE Proceedings-Vision Image and Signal Processing*, **149**(5), 253–262.

4

Evaluation Criteria for the BSS Algorithm

4.1 Evaluation Criteria for Objective Functions

The independence criterion of blind source separation (BSS) based on independent component analysis (ICA) is applied to joint probability density function (PDF) factorization. Researchers have presented various criteria, such as information theory, system theory, and neural network coding, amongst others. The central limit theorem proved that, generally, when the number of independent random variables increases, their sum tends towards a Gaussian distribution. So, non-Gaussianity could be used to estimate independence. The common measures of non-Gaussianity are kurtosis (the fourth-order cumulant) and negentropy.

For zero-mean real signals, kurtosis is just fourth-order statistics. The classic definition is given as (Hyvärinen and Oja, 2000)

$$\mathrm{kurt}(\boldsymbol{y}) = E\{\boldsymbol{y}^4\} - 3(E\{\boldsymbol{y}^2\})^2 \tag{4.1}$$

The normalization of Equation 4.1 is

$$\mathrm{kurt}(\boldsymbol{y}) = E\{\boldsymbol{y}^4\}/(E\{\boldsymbol{y}^2\})^2 - 3 \tag{4.2}$$

When the variance is 1, Equation 4.2 can be written as

$$\mathrm{kurt}(\boldsymbol{y}) = E\{\boldsymbol{y}^4\} - 3 \tag{4.3}$$

This illustrates that kurtosis can be formed as a standard fourth moment. The fourth moment of a Gaussian random variable is equal to $3(E\{\boldsymbol{y}^2\})^2$, and thus its kurtosis is zero.

Kurtosis as a measure of statistical independence is widely used in ICA. Its outstanding advantage can easily be realized. However, kurtosis is sensitive to singular values and is not a robust measure of non-Gaussianity.

Other measures of non-Gaussianity may be better than kurtosis in certain situations. Negentropy is the second important measure of non-Gaussianity, and thus various

Blind Source Separation: Theory and Applications, First Edition. Xianchuan Yu, Dan Hu and Jindong Xu.

objectives, such as mutual information minimization, negentropy maximization, and maximum likelihood, among others, are considered. These are introduced in the following subsections.

4.1.1 Mutual Information Minimization

Given $y(t) = \boldsymbol{W}x(t)$, assume that the PDF of $\boldsymbol{y}$ is $p_y(\boldsymbol{y})$ and every element's PDF is $p_{y_i}(y_i)$. The Kullback–Leibler (KL) divergence between $p_y(\boldsymbol{y})$ and $\prod_{i=1}^{N} p_{y_i}(y_i)$ can be used as the measure of independence between each element. Here, KL divergence is called mutual information $I(p_y(\boldsymbol{y}))$, that is,

$$\begin{aligned} I(p_{\mathbf{y}}(\boldsymbol{y})) &= KL\left[p_{\mathbf{y}}(\boldsymbol{y}), \prod_{i=1}^{N} p_{y_i}(y_i)\right] \\ &= \int_{\mathbf{y}} p_{\mathbf{y}}(\boldsymbol{y}) \lg \left[\frac{p_{\mathbf{y}}(\boldsymbol{y})}{\prod_{i=1}^{N} p_{y_i}(y_i)}\right] \mathrm{d}\boldsymbol{y} \end{aligned} \tag{4.4}$$

Obviously, if and only if $p_{\mathbf{y}}(\boldsymbol{y}) = \prod_{i=1}^{N} p_{y_i}(y_i)$, the mutual information is zero and every element is statistically independent.

For the invertible linear transformation $\boldsymbol{y}(t) = \boldsymbol{Wx}(t)$, we have another expression of mutual information (Basseville, 1989; Xie and Zhang, 2002):

$$I(p_{\mathbf{y}}(\boldsymbol{y})) = -H(\boldsymbol{x}) - \lg|\det(\boldsymbol{W})| + \sum_{i=1}^{N} H_i(y_i) \tag{4.5}$$

where $H_i(y_i) = -\int_{y} p_{\mathbf{y}}(\boldsymbol{y}) \lg[p_i(y_i)]\mathrm{d}\boldsymbol{y}$.

4.1.2 Negentropy Maximization

Negentropy is an important measure of non-Gaussianity. Under the conditions of a high noise-to-signal ratio, mutual information maximization between input and output means that the information redundancy between input and output is at a minimum. Then, the mutual information of the outputs is also at a minimum; thus, entropy is at a maximum and the outputs are statistically independent.

Assume input $\boldsymbol{x} = (x_1, x_2, \ldots, x_N)^{\mathrm{T}}$, the PDF thereof p_1, and p_g, which denotes a Gaussian PDF with the same mean and variance. Using KL divergence, one of the negentropy definitions is given as

$$J(\boldsymbol{x}) = KL[p_1(\boldsymbol{x}), p_g(\boldsymbol{x})] = \int_x p_1(\boldsymbol{x}) \lg \left[\frac{p_1(\boldsymbol{x})}{p_g(\boldsymbol{x})}\right] \mathrm{d}\boldsymbol{x} \tag{4.6}$$

Thus, we can see the similarity between $p_1(\boldsymbol{x})$ and the Gaussian PDF. Negentropy, which is always nonnegative, is zero when $\boldsymbol{x}$ is a Gaussian distribution. So we can separate signals using maximum negentropy.

Negentropy is the best criterion for non-Gaussianity. However, it needs to compute the PDF and has high computational complexity; thus an approximate calculation of negentropy is needed.

4.1.3 Maximum Likelihood

The aim of BSS is to estimate the separation matrix $\boldsymbol{W}$ to calculate $\boldsymbol{WA} = \boldsymbol{I}$, and then obtain $\boldsymbol{y}(t) = \boldsymbol{s}(t)$ to realize signal separation. The objective of maximum likelihood estimation is to estimate the real PDF of the source signals via the observed data samples. This has the advantage of uniformity, minimum variance, and a global optimum. However, prior knowledge of the PDF of the input signals is needed. The existing method uses non-parameter estimation of the kernel function to identify the PDF of the signals.

$$\boldsymbol{x}(t) = \boldsymbol{As}(t) = \boldsymbol{W}^{-1}\boldsymbol{s}(t) \tag{4.7}$$

If the PDF of the signal source is $p_s(\boldsymbol{s}) = \prod_{i=1}^{N} p_{s_i}(s_i)$, the likelihood PDF $\widehat{p}_x(x)$ of $\boldsymbol{x}(t)$ is

$$\widehat{p}_{\mathbf{x}}(\boldsymbol{x}) = |\det \boldsymbol{W}| p_s(\boldsymbol{s})|_{s=\boldsymbol{W}x} = |\det \boldsymbol{W}| \prod_{i=1}^{N} p_{s_i}(s_i)|_{s=\boldsymbol{W}x} \tag{4.8}$$

$$L_{ML} = \int_x p_x(\boldsymbol{x}) \lg \widehat{p}_x(\boldsymbol{x}) \mathrm{d}\boldsymbol{x} \tag{4.9}$$

On the other hand, we have another definition of maximum likelihood that uses minimization of the KL divergence.

$$\begin{aligned} L_{KL} &= \mathrm{KL}(\boldsymbol{W}) = \mathrm{KL}(p_x(\boldsymbol{x}), \widehat{p}_x(\boldsymbol{x})) \\ &= -\int_x p_x(\boldsymbol{x}) \lg \frac{\widehat{p}_x(\boldsymbol{x})}{p_x(\boldsymbol{x})} \mathrm{d}\boldsymbol{x} = -H(\boldsymbol{x}) - L_{ML} \end{aligned} \tag{4.10}$$

where $p_x(\boldsymbol{x})$ is the real PDF of $\boldsymbol{x}$. If the KL divergence is smaller, $p_x(\boldsymbol{x})$ and $\widehat{p}_x(\boldsymbol{x})$ are more similar.

When using maximum likelihood as the objective function, we first need to know the PDF of the signal source to compute the likelihood function of $\boldsymbol{x}(t)$. If this is not known, it should be assumed or confirmed during the learning process.

4.1.4 *Information Maximization*

Information maximization is just entropy maximization and, therefore, it is also called maximum entropy. Entropy is the measure of uncertainty of random variables. The higher the statistical independence of each element of $\boldsymbol{y}$, the greater the entropy. Thus, entropy can be used as the measure of statistical independence of each element.

Easy entropy maximization is to make the output entropy divergence infinite. So, the entropy of $z = g(\boldsymbol{y}) = g(\boldsymbol{Wx})$ passing through a nonlinear node is the greatest and the objective function is defined as

$$L_H(\boldsymbol{W}) = H(z) = H(g(\boldsymbol{y})) \tag{4.11}$$

where g is a nonlinear function.

The above four objective functions all aim to obtain statistically independent outputs from the perspective of information theory and they are equivalent under certain conditions. Lee, Girolami, and Bell (2000) unified minimum mutual information, information maximization, and maximum likelihood in the information theory framework.

4.1.5 *Permanent Model Objective Function*

Using the permanent model property of binary phase-shift keying (BPSK) and quadrature phase-shift keying (QPSK) modulation signals in a communication system, the following objective function can be used in BSS (Zhang and Bao, 2000).

$$J(y, W) = -\sum_{i=1}^{n} E\left\{\left[|y_i|^2 - 1\right]^2\right\} - \rho \sum_{i=1}^{n} \sum_{j=1, j \neq i}^{n} |E\{y_i \cdot y_j^*\}|^2 \tag{4.12}$$

where ρ is a positive real constant, whose function is to adjust the proportion between the first part (permutation model cost function) and second part (decorrelation cost function). The superscript * denotes the complex conjugate operation, while $E\{\cdot\}$ denotes the calculation of mathematical expectation of the variables.

4.1.6 *Higher-Order Cumulant Objective Function*

$$J(y, W) = \sum_{i=0}^{n} [\text{cum}_4(y_i)]^2 \tag{4.13}$$

where $\text{cum}_4(y_i)$ is a fourth-order cumulant. Moreau *et al.* proved that the square operation in the above equation can be simplified as an absolute value operation (as shown in Equation 4.14). Thus, a higher-order cumulant objective function can be obtained.

$$J(y, W) = \sum_{i=0}^{n} |\text{cum}_4(y_i)| \tag{4.14}$$

If all the source signals' fourth-order cumulants have the same sign $\varepsilon = \text{sgn}[\text{cum}_4(s_i)]$, the higher-order objective function can be obtained by simplifying the contrast function.

4.2 Evaluation Criteria for Correlations

The correlation coefficient between the output signal y_i and source signal s_j is used as the evaluation criterion, the definition of which is given as

$$\xi_{ij} = \xi(y_i, s_j) = \frac{\left|\sum_{i=1}^{N} y_i(t)\, s_j(t)\right|}{\sqrt{\sum_{i=1}^{N} y_i^2(t) \sum_{j=1}^{M} s_j^2(t)}} \tag{4.15}$$

If $\xi_{ij} = 1$, it means that output signal y_i is the same as the source signal s_j up to scale. If ξ_{ij} is close to 0 or far from 1, it means that the separation is not complete. If the source signals do not satisfy the independence conditions, a correlation criterion is usually used, such as sparse component analysis (SCA).

4.3 Evaluation Criteria for Signal-to-Noise Ratio

Peak signal-to-noise-ratio (PSNR) is usually used as the criterion for evaluating separation performance, and is calculated as

$$PSNR = 10 \times \lg\left(\frac{MN \max_{m,n} s_i(m,n)^2}{\sum_{m=1}^{M}\sum_{n=1}^{N}(s_i(m,n) - y_i(m,n))^2}\right) \tag{4.16}$$

where $s_i(m, n)$ and $y_i(m, n)$, respectively, denote the gray values of images s_i and y_i in (m, n), and M and N are the sizes of the images.

In addition, the number of iterations can also be used to measure the performance of the algorithm.

References

Basseville, M. (1989) Distance measures for signal processing and pattern recognition. *Signal Processing*, **18**(4), 349–369.

Hyvärinen, A., and Oja, E. (2000) Independent component analysis: algorithms and applications. *Neural Networks*, **13**(4), 411–430.

Lee, T.W., Girolami, M., and Bell, A.J. (2000) A unifying information-theoretic framework for independent component analysis. *Computers and Mathematics with Applications*, **31**(11), 1–21.

Xie, S.L., and Zhang, J.L. (2002) Blind separation algorithm of minimal mutual information based on rotating transform. *Acta Electronica Sinica*, **30**(5), 628–631, (in Chinese).

Zhang, X.D., and Bao, Z. (2000) *Communication Signal Processing*, Beijing: National Defense Industry Press, pp. 383–384, (in Chinese).

Part II

Independent Component Analysis and Applications

5

Independent Component Analysis

5.1 History of ICA

ICA (independent component analysis) is a BSS (blind source separation) method that finds statistically independent and non-Gaussian factors or components from multivariate data. The research target of this method is a non-Gaussian signal.

The development of ICA theory can be traced back to the early 1990s, when the French scholars Jutten and Hérault (1991) first proposed the concept of ICA. At the time, research on neural networks was increasing, but ICA theory had been investigated only in a small area. Till the mid-1990s, the theory of ICA and several algorithms were developed and the technique quickly expanded to other research areas.

While studying BSS problems, scholars proposed a number of classic methods: principal component analysis (PCA), factor analysis (FA), projection pursuit (PP), redundancy reduction (RR), and blind deconvolution (BD), among others. Considering the shortcomings of these methods, in 1986 Hérault and Jutten presented a research report (Herault and Jutten, 1986) in which they proposed a recursive artificial neural network model and learning rules based on the Hebbian learning algorithm, called the H–J algorithm, which realizes separation of two independent mixed source signals (Jutten and Hérault, 1991). Thereafter, Sorouchyari (1991), Comon, Jutten, and Hérault (1991), and Deville (1996, 1999) analyzed the convergence and stability of the H–J algorithm. Through the efforts of many scholars, the convergence of the H–J algorithm with two source signals and two mixed signals has been resolved.

In 1988 Giannakis and Swami (1988) proposed a BSS identification method that introduced third-order cumulants. However, this method requires an exhaustive search.

Shortly thereafter, Linsker (1988, 1989) published a few articles that had an important impact on BSS studies. He proposed the maximum mutual information criterion, the basic form of which is most suitable for the establishment of self-organizing and feature models.

In the early 1990s, Tong *et al.* (1991) and Tong, Inouye, and Liu (1993) conducted a more systematic study of the uncertainty of the BSS problem solution, that is, the

Blind Source Separation: Theory and Applications, First Edition. Xianchuan Yu, Dan Hu and Jindong Xu.

identifiability. They pointed out that the BSS problem is actually an unsolvable problem or, more precisely, a multi-solution problem, and that the unsolvability of the problem can be parametrized with an arbitrary full rank matrix. They also pointed out that by decomposing an unsolvable full rank matrix of the parametric BSS problem into the product of a full rank diagonal matrix and an elementary matrix, the uncertainty of the BSS problem can be reflected by the unsolvability of the true amplitude of the source signal and that of the original order of the source signals. The source signal waveform can be restored, thereby defining the solvability of the BSS problem. Meanwhile, Tong *et al.* proved that if all vector components of the source signal are statistically independent of each other and, when implementing matrix transformation on the respective components of the source signals, if only the transformed components are statistically independent of each other, the source signal vector matrix can be decomposed into the product of a diagonal matrix and an elementary matrix of full rank. In other words, the transformation only changes the amplitude and the order of the respective components of the source signal vector, and does not change the waveform of the signal.

Burel (1992) proposed using a BP neural network to implement linear and nonlinear mixed-signal BSS.

In 1994, Comon (1994) expanded the data processing and compression PCA into ICA, and elaborated on the concept of ICA by defining a basic assumption of the ICA method for solving BSS problems, namely, that the components recovered from the mixed signals using appropriate linear transformations are statistically independent of each other. This finding was of great importance. Comon also provided a strict mathematical definition of ICA in the literature, using the concept of the contrast function of BD in the ICA definition. At the same time, Comon defined an ICA contrast function. Simply put, the ICA contrast function is a functional function used to measure the PDF (probability density function) of random variables; if and only if the random variables measured by the contrast function are statistically independent of each other, the function obtains the maximum or minimum value. With the contrast function maximized or minimized, a random variable ICA problem is changed into an optimization problem. Comon also pointed out that ICA is an expansion and extension of PCA.

Owing to the complexity of the BSS problem itself, early BSS algorithms were only applicable if the number of source signals was equal to the number of mixed signals. After the landmark publication by Comon, research on BSS increased rapidly. In fact, several BSS algorithms were based on the results of Tong and Comon. Comon provided a clear context for the BSS problem, creating an algorithm design with a clear theoretical basis. Cao and Liu (1996) studied the separability condition and solving principle of the BSS problem. In addition to the case where there is the same number of source signals and mixed signals, they also studied the case where the number of source signals is less than or greater than the number of mixed signals. They investigated the situation where the number of source signals obeying the Gaussian distribution was greater than one. The results of Tong and Comon were widely incorporated in their research, with their results. Cao *et al.* completely solved the solvability of the blind source problem and theoretically proposed a solution of BSS, by defining the principle to use for BSS, including the case when the number of sources is greater than the number of mixed signals.

In fact, the vast majority of BSS algorithms developed after publication of Comon's research, were implemented using the ICA method. A variety of the same or different corresponding comparison functions and optimization algorithms were proposed according to the different principles and methods. It can be said that the work of Comon actually transformed research on the BSS algorithm into research on the control functions of ICA and its optimization algorithm. Based on the principle of ICA, a variety of BSS algorithms have been developed, including the information maximization BSS algorithm, robust neural network BSS, nonlinear PCA BSS algorithm, vastly similar contingent BSS algorithm, natural gradient algorithm for BSS, and negentropy maximization BSS algorithm.

In 1995, Bell and Sejnowski (1995) presented a landmark paper on the history of the development of ICA. In this paper, they proposed a neural network based ICA algorithm. They studied the transfer of information in a network composed of nonlinear elements, and proposed a self-organizing learning gradient algorithm. The algorithm can maximize the information transmitted in a nonlinear network. They illustrated that for a super-Gaussian source signal, that is, a source signal with fourth-order cumulants greater than zero, maximizing the transmission information of the signal through the nonlinear network is equivalent to minimizing the mutual information of the network output. Therefore, the information maximization algorithm is able to recover the statistical independence between the respective components of the mixed signal in order to achieve the signal source. The neural network-type optimal weight matrix iterative learning algorithm is given in the literature, and the algorithm has become the basis of a variety of follow-up algorithms. Simulation experiments on a "cocktail party conversation" among 10 people show that good separation effects can be achieved. However, there are also defects in the algorithm. It can only separate a source signal with a super-Gaussian distribution; for sub-Gaussian signals with fourth-order cumulant source signals less than zero, the algorithm cannot realize successful BSS.

The robustness of various algorithms, that is, the H–J algorithm of Hérault and Jutten, the maximization algorithm of Bell and Sejnowski, and many other BSS algorithms, and their blind separation ability depends on the specific mixed signals. If the mixing matrix approximates a singular matrix or a certain component of the source signal is much weaker than the other components, these algorithms cannot successfully implement signal source separation and may not converge. Cichocki, Unbehauen, and Rummert (1994) and Cichocki and Unbehauen (1996) proposed a robust neural network BSS algorithm based on the H–J algorithm that includes a forward neural network and a recursive reverse neural network. These two networks have the ability to automatically adjust the signal amplitude of the output, and it is possible to adjust all weights of the neural network so that the convergence characteristic of the neural network algorithm does not depend on the initial conditions of the algorithm or the specific mixed matrix and other factors. Thus, the two algorithms are called the robust neural network method.

Oja (1997), Karhunen and Joutsensalo (1994); Karhunen *et al.* (1997), Karhunen, Wang, and Vigario (1995), and Karhunen, Pajunen, and Oja (1998) proposed a nonlinear PCA BSS algorithm using higher-order statistics. This is an extension of the linear PCA method, and achieves separation of signals under the condition of an orthogonal constraint. The maximum likelihood estimation (MLE) ICA method was first proposed by Gaeta and Lacoume (1990) and later improved by Pearlmutter and Parra (1996). Amari studied the neural network BSS algorithm using a natural gradient method. In Euclidean space, the negative gradient of the function represents the steepest descent direction of the function value, and the gradient descent method is widely used in the optimization method. If the algorithms are involved in non-Euclidean space, the negative gradient direction no longer represents the steepest descent direction of the function. To solve this problem, Amari proposed the concept of the natural gradient using differential geometry and the Riemannian structure of the parameter space. The authors proved that, in Riemann space, the natural gradient of the function represents the true direction of steepest descent, while the natural gradient of Euclidean space is the same as the conventional gradient. Yang and Amari (1997) and Amari and Cichocki (1998) proposed a class of online learning algorithms using the natural gradient to minimize the mutual information between the respective output signals of the BSS network. Amari and Cardoso (1997) noted that the natural gradient and relative gradient are the same for the BSS problem. Because of the equivalence between the natural gradient and the relative gradient for BSS problems, the natural gradient and relative gradient algorithms are essentially equivalent. Both use a similar iterative rule to update the separation of the network weights matrix and have good varying characteristics. Amari and Cardoso (1997) and Cardoso (1998) studied these algorithms from the perspective of semi-parametric statistical estimation theory, summarized this type of BSS algorithm as an iterative algorithm that estimates equations, and analyzed the effectiveness of this type of blind separation algorithm. Hyvärinen and Oja (1997) proposed an ICA algorithm based on the negentropy maximization of random variables method. The negentropy concept, which was first proposed by Comon, is defined as the difference in the differential entropy of a random vector and Gaussian random vector with the same covariance matrix. Negentropy actually reflects the degree to which a random variable deviates from the Gaussian distribution. According to the central limit theorem, the sum of statistically independent random variables tends to be a Gaussian distribution. By transforming the source signals into vectors with a non-Gaussian distribution, random vector ICA can be implemented to achieve blind separation. Since negentropy is difficult to calculate, Hyvärinen (1997a, 1997b) used an approximate of the negentropy to obtain a class of ICA contrast functions. The first step in BSS using the negentropy contrast function is whitening the mixed signal, to ensure that each column of the matrix mold is a constant when maximizing the contrast function. Using K-T conditions of the constrained optimization problem, Hyvärinen (1997a, 1997b,

1999a, 1999b) obtained a unary equation and used Newton's method to achieve an iterative solution of this equation yielding a fixed ICA algorithm. Compared with some of the other ICA algorithms, this algorithm has a great advantage in terms of calculation speed.

The nonlinear PCA method proposed by Karhunen and Joutsensalo (1994) can be considered as an algorithm that minimizes the information of the network output vector. Girolami and Fyfe used marginal negentropy as the projection index, and extracted the main component from the linear mixture; the multiple output exploratory projection pursuit (EPP) can extract all the components (Girolami and Fyfe, 1996a,b, 1997a,b). At the 2002 ICA International Conference, Bach and Jordan (2003) proposed the KICA (kernel independent component analysis) algorithm, which gives a new typically associated contrast function based on a regenerate kernel Hilbert space (RKHS), and two new algorithms based on the RKHS.

Theis (2004) calculated the uncertainty of a problem with complex random variables and coefficient ICA using the Darmois–Skitovitch Theorem in the complex field, and studied the uniqueness and separability of a complex-valued ICA problem.

Tsai *et al.* (2006) applied a random evolutionary algorithm, namely, particle swarm optimization, to the constrained ICA model. The results showed that the ICA algorithm can effectively detect texture image defects in the glass substrate of an LCD display.

Masnadi-Shirazi and Rao (2009) proposed a state-based ICA algorithm aimed at the non-stationarity of the signal, which achieves very good results when processing voice signals.

After nearly two decades of development, the theoretical framework and algorithms for ICA have basically been perfected, although there are still some unsolved problems, such as the nonlinear mixing problem, underdetermined problem, and noisy ICA problem. Although these issues are very complex, they tend to be more applicable to realistic models and thus need to be investigated. ICA has also been combined with other theories, yielding several new theories and new methods based on ICA, such as the new BSS method that combines ICA with wavelet theory.

5.2 Principle of ICA

ICA is mainly used to solve the problem of BSS. In cases where the source signal and mixing matrix are unknown, simply assuming that the source signals are statistically independent ensures that ICA can separate the source signals from the mixed signals well. In practical problems, the basic assumption of independence is reasonable, and therefore ICA can be widely applied in many fields.

5.2.1 Concept of Independence of Random Variables

In ICA theory, an important concept is that of statistical independence. Simply stated, for any two random variables, x and y, if no information about x can be obtained from y, and vice versa, we say that random variables x and y are independent of each other. For example, two random signals generated by different physical processes are independent of each other.

From a mathematical point of view, the PDF of the random variable is used to define the statistical independence concept. Random variables x and y are statistically independent, if and only if their joint probability density can be decomposed into the product of the marginal densities, that is,

$$p_{x,y}(x, y) = p_x(x)p_y(y) \tag{5.1}$$

This definition can be extended to the case of n random variables.

5.2.2 ICA Definition

The essence of the ICA algorithm is based on statistical independence and trying to express a set of random variables as a linear combination of statistically independent variables when there is no information about the source signals and mixing matrix. The ICA method minimizes the statistical dependence between each component of the signal, highlighting the essential structure of the source signal. Currently, two of the most promising applications of ICA are BSS and feature extraction.

Next we illustrate how ICA implements a linear representation of multidimensional data. Let there be a set of n-dimensional observation signals generated by a number of implicit variables, where $i = 1, \ldots, n$ represents the ordinal number of the observation signal, $t = 1, \ldots, T$ represents the length of the observation data, and T is the number of sample points. We hope to find hidden variables $\boldsymbol{Y}(t) = (y_j(t))$ from the n-dimensional observed signals $\boldsymbol{X}(t)$, such that these hidden variables can describe the essential characteristics of the observed signals $\boldsymbol{X}(t)$. At the same time we also hope that these hidden variables have some kind of necessary connection (or physical source) with the actual physical source of the observed signal. To avoid the problem of multivariate data analysis, we usually reduce it to a linear problem for processing, that is,

$$y_j(t) = \sum_{i=1}^{n} \omega_{ji} x_i(t) \quad i = 1, 2, \ldots, n; j = 1, 2, \ldots, m \tag{5.2}$$

Such data can be written in the form of a matrix

$$\begin{pmatrix} y_1(t) \\ \vdots \\ y_m(t) \end{pmatrix} = \begin{pmatrix} \omega_{11} & \cdots & \omega_{1n} \\ \vdots & \ddots & \vdots \\ \omega_{m1} & \cdots & \omega_{mn} \end{pmatrix} \begin{pmatrix} x_1(t) \\ \vdots \\ x_n(t) \end{pmatrix}, \tag{5.3}$$

$$\boldsymbol{Y} = \boldsymbol{W} \cdot \boldsymbol{X} \tag{5.4}$$

From a statistical analysis, the time-series $x_i(t)$, $y_j(t)$, $t = 1, 2, \ldots, T$ are seen as finding a random variable T times, while the time characteristics of the signal are ignored. We abbreviate the $x_i(t)$, $y_j(t)$ time series (or signals) as x_i and y_j. This approach can largely compensate for the deficiencies brought about by the limitations of the time-domain analysis methods. Without loss of generality, we assume that the values of these random variables are continuous. Let all components have a zero mean, that is, $E\{x_i\} = \overline{0}$. Therefore, the multivariate data linear representation problem is briefly described as finding a mapping from element x in n-dimensional space to element y in m-dimensional space, such that the transformed vectors are capable of expressing the statistical information of the data space. These transformed vectors $\boldsymbol{y}$ are the so-called "factors" or "components". Thus, in Equation 5.4, the data representation problem is transformed into solving the parameter matrix $\boldsymbol{W}$.

5.2.3 ICA Estimation Principle

5.2.3.1 ICA Assumptions

Figure 5.1 depicts a schematic block diagram of ICA.

Assumptions and constraints of the ICA problem:

1. Source signals $s_i(i = 1, \ldots, m)$ are real random variables with a zero mean, and are statistically independent of each other at any arbitrary time. If the PDF of s_i is $p_i(s_i)$, the PDF of the source of signal vector $\boldsymbol{s}$ is

$$p(\boldsymbol{s}) = \prod_{i=1}^{n} p_i(s_i) \tag{5.5}$$

2. The number of source signals m should be less than or equal to the number of observed signals n $(m \leq n)$. The mixing matrix $\boldsymbol{A}$ is an $n \times n$ unknown phalanx. If $\boldsymbol{A}$ is assumed to be full rank, the inverse matrix $\boldsymbol{A}^{-1}$ exists.
3. Only one source signal s_i $(i = 1, \ldots, m)$ is allowed to satisfy the Gaussian distribution. Since a mixed signal with two statistically independent white Gaussian signals is still a white Gaussian signal, and its PDF involves only second-order statistics, there are no higher-order statistical properties that can be used, and their independence is equivalent to being unrelated. It can be proved that by any rotation $\boldsymbol{y} = \boldsymbol{Wx}$ ($\boldsymbol{W}$ is the rotation matrix with $\boldsymbol{WW}^{\mathrm{T}} = \mathrm{I}$), the separation results do not change the second correlation property of the Gaussian vector; that is, it is always in line with the requirement of statistical independence. Obviously, the source signals do not

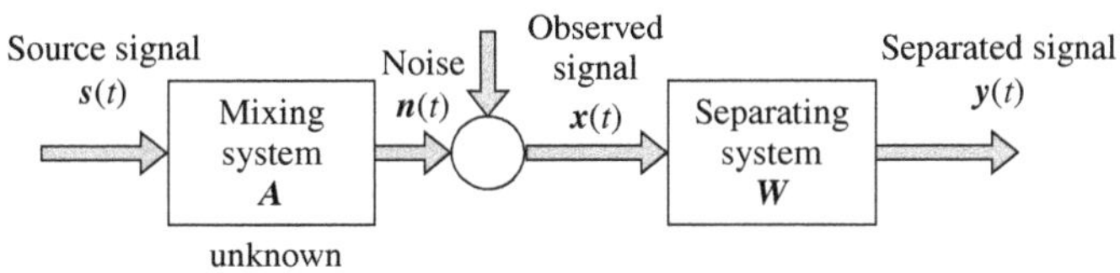

Figure 5.1 Block diagram of ICA

always satisfy this requirement. If the number of source signals with a Gaussian distribution is greater than one, the source signals cannot be separated.

4. The noise introduced by each of the sensors is very small and, therefore, negligible. The output mutual information of the information maximization method can be minimized only in low noise conditions. However, we can consider the noise itself as a source signal that can be separated by blind separation processing from the mixed signal, in the same way as the other "real" source signals. This makes the algorithm more robust with a broader application area.
5. We should have some prior knowledge about the probability distribution function of the source signals. For example, natural speech and music signals have super-Gaussian characteristics (such as the Laplace distribution); most image signals have sub-Gaussian characteristics (such as the uniform distribution); and noise typically has Gaussian characteristics. In addition, if s_i is the sum of multiple random variables, its PDF $p_i(s_i)$ is also close to a Gaussian distribution (by the central limit theorem).

5.2.3.2 ICA Mathematical Model

The ICA model estimate usually first selects a suitable objective function and then minimizes or maximizes it. Therefore, the nature of the ICA method depends on the objective function and optimization algorithm. In general, the selection of the objective function determines the statistical properties of the ICA method, such as consistency, gradient (asymptotic variance), and robustness, while the choice of optimization algorithm determines the nature of the algorithm, such as convergence speed, memory requirements, and stability. We can use different algorithms with the same objective to implement function optimization and an optimization algorithm can be used with different objective functions.

Let $\boldsymbol{x} = (x_1, x_2, \dots, x_n)^{\mathrm{T}}$ be the n-dimensional random observed vector, where $\boldsymbol{x}$ is the linear combination of n unknown source signals (or independent sources) $\boldsymbol{s} = (s_1, s_2, \dots, s_n)^{\mathrm{T}}$. Without loss of generality, we assume that the mean value of the random variables and independent sources of the mixed signal is zero. Below we use a linear representation of the multidimensional data in matrix form to define the ICA linear model:

$$\boldsymbol{x} = \boldsymbol{A}\boldsymbol{s} = \sum_{i=1}^{n} a_i s_i, \quad i = 1, 2, \dots, n \tag{5.6}$$

where s_i is called the independent component (IC), $\boldsymbol{A} = (\boldsymbol{a}_1, \boldsymbol{a}_2, \dots, \boldsymbol{a}_n) \in \boldsymbol{R}^{n\times n}$ is a full-rank matrix called the mixing matrix, and $\boldsymbol{a}_i$ is the base vector of the mixing matrix. This shows that each observed datum x_j is obtained by the linear weight of the different a_{ij} in the independent source signals s_i. The purpose of ICA processing is to find a linear transformation matrix $\boldsymbol{W}$ of mixed signal $\boldsymbol{x}$, so that the output is as independent as possible:

$$\boldsymbol{y}(t) = \boldsymbol{W}\boldsymbol{x}(t) = \boldsymbol{W}\boldsymbol{A}\boldsymbol{s}(t) \tag{5.7}$$

The independent source s_i is also known as the hidden variable, that is, these independent components cannot be directly observed. Since mixing matrix $\boldsymbol{A}$ is also an unknown matrix, the only available information is the observed random vector $\boldsymbol{x}$. If there are no limiting conditions, and if we only use $\boldsymbol{x}$ to estimate $\boldsymbol{s}$ and $\boldsymbol{A}$, the solution of Equation 5.6 will contain multiple solutions. While ICA has certain limiting conditions, it gives only one solution of the equation, and realizes independent component extraction according to the statistical properties of $\boldsymbol{x}$.

5.2.3.3 ICA Unmixing Model

In order to separate source signal $\boldsymbol{s}$ from the mixed signal in as realistic a way as possible using only the ICA assumptions and sensor detected signal $\boldsymbol{x}$ (referred to as the sensor signal), under the condition that mixed matrix $\boldsymbol{A}$ and source signal $\boldsymbol{s}$ are unknown, it is necessary to build an unmixing matrix (or separation matrix) $\boldsymbol{W} = (w_{ij})$. After the transformation of the separation matrix $\boldsymbol{W}$ and $\boldsymbol{x}$, the output of the n-dimensional column vector $\boldsymbol{y} = (y_1, y_2, \cdots, y_n)^{\mathrm{T}}$ is obtained, and the ICA unmixing model can be expressed as

$$\boldsymbol{y} = \boldsymbol{Wx} = \boldsymbol{WAs} = \boldsymbol{Gs} \tag{5.8}$$

where $\boldsymbol{G}$ refers to the global/system matrix. $\boldsymbol{G} = \boldsymbol{I}$ ($\boldsymbol{I}$ is an $n \times n$ unit matrix) can be learned, and $\boldsymbol{y} = \boldsymbol{s}$, in order to achieve the separation (or restoration) of the source signal. In fact, as long as only one element in each column of $\boldsymbol{G}$ is close to 1, while the other elements are close to zero, the separation can be considered successful.

Two performance indicators that measure the performance of the unmixing algorithm more accurately are the similarity (related) coefficient and performance index (PI).

1. **Similarity coefficient** (in Equation 4.15). If $y_i = cs_j$ (c is a constant), $\zeta_{ij} = 1$; if y_i and s_j are mutually independent, $\zeta_{ij} = 0$. If each row and each column of the similarity coefficient matrix has one and only one element that is close to 1, with the other elements close to zero, the separation algorithm can be considered to be ideal.
2. **PI**

$$PI = \frac{1}{n(n-1)} \sum_{i=1}^{n} \left\{ \left(\sum_{i=1}^{n} \frac{|g_{ik}|}{\max_j |g_{ij}|} - 1 \right) + \left(\sum_{k=1}^{n} \frac{|g_{ki}|}{\max_j |g_{ji}|} - 1 \right) \right\} \tag{5.9}$$

where g_{ij} is an element of the global matrix $\boldsymbol{G}$; $\max_j |g_{ij}|$ represents the maximum absolute values in the i-th row element of $\boldsymbol{G}$; and $\max_j |g_{ji}|$ represents the maximum absolute values in the ith column element of $\boldsymbol{G}$. If the separated estimated signal $\boldsymbol{y}(t)$ has the same waveform as the source signal $\boldsymbol{s}(t)$, $PI = 0$. In fact, when PI approaches 10^{-2}, the separation performance of the algorithm is considered to be quite good.

The assumptions of ICA, the ICA mixture model (given by Equation 5.6), and the ICA unmixing model (given by Equation 5.8) constitute a complete definition of the ICA problem. In fact, the ICA problem aims to solve separation system $\boldsymbol{W}$ so that output y_i approximates the approximated independent source signal s_i as closely as possible.

5.2.4 Uncertainty of ICA

An ICA solution has two internal uncertainties (uncertainty here does not refer to fuzziness or randomness): uncertainty of the order of the separation signal permutation (referred to as the permutation), that is, not knowing which component in $\boldsymbol{s}$ corresponds to signal component y_i, and uncertainty of signal scaling (referred to as scaling) (the multiplexing amplitude of the waveform), that is, the true waveform of source signal $\boldsymbol{s}$ cannot be recovered from $\boldsymbol{x}$, or we cannot determine the variance (energy) of each of the independent source signals s_i. The scalar product of an arbitrary source can be obtained by dividing the corresponding column α_i of matrix $\boldsymbol{A}$ by the scalar α_i,

$$\boldsymbol{x} = \sum_i \left(\frac{1}{\alpha_i} a_i \right) (\alpha_i s_i) \tag{5.10}$$

It can be seen that, if component s_i in $\boldsymbol{s}$ is multiplied by any non-zero complex factor α_i, and each corresponding element of the jth column in $\boldsymbol{A}$ is multiplied by $1/\alpha_i$, regardless of the value of α_i, $\boldsymbol{x}$ remains unchanged. Therefore, when trying to obtain each of the source signals from $\boldsymbol{x}$, scale uncertainty exists. The most natural method for eliminating this is to convert each of the source signals using unit variance (i.e., $E\{s_i^2\} = 1, \forall i$), such that the autocorrelation matrix of $\boldsymbol{s}$ is the unit matrix $\boldsymbol{R}_{ss} = E\{\boldsymbol{s}\boldsymbol{s}^{\mathrm{T}}\}$. After processing the digitized variance, for real amplitude, signal uncertainty may also exist. However, this is not a major problem, as it can be handled without difficulty. We only need to multiply the amplitude of the component signal by -1, although for complex amplitude values, there would still be uncertainties on the phase.

5.2.5 Relationship between ICA and BSS, PCA, and Whitening

5.2.5.1 ICA and BSS

BSS is the process of recovering an original source signal that cannot be observed directly from several observed mixed signals. The method combines artificial neural networks, statistical signal processing, and information theory. BSS is a challenging task in signal processing. "Blind" refers to two aspects: first, the source signal is unknown, and second, the hybrid system is also unknown. In scientific and engineering applications, many of the observed signals can be seen as a mix of blind sources, with the so-called "cocktail party" problem a typical example. ICA has developed over time to solve BSS problems, and has become a powerful tool for signal processing and data analysis. When ICA is applied to BSS, the unobserved source signal components can be reproduced from the mixed signals, with the only requirement being the simple a priori condition that the source signals are statistically independent. Therefore, ICA is a BSS method and BSS is one of the applications of ICA. Although most of the ICA algorithms can be used to solve BSS, there are many other theories that can also be used, and thus, the solution of BSS is not limited to ICA.

Under the assumption that the source signals are independent of each other, ICA has the same or a similar model as BSS, and uses the same or a similar algorithm to separate signals. Thus, in some literature, BSS is indiscriminately interchanged with ICA. However, under normal circumstances (especially for practical problems), the target of ICA is slightly different from that of BSS. In fact, the purpose of BSS is to estimate the original source signals, even if they are not completely mutually statistically independent, whereas the purpose of ICA processing is to identify a transformation, to ensure that each of the output signals is as independent of each other as possible so that the non-Gaussian property can be maximized.

5.2.5.2 ICA and PCA

The PCA process aims to reduce vector dimensions and remove the linear correlation between the random vectors to ascertain the intrinsic information implicit in the original signal. However, owing to the PCA method using only the covariance matrix of the signals in its computation, the actual calculation involves only second-order statistical properties of the probability distribution function of the input data. The decomposed principal components are only orthogonal to each other (owing to the assumption that the mean of the signals is zero, the principal components are actually not related), which does not satisfy the independence requirements of the ICA output. From a statistical point of view, the most important information of the actual signal is often included in the higher-order statistics characteristics. Therefore, only if the multi-variable observed data are constituted by source signals with Gaussian distribution, can the PCA method be used to achieve separation of the signal (since the uncorrelation property of a normal distribution random process is equivalent to statistical independence, and the signal can be described by second-order statistics). However, in reality, only a few random variables have a normal (Gaussian) distribution.

In addition, ICA often requires higher-order statistics; that is, a nonlinear function is needed in the learning phase. The use of a nonlinear function, however, does not affect the final result of the linear mapping of input and output. In many cases, ICA can provide a more meaningful data representation than PCA, and PCA can only achieve ICA separation results when the source signal has a Gaussian distribution. Nevertheless, in the signal preprocessing stage, the PCA method can be used to achieve data pre-whitening, which can simplify the follow-up stage.

5.2.5.3 ICA and Whitening

A zero-mean random vector $\boldsymbol{y}$ is a white signal, which refers to the fact that each component has unit variance and is not related to the others, that is, the covariance matrix of $\boldsymbol{y}$ is a unit matrix. The process of applying a linear transformation to any multidimensional signal so that it becomes a white signal is called whitening (also known as "sphering" of the mixing vector to provide uncorrelated components), and the corresponding transformation matrix is called the whitening matrix. Let $\boldsymbol{Q}$ be the whitened matrix of observed signal $\boldsymbol{x}(t)$,

$$\overline{\boldsymbol{x}}(t) = \boldsymbol{Q}\boldsymbol{x}(t) \tag{5.11}$$

where $\overline{\boldsymbol{x}}(t)$ is the whitened mixed signal, and thus $\mathrm{cov}\{\overline{\boldsymbol{x}}\} = \boldsymbol{I}$. Then, by substituting $\boldsymbol{x} = \boldsymbol{A}\boldsymbol{s}$ into Equation 5.11, and letting $\boldsymbol{A}_{\mathrm{G}} = \boldsymbol{Q}\boldsymbol{A}$ ($\boldsymbol{A}_{\mathrm{G}}$ is called the global mixing matrix),

$$\overline{\boldsymbol{x}}(t) = \boldsymbol{Q}\boldsymbol{A}s(t) = \boldsymbol{A}_{\mathrm{G}}s(t) \tag{5.12}$$

Since $\overline{\boldsymbol{x}}(t)$ and $\boldsymbol{s}(t)$ connected by the linear transformation $\boldsymbol{A}_{\mathrm{G}}$ are two white random vectors, matrix $\boldsymbol{A}_{\mathrm{G}}$ must be an orthogonal matrix. If $\overline{\boldsymbol{x}}(t)$ is seen as the new observed signal, the whitening process simplifies the original mixing matrix $\boldsymbol{A}$ as the new orthogonal matrix $\boldsymbol{A}_{\mathrm{G}}$, which can be proved as follows:

$$E\{\overline{\boldsymbol{x}}\overline{\boldsymbol{x}}^{\mathrm{T}}\} = E\{\boldsymbol{A}\boldsymbol{s}\boldsymbol{s}^{\mathrm{T}}\boldsymbol{A}^{\mathrm{T}}\} = \mathbf{A}E\{\boldsymbol{s}\boldsymbol{s}^{\mathrm{T}}\}\boldsymbol{A}^{\mathrm{T}} = \boldsymbol{A}\boldsymbol{A}^{\mathrm{T}} = \boldsymbol{I} \tag{5.13}$$

Similarly, if the separating matrix $\boldsymbol{W}$ is for a whitened mixed signal $\overline{\boldsymbol{x}}$, if the separated output $\boldsymbol{y}$ satisfies $E\{\boldsymbol{y}\boldsymbol{y}^{\mathrm{T}}\} = \mathbf{I}$ (eliminating the uncertainty scales), then

$$E\{\boldsymbol{y}\boldsymbol{y}^{\mathrm{T}}\} = E\{\boldsymbol{W}\overline{\boldsymbol{x}}\overline{\boldsymbol{x}}^{\mathrm{T}}\boldsymbol{W}^{\mathrm{T}}\} = \boldsymbol{W}\boldsymbol{W}^{\mathrm{T}} = \boldsymbol{I} \tag{5.14}$$

The above formula shows that when whitened data are processed using blind separation, the separation matrix $\boldsymbol{W}$ is an orthogonal matrix.

Figure 5.2 shows the implementation principle of the algorithm. Obviously, using a whitening method as ICA preprocessing can effectively reduce the complexity of the

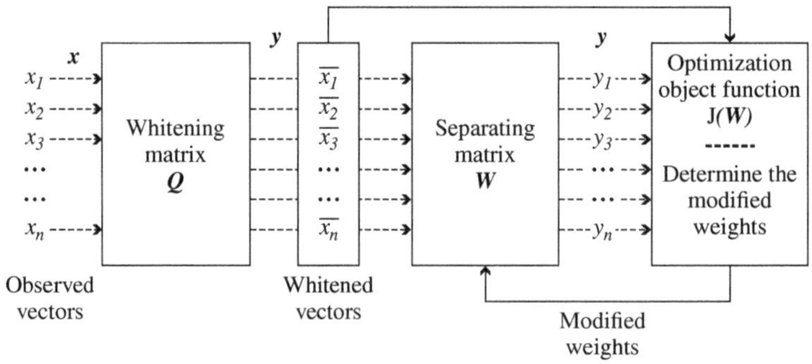

Figure 5.2 Implementation principle of the algorithm

problem, since the whitening algorithm is simple and varied. PCA with singular value decomposition (SVD) can accurately calculate the results from the observed data. Some ICA algorithms do not need whitening preprocessing as the whitening process is included in the separation algorithm; instead, they can directly and iteratively estimate the separation matrix.

5.2.6 *Basic Methods for ICA*

ICA is actually an optimization process: separating independent components such that they maximally approximate the source signal. Therefore, ICA includes two key aspects: the optimization criterion (objective function) and the optimization algorithm.

The optimization criterion involves selecting the appropriate objective function such that the criterion measure of the independence of each component can be found from different angles. The basic criterion of independence should be obtained from the PDF. Since the PDF is generally unknown and difficult to estimate, we need some way of circumventing this difficulty. The learning process of the ICA algorithm first establishes an objective function $J(\boldsymbol{W})$ (used as the optimization criterion function) of the $\boldsymbol{W}$ argument, based on methods from information and statistical theory. The value of $J(\boldsymbol{W})$ indicates the degree of independence between each of the output components of $\boldsymbol{y}(t)$. If a particular $\hat{\boldsymbol{W}}$ can maximize (or minimize) the value of $J(\boldsymbol{W})$, $\boldsymbol{W}$ is the solution we are seeking. Two commonly used methods are discussed below.

1. The PDF is implemented with series expansion so as to translate the estimates of the PDF into estimates of higher-order statistics.
2. A nonlinear link is introduced at the output to establish the optimization criterion, with the practice actually implying the introduction of higher-order statistics.

The optimization algorithm finds an efficient method to solve the $\hat{\boldsymbol{W}}$, generally using the conventional (or standard) gradient method, the stochastic gradient method proposed by Bell and Sejnowski (1995), the relative gradient method of Cardoso (1997), the natural gradient method proposed by Cichocki and Unbehauen (1996), or another heuristic learning algorithm. It has already been proved that the two descent methods, namely, the relative gradient method and the natural gradient method, are equivalent.

5.2.6.1 Objective Function of ICA

The objective function is an operator that maps the probability density distribution collection of vector $\boldsymbol{y}(t)$ to a real-valued function, referred to as $\psi(\boldsymbol{y})$, and satisfies the following conditions:

1. When the sequence of each component of $\boldsymbol{y}(t)$ changes, the value of $\psi(\boldsymbol{y}(t))$ remains constant, that is, $\psi(\boldsymbol{y}(t)) = \psi(\boldsymbol{P}\boldsymbol{y}(t))$, where $\boldsymbol{P}$ is a switching matrix.

2. When the amplitude of each component of $\boldsymbol{y}(t)$ changes, the value of $\psi(\boldsymbol{y}(t))$ remains constant, that is, $\psi(\boldsymbol{y}(t)) = \psi(\boldsymbol{D}\boldsymbol{y}(t))$, where $\boldsymbol{D}$ is a real diagonal matrix.
3. When $\boldsymbol{y}(t)$ has independent components, $\psi(\boldsymbol{A}\boldsymbol{y}(t)) \leq \psi(\boldsymbol{y}(t))$, where $\boldsymbol{A}$ is the mixing matrix.

The objective function is actually the functional of the PDF, and only if the random variables are independent of each other, can the objective function achieve its maximum or minimum value. During ICA processing, selecting different objective functions can produce different algorithms, including the minimum mutual information (MMI) method, the maximum entropy (ME) method, and MLE as examples of three of the categories.

Maximum Non-Gaussian

The non-Gaussian is most important in ICA estimation; if there is no non-Gaussian, ICA estimation cannot be carried out. The central limit theorem states that, under certain conditions, the sum of the independent random variables is closer to a Gaussian distribution. That is, the sum of two independent random variables is closer to a Gaussian distribution than the distribution of any one component.

Assuming that vector $\boldsymbol{X}$ is the observation vector, $\boldsymbol{X} = \boldsymbol{A}\boldsymbol{S}$. If one wishes to estimate a separate component, the linear combination $\boldsymbol{y} = \boldsymbol{b}^{\mathrm{T}}\boldsymbol{X} = \sum_i b_i x_i$ of each component of $\boldsymbol{X}$ must be considered. According to the central limit theorem, we only need to change $\boldsymbol{b}$ so that it reaches the greatest non-Gaussian $\boldsymbol{b}$ from $\boldsymbol{b}^{\mathrm{T}}\boldsymbol{X}$, to ensure that $\boldsymbol{y}$ is an independent component.

Nonlinear Correlation

Some researchers (Cichocki, Unbehauen, and Rummert, 1994; Everson and Roberts, 1999) have obtained independent components by removing the nonlinear correlation in the observation vectors. Generally, the nonlinear correlation takes the form $E\{g_1(y_i)g_2(y_j)\}$, where g_1 and g_2 are suitable nonlinear functions. If y_i and y_j are independent, the nonlinear correlation is 0, and the premise is that the density distributions of y_i and y_j are symmetrical. An exact objective function does not exist, because the nonlinear function should be selected according to the PDF of the independent sources in an actual situation.

Higher-Order Cumulants

For multi-vectors, ICA estimation theory that has no direct contact with the objective function is the eigenmatrix decomposition of higher-order cumulants. Many methods make use of fourth-order cumulants. The linear operator for fourth-order cumulants can be defined in the space of an $m \times m$ matrix as

$$T(\boldsymbol{K})_{ij} = \sum_{k,l} \mathrm{cum}(x_i, x_j, x_k, x_l)\boldsymbol{K}_{kl} \tag{5.15}$$

where $\boldsymbol{K}$ is an $m \times m$ matrix. Since it is a linear operator, there are m^2 characteristic values associated with the eigenmatrix. By solving the eigenvectors of the eigenmatrix, the ICA model can be estimated.

To estimate a single vector, the objective function usually uses higher-order statistics such as kurtosis. To estimate the source vector, a linear transformation $\boldsymbol{w}^{\mathrm{T}}\boldsymbol{x}$ of the observation vector $\boldsymbol{x}$ must be found, such that its kurtosis is at a maximum or minimum. Let $z = \boldsymbol{w}^{\mathrm{T}}\boldsymbol{x}$, then

$$\mathrm{kurt}(\boldsymbol{w}^{\mathrm{T}}\boldsymbol{x}) = \mathrm{kurt}(\boldsymbol{w}^{\mathrm{T}}\boldsymbol{A}\boldsymbol{s}) = \mathrm{kurt}(z^{\mathrm{T}}\boldsymbol{s}) = \sum_{i=1}^{m} z_i^4 \mathrm{kurt}(s_i) \tag{5.16}$$

Under the given conditions, by maximizing or minimizing kurtosis in the above formula, we can obtain the independent component $\boldsymbol{w}^{\mathrm{T}}\boldsymbol{x} = \pm s_j$. As the accumulated value is easy to calculate, and particularly since it is possible to obtain a global convergence result, cumulants are widely used in ICA processing. However, a higher-order cumulant is only impacted by both ends (tails) and has nothing to do with the middle distribution. Moreover, since it is very sensitive to both ends, from a statistical significance perspective, it is not a good solution for the ICA problem.

Mutual Information Minimization Objective Function

Since the mutual information function $I(\boldsymbol{y})$ satisfies the three conditions in the definition of an objective function, it is an objective function. It is difficult to use the original formula of $I(\boldsymbol{y})$ directly to implement the optimization calculation, because the joint PDF of $\boldsymbol{y}(t)$ and the marginal PDF are difficult to obtain. Thus, there is a need to introduce higher-order statistics. First, using higher-order statistics we implement a polynomial expansion of the probability distribution function of each component of output $\boldsymbol{y}(t)$ to estimate the PDF (this estimation generally uses either the Edgeworth or Gram-Charlier expansion method), and then we substitute this estimate into Equation 4.4 to expand and obtain the higher-order statistics $I(\boldsymbol{y})$. The obtained $I(\boldsymbol{y})$ is a more complex form of the polynomial, and using it to carry out the optimization calculation is still relatively difficult. But we can choose a relatively simple separate part from the expression of $I(\boldsymbol{y})$ according to the circumstances, and its maximization or minimization enables $I(\boldsymbol{y})$ to be minimized. Thus, we can use it as the criterion function $J(\boldsymbol{W})$ to carry out the optimization calculation. Finally, Equation 5.17 can be obtained by a gradient method. By substituting it into the iterative process, we obtain the optimization algorithm of the separating matrix

$$\frac{\mathrm{d}\boldsymbol{W}}{\mathrm{d}t} = \mu(t)\frac{\partial J(\boldsymbol{W})}{\partial \boldsymbol{W}} \tag{5.17}$$

Information Transmission Maximization or Negentropy Maximization Objective Function

Mutual information maximization (or information transfer maximization) between the input and output in the case of low noise means that the amounts of redundancy between the information inputs and outputs are at their minimum, thereby minimizing the mutual information between each of the outputs and ensuring that each of the output components is statistically independent of the others. From the definition of mutual information, $I(\boldsymbol{x}|\boldsymbol{y})$ is the mutual information expression between system output $\boldsymbol{y}$ and input $\boldsymbol{x}$. $I(\boldsymbol{x}|\boldsymbol{y})$ reaches its maximum when both forms of redundant information are minimized.

Negentropy maximization of the output of the system can also lead to separation of the signal. The characteristics of negentropy are that it remains unchanged with any linear transformation of $\boldsymbol{y}$, and it is always non-negative; only when $\boldsymbol{y}$ is a Gaussian distribution is it 0. Thus we can use negentropy to describe the degree to which a distribution differs from the Gaussian distribution. These characteristics imply that negentropy is a good objective function. It can be proved that the relation between negentropy and mutual information is

$$I(\boldsymbol{y}) = J(\boldsymbol{y}) - \sum_{i=1}^{n} J_i(y_i) + \frac{1}{2}\log\frac{\prod_{i=1}^{n} C_{ii}}{\det(\boldsymbol{C})} \tag{5.18}$$

where $J_i(y_i)$ is the negentropy of the y_i, $\boldsymbol{C}$ is the covariance matrix of $\boldsymbol{y}$, and C_{ii} is a matrix of the diagonal elements. If the respective components of $\boldsymbol{y}$ are not relevant, the third term on the right-hand side of the equation is 0, and the formula can be simplified as

$$I(\boldsymbol{y}) = J(\boldsymbol{y}) - \sum_{i=1}^{n} J_i(y_i) \tag{5.19}$$

From the above formula it is not difficult to see that minimizing the mutual information $I(\boldsymbol{y})$ of the respective components of the output signal $\boldsymbol{y}$ is equivalent to maximizing the sum of negentropy $\sum_{i=1}^{n} J_i(y_i)$ of each component. Therefore, an objective function based on negentropy is given as

$$\rho(\boldsymbol{y}) = \sum_{i=1}^{n} J_i(y_i) \tag{5.20}$$

5.2.6.2 ICA Optimization Algorithm

After establishing the target function, it is necessary to choose a learning algorithm (also known as an optimization algorithm), with the objective function reaching its maximum or minimum value through iterative calculation. The learning algorithms that are normally used are the stochastic gradient method and the natural gradient method.

Stochastic Gradient Method

The stochastic gradient method is a classic method that aims to find the minimum (or maximum) of the objective function $L(\boldsymbol{W})$ (the likelihood function). The basic idea is first to determine an initial value $\boldsymbol{W}_0$ of the separation matrix $\boldsymbol{W}$, calculate the gradient at $\boldsymbol{W}(0)$ of the objective function $L(\boldsymbol{W})$, and then obtain a new separation matrix $\boldsymbol{W}(1)$ by moving a proper step size in the negative gradient direction of the gradient (when seeking the maximum value, we use the positive gradient direction). By repeating the above process, the iterative formula for solving $\boldsymbol{W}$ can be obtained:

$$\boldsymbol{W}(k+1) = \boldsymbol{W}(k) + \alpha(k)\frac{\partial L(\boldsymbol{W})}{\partial \boldsymbol{W}}|_{W=W(k)} \tag{5.21}$$

where k is the number of iterations and $\alpha(k)$ is the step, otherwise known as the learning rate. Let $\boldsymbol{W}(k+1) - \boldsymbol{W}(k) = \Delta\boldsymbol{W}$. If the number of iteration steps does not need to be considered, the above equation can be rewritten as

$$\Delta\boldsymbol{W} = \alpha\frac{\partial L(\boldsymbol{W})}{\partial \boldsymbol{W}} \tag{5.22}$$

This is the common iteration formula for the gradient descent method, and different objective functions can yield different iteration formulas. In the following example we use the maximum logarithmic likelihood function to derive the iterative formula of the stochastic gradient method. Substituting $\boldsymbol{y} = \boldsymbol{W}\boldsymbol{x}$ into the ICA logarithmic likelihood function (considering only the continuous case) yields

$$\log L(\boldsymbol{W}) = \int p_x(\boldsymbol{x}) \log p_s(\boldsymbol{y})\mathrm{d}\boldsymbol{x} + \log|\det(\boldsymbol{W})| \tag{5.23}$$

Derivating $\boldsymbol{W}$ on the right-hand side of the equal sign in Equation 5.23 gives the gradient of the last part of the equation as $\frac{\partial \log|\det \boldsymbol{W}|}{\partial \boldsymbol{W}} = \boldsymbol{W}^{-T}$, and the partial derivative of $\boldsymbol{W}$ in the first part as

$$\frac{\partial \int p_x(\boldsymbol{x}) \log p_s(\boldsymbol{y})\mathrm{d}\boldsymbol{x}}{\partial w_{ij}} = \int p_x(\boldsymbol{x})\frac{\partial \log p_s(\boldsymbol{y})}{\partial w_{ij}}\mathrm{d}\boldsymbol{x} = \int p_x(\boldsymbol{x})\frac{p'_s(\boldsymbol{y})}{p_s(\boldsymbol{y})}\cdot\frac{\partial \boldsymbol{y}}{\partial w_{ij}}\mathrm{d}\boldsymbol{x} \tag{5.24}$$

where operator "·" represents the dot product of two vectors. Next, we define the non-linear function $\varphi_i(\boldsymbol{x}) = -p'_s(\boldsymbol{x})/p_s(\boldsymbol{x})$, substitute the gradient formula of the matrix

$$\frac{\partial \boldsymbol{y}}{\partial \boldsymbol{W}} = \frac{\partial(\boldsymbol{W}\boldsymbol{x})}{\partial \boldsymbol{W}} = \boldsymbol{x}^{\mathrm{T}} \tag{5.25}$$

into Equation 5.24, and write it in the form of the derivative of matrix $\boldsymbol{W}$ as

$$\frac{\partial \int p_x(\boldsymbol{x}) \log p_s(\boldsymbol{y})\mathrm{d}\boldsymbol{x}}{\partial w_{ij}} = -\int \varphi(\boldsymbol{y})\boldsymbol{x}^{\mathrm{T}} p_x(\boldsymbol{x})\mathrm{d}\boldsymbol{x} = -E\{\varphi(\boldsymbol{y})\boldsymbol{x}^{\mathrm{T}}\} \tag{5.26}$$

Then, the gradient of the likelihood function on separation matrix W is

$$\frac{\partial L(\boldsymbol{W})}{\boldsymbol{W}} = \boldsymbol{W}^{-\mathrm{T}} - E\{\varphi(\boldsymbol{y})\boldsymbol{x}^{\mathrm{T}}\} \tag{5.27}$$

Substituting Equation 5.27 into Equation 5.21 we get the iterative formula of separation matrix $\boldsymbol{W}$

$$\boldsymbol{W}(k+1) = \boldsymbol{W}(k) + \alpha(k)[\boldsymbol{W}^{-\mathrm{T}}(k) - E\{\varphi(\boldsymbol{y})\boldsymbol{x}^{\mathrm{T}}\}] \tag{5.28}$$

where nonlinear function $\varphi(\boldsymbol{y})$ is related to the PDF of the source signal. However, since the probability density of the source signal is unknown, we usually use different nonlinear functions for different distributions. For a sub-Gaussian signal, we can use $\varphi(\boldsymbol{y}) = \boldsymbol{y}^3$, and for a super-Gaussian signal we can use $\varphi(\boldsymbol{y}) = \tanh(\boldsymbol{y})$.

Natural Gradient Method

The stochastic gradient method requires obtaining the inverse of the separation matrix $\boldsymbol{W}$. If the dimension of W is large, the amount of calculation will also be large, resulting in deterioration of the stability of the algorithm. To address this problem, Amari (1998) started in Riemann space and proposed the natural gradient method. The relationship between the natural gradient and stochastic gradient is

$$\frac{\partial L(\boldsymbol{W})}{\partial \boldsymbol{W}_{\mathrm{nat}}} = \frac{\partial L(\boldsymbol{W})}{\partial \boldsymbol{W}}\boldsymbol{W}^{\mathrm{T}}\boldsymbol{W} \tag{5.29}$$

Substituting this into the iterative formula of the stochastic gradient method, we can obtain the iteration formula for the natural gradient method,

$$\begin{aligned}\boldsymbol{W}(k+1) &= \boldsymbol{W}(k) + \alpha(k)[\boldsymbol{W}^{-\mathrm{T}}(k) - E\{\varphi(\boldsymbol{y})\boldsymbol{x}^{\mathrm{T}}\}]\boldsymbol{W}^{\mathrm{T}}(k)\boldsymbol{W}(k) \\ &= \boldsymbol{W}(k) + \alpha(k)[\boldsymbol{I} - E\{\varphi(\boldsymbol{y})\boldsymbol{y}^{\mathrm{T}}\}]\boldsymbol{W}(k)\end{aligned} \tag{5.30}$$

Practice has shown that separation by the natural gradient method works well, and it has been used as a standard learning algorithm. The algorithm does not require whitening preprocessing of the observed data. Cardoso (1998) included the whitening process in the blind separation algorithm and proposed the EASI algorithm:

$$\boldsymbol{W}(k+1) = \boldsymbol{W}(k) + \alpha(k)[\boldsymbol{I} - E\{\boldsymbol{y}\boldsymbol{y}^{\mathrm{T}}\} + \boldsymbol{I} - E\{\varphi(\boldsymbol{y})\boldsymbol{y}^{\mathrm{T}}\}]\boldsymbol{W}(k) \tag{5.31}$$

5.2.7 Impact of the Statistical Properties of the Signal on the Algorithm

Another idea for the independence measure comes from the central limit theorem. The central limit theorem indicates that the result of the superposition of a group of random variables with means and variances of the same order of magnitude is close to a Gaussian distribution. Therefore, if the observed signal is a linear combination of a number of separate sources, it will be closer to a Gaussian distribution than the source signals, or the non-Gaussian nature of the source signals will be stronger than the non-Gaussian nature of the observed signal. According to this idea, we can use the non-Gaussian nature of the separation results as a measure: when the non-Gaussianity reaches its maximum, the best separation results are achieved.

In the actual calculation, the degree of non-Gaussianity is usually expressed as kurtosis of the fourth-order cumulants. The kurtosis of random variables $y_i(t)$ is defined as

$$\text{kurt}(y_i) = \text{cum}(y_i^{\,4}(t)) = E\{y_i^{\,4}(t)\} - 3(E\{y_i^{\,2}(t)\})^2 \tag{5.32}$$

For a Gaussian signal $\text{cum}(y_i^{\,4}(t)) = 0$, if $\text{cum}(y_i^{\,4}(t)) < 0$, $y_i(t)$ is called a sub-Gaussian signal, and if $\text{cum}(y_i^{\,4}(t)) > 0$, it is called a super-Gaussian signal. Figuratively speaking, random variables that obey the super-Gaussian are more spiky than those obeying the Gaussian distribution, with the Laplace distribution being a typical density function obeying the super-Gaussian distribution. On the other hand, random variables that obey the sub-Gaussian distribution will be flatter than those obeying the Gaussian distribution, with the uniform distribution being a typical density function obeying the sub-Gaussian distribution. Both of these distributions are shown in Figure 5.3.

In Figure 5.3, the dashed line represents a Laplace super-Gaussian distribution, the dotted line represents a Pearson's sub-Gaussian distribution, while the solid line represents the Gaussian distribution. The means of the three density functions are all zero with a variance of 1.

The fact that the source signal is divided into sub-Gaussian and super-Gaussian signals is very important, because in the real world, signals satisfying the Gaussian distribution are not common. Usually signals are only very close to the Gaussian distribution (kurtosis close to 0), although super-Gaussian and sub-Gaussian signals are common. For example, there are both super-Gaussian and sub-Gaussian distributions

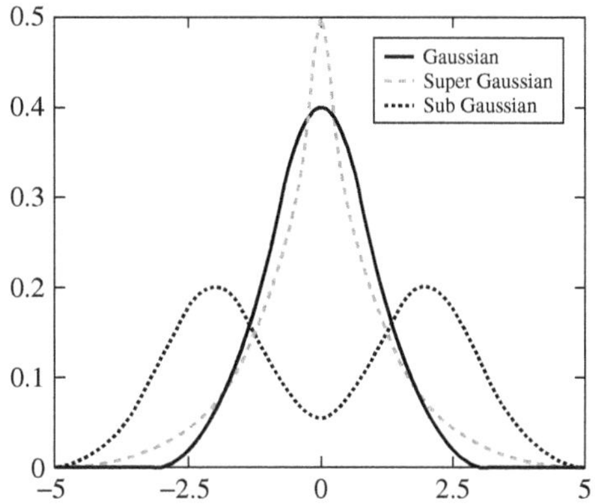

Figure 5.3 Super-Gaussian and sub-Gaussian distribution density functions of a Gaussian distribution

among biomedical signals. As another example, voice signals usually have a super-Gaussian distribution, while image data of natural features generally have a sub-Gaussian distribution. At the same time, the kurtosis of the source signal has a very important effect on the separation results of many algorithms.

5.2.7.1 Symmetrical Distribution Assumptions of Traditional ICA

As explained above, most of the existing ICA algorithms in the separation process involve only second and fourth-order statistics of the signals, which shows that the algorithm has an implicit assumption of the symmetrical distribution of the source signal. Traditional ICA assumes that the source signals are strictly statistically independent, and that the PDF is symmetric. The characteristics of the signal extracted by ICA are determined by the PDF of the corresponding model. For example, the symmetry of the PDF of the signal extracted through application of the ICA standard model is generally high; natural source signals, such as voice signals and the like, typically have this feature. Figure 5.4 shows histograms of the PDFs of a few sound source signals. These histograms show that speech signals obey a symmetric distribution, and hence the symmetric model of the conventional ICA algorithm can be applied to the separation of natural source signals, which can yield a better separation effect.

However, in actual biomedicine, engineering construction, signal processing, environmental science, information management, and economic forecasts, as well as many other applications, there are many very important asymmetrically distributed signals, which cannot guarantee that these assumptions are met. Moreover, use of the

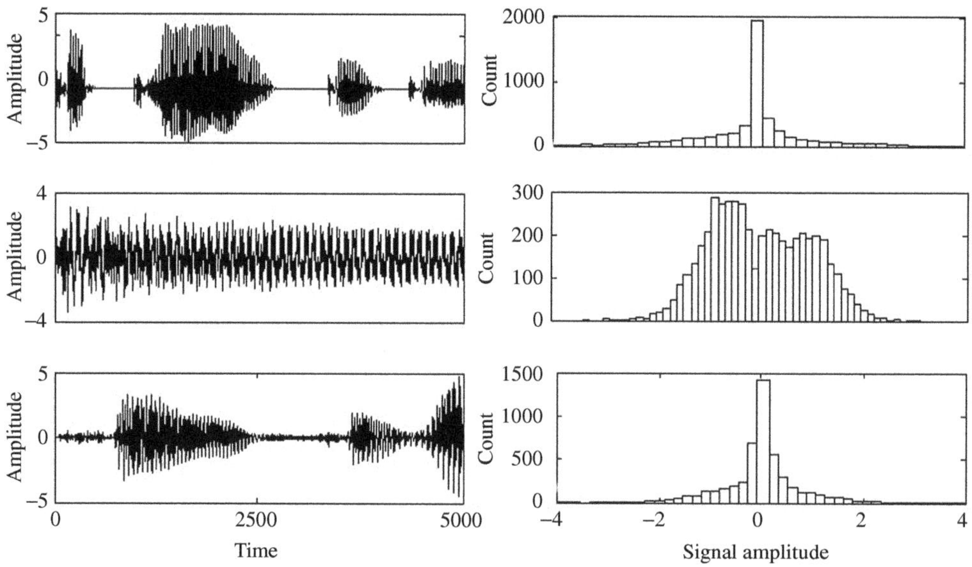

Figure 5.4 PDF of the speech signal histogram

traditional method is usually for conventional thinking or for the sake of calculation convenience. Taking into account the case of a non-symmetrical distribution, the ICA algorithm based on fourth-order statistics may not produce the desired results, and the traditional ICA model cannot be applied to solving BSS problems. The optimization ICA method described in Chapter 13 tries to solve the space source signal decomposition problem of functional magnetic resonance imaging data that obey a skewed distribution.

5.2.7.2 Non-symmetrical Distribution and Skewness

In statistics, distributions can be divided into symmetric and non-symmetric distributions according to the shape of the histogram. In the symmetric distribution in Figure 5.5a, the right and left halves of the histogram mirror each other, such as in a

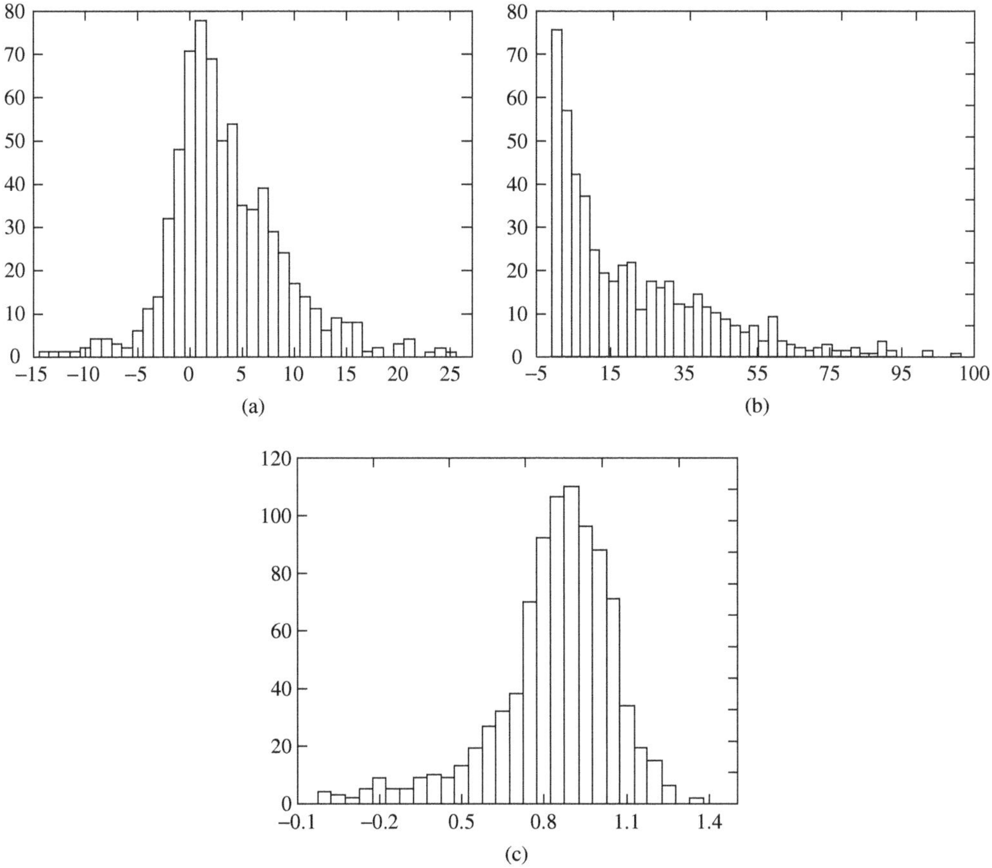

Figure 5.5 Histograms of a symmetric distribution and the PDFs of asymmetric distributions (a) symmetric distribution, (b) skewed right distribution, and (c) skewed left distribution

normal distribution. However, the distribution that is more prevalent in reality is the asymmetric distribution, also referred to as a skewed distribution, either positively skewed or negatively skewed. Only a small number of data in a positively skewed distribution have very large values, and thus after the peak in the histogram, there is a long "tail" to the right, which is also called skewed right, as shown in Figure 5.5b. Conversely, the histogram curve of data in a negatively skewed distribution is exactly the opposite. Because there are only a few small values, the "tail" appears before the peak; this is also called skewed left, as shown in Figure 5.5c.

Generally skewness can be used to measure numerically the skew characteristics of the data. Similar to the introduced kurtosis (fourth-order statistics), skewness also reflects the characteristics of higher-order statistics. Skewness, also referred to as the skewness coefficient, is a measure of the skewness direction and the extent thereof in a statistical data distribution, and is a numeric characteristic that determines the asymmetric extent of the statistical data distribution. A formula to calculate skewness that uses the third-order cumulant is given as

$$\text{skew}(y_i) = \frac{\text{cum}(y_i{}^3(t))}{(\text{cum}(y_i{}^2(t)))^{3/2}} = \frac{E\{y_i{}^3(t)\}}{\sigma^3} \tag{5.33}$$

where σ is the standard deviation. Under normal circumstances, if the statistical data have a right-skewed distribution, $\text{skew}(\cdot) > 0$, where a greater value implies a higher degree of right-skewness. Conversely, if the statistical data have a left-skewed distribution, $\text{skew}(\cdot) < 0$, with a smaller value implying a higher degree of left-skewness. Clearly if the statistical data have a symmetrical distribution, $skew(\cdot) = 0$.

5.3 Chapter Summary

ICA is a very common statistical method that can represent random observed data as a linear transformation of various statistically independent components. In this chapter, we formalized ICA as an estimate of the independent latent variable generative mode. This decomposition is identifiable or adaptable in the case where the independent components are non-Gaussian (or there is only a single Gaussian component). To simplify the estimation problem, we can carry out whitening of the data. The whitening process estimates some parameters, leaving only an undetermined orthogonal transform. Moreover, by using higher-order information contained in the non-Gaussian variables, we can estimate the corresponding orthogonal transform.

We detailed the origin and development of the ICA algorithm and its application in various fields; briefly summarized the basic principles of the ICA algorithm and the main problems in ICA; and provided a more detailed description of several typical cost functions for the BSS algorithm based on ICA, with an analysis and derivation of several typical learning algorithms of ICA based on the cost functions. Finally, we discussed the impact of the statistical properties of the signal on the ICA algorithm.

References

Amari, S.I. (1998). Natural gradient works efficiently in learning. *Neural Computation*, **10**(2), 251–276.

Amari, S.I., and Cardoso, J.F. (1997). Blind source separation-semiparametric statistical approach. *IEEE Transactions on Signal Processing*, **45**(11), 2692–2700.

Amari, S.I., and Cichocki, A. (1998). Adaptive blind signal processing-neural network approaches. *Proceedings of the IEEE*, **86**(10), 2026–2048.

Bach, F.R., and Jordan, M.I. (2003). Kernel independent component analysis. *The Journal of Machine Learning Research*, **3**, 1–48.

Bell, A.J., and Sejnowski, T.J. (1995). An information-maximization approach to blind separation and blind deconvolution. *Neural Computation*, **7**(6), 1004–1034.

Burel, G. (1992). Blind separation of sources: a nonlinear neural algorithm. *Neural Networks*, **5**(6), 937–947.

Cao, X.R., and Liu, R.W. (1996). General approach to blind source separation. *IEEE Transactions on Signal Processing*, **44**(3), 562–571.

Cardoso, J.F. (1997). Estimating equations for source separation. Proceedings of ICASSP-97, IEEE, pp. 3449–3452.

Cardoso, J.F. (1998). Blind signal separation: statistical principles. *Proceedings of the IEEE*, **86**(10), 2009–2025.

Cichocki, A., and Unbehauen, R. (1996). Robust neural networks with on-line learning for blind identification and blind separation of sources. *IEEE Transactions on Circuits and Systems I: Fundamental Theory and Applications*, **43**(11), 894–906.

Cichocki, A., Unbehauen, R., and Rummert, E. (1994). Robust learning algorithm for blind separation of signals. *Electronics Letters*, **30**(17), 1386–1387.

Comon, P. (1994). Independent component analysis, a new concept? *Signal Processing*, **36**(3), 287–314.

Comon, P., Jutten, C., and Hérault, J. (1991). Blind separation of sources, Part II: Problems statement. *Signal Processing*, **24**(1), 11–20.

Deville, Y. (1996). A unified stability analysis of the Hérault-Jutten source separation neural network. *Signal Processing*, **51**(3), 229–233.

Deville, Y. (1999). Analysis of the convergence properties of self-normalized source separation neural networks. *IEEE Transactions on Signal Processing*, **47**(5), 1272–1287.

Everson, R., and Roberts, S. (1999). Independent component analysis: a flexible nonlinearity and decorrelating manifold approach. *Neural Computation*, **11**(8), 1957–1983.

Gaeta, M., and Lacoume, J.L. (1990). Source separation without a priori knowledge: the maximum likelihood solution. Proceedings of EUSIPCO, IEEE, pp. 621–624.

Giannakis, G.B., and Swami, A. (1988). New results on state-space and input-output identification of non-Gaussian processes using cumulants. 31st Annual Technical Symposium, International Society for Optics and Photonics, pp. 199–204.

Girolami, M., and Fyfe, C. (1996a). Blind separation of sources using exploratory projection pursuit networks. International Conference on Engineering Applications of Neural Networks, IEEE, pp. 249–252.

Girolami, M., and Fyfe, C. (1996b). Higher order cumulant maximization using non-linear Hebbian and anti-Hebbian learning for adaptive blind separation of source signals. 3rd International Workshop in Signal and Image Processing, IEEE, pp. 141–144.

Girolami, M., and Fyfe, C. (1997a). Multivariate density factorization for independent component analysis: an unsupervised artificial neural network approach. Proceedings of 3rd International Workshop on Artificial Intelligence and Statistics, IEEE, pp. 223–230.

Girolami, M., and Fyfe, C. (1997b). Kurtosis extrema and identification of independent components: a neural network approach. IEEE International Conference on Acoustics, Speech, and Signal Processing, IEEE, pp. 3329–3332.

Herault, J., and Jutten, C. (1986). Space or time adaptive signal processing by neural network models. AIP Conference Proceedings, IEEE, p. 206.

Hyvärinen, A. (1997a). A family of fixed-point algorithms for independent component analysis. Proceedings of ICASSP-97, IEEE, pp. 3917–3920.

Hyvärinen, A. (1997b). One-unit contrast functions for independent component analysis: a statistical analysis. Neural Networks for Signal Processing VII, IEEE, pp. 388–397.

Hyvärinen, A. (1999a). Fast and robust fixed-point algorithms for independent component analysis. *IEEE Transactions on Neural Networks*, **10**(3), 626–634.

Hyvärinen, A. (1999b). The fixed-point algorithm and maximum likelihood estimation for independent component analysis. *Neural Processing Letters*, **10**(1), 1–5.

Hyvärinen, A., and Oja, E. (1997). A fast fixed-point algorithm for independent component analysis. *Neural Computation*, **9**(7), 1483–1492.

Jutten, C., and Hérault, J. (1991). Blind separation of sources, part I: an adaptive algorithm based on neuromimetic architecture. *Signal Processing*, **24**(1), 1–10.

Karhunen, J., and Joutsensalo, J. (1994). Representation and separation of signals using nonlinear PCA type learning. *Neural Networks*, **7**(1), 113–127.

Karhunen, J., Oja, E., Wang, L. *et al.* (1997). A class of neural networks for independent component analysis. *IEEE Transactions on Neural Networks*, **8**(3), 486–504.

Karhunen, J., Pajunen, P., and Oja, E. (1998). The nonlinear PCA criterion in blind source separation: relations with other approaches. *Neurocomputing*, **22**(1), 5–20.

Karhunen, J., Wang, L., and Vigario, R. (1995). Nonlinear PCA type approaches for source separation and independent component analysis. IEEE International Conference on Neural Networks, IEEE, pp. 995–1000.

Linsker, R. (1988). Self-organization in a perceptual network. *Computer*, **21**(3), 105–117.

Linsker, R. (1989). An application of the principle of maximum information preservation to linear systems. *Advances in Neural Information Processing Systems*, Vol. **1**, Morgan Kaufman Publishers Inc., San Mateo, CA, pp. 186–194.

Masnadi-Shirazi, A., and Rao, B. (2009). Independent vector analysis incorporating active and inactive states. IEEE International Conference on Acoustics, Speech and Signal Processing, IEEE, pp. 1837–1840.

Oja, E. (1997). The nonlinear PCA learning rule in independent component analysis. *Neurocomputing*, **17**(1), 25–45.

Pearlmutter, B.A., and Parra, L.C. (1996). A context-sensitive generalization of ICA. International Conference on Neural Information Processing, IEEE, pp. 151–156.

Sorouchyari, E. (1991). Blind separation of sources, part III: stability analysis. *Signal Processing*, **24**(1), 21–29.

Theis, F.J. (2004). Uniqueness of complex and multidimensional independent component analysis. *Signal Processing*, **84**(5), 951–956.

Tong, L., Inouye, Y., and Liu, R.W. (1993). Waveform-preserving blind estimation of multiple independent sources. *IEEE Transactions on Signal Processing*, **41**(7), 2461–2470.

Tong, L., Liu, R.W., Soon, V.C., and Huang, Y.F. (1991). Indeterminacy and identifiability of blind identification. *IEEE Transactions on Circuits and Systems*, **38**(5), 499–509.

Tsai, D.M., Tseng, Y.H., Chao, S.M., and Yen, C.H. (2006). Independent component analysis based filter design for defect detection in low-contrast textured images. 18th International Conference on Pattern Recognition, IEEE, pp. 231–234.

Yang, H.H., and Amari, S.I. (1997). Adaptive online learning algorithms for blind separation: maximum entropy and minimum mutual information. *Neural Computation*, **9**(7), 1457–1482.

6

Fast Independent Component Analysis and Its Application

6.1 Overview

Fast Independent Component Analysis (FastICA), also known as the fixed-point algorithm, was proposed and developed by Hyvärinen (1998, 1999), Hyvärinen, Hoyer, and Oja (1998), Hyvärinen and Oja (1997, 2000), and Hyvärinen and Karhunen (2001) in the Computer and Information Science Laboratory at Finland's Aalto University's Institute of Technology (formerly known as the Helsinki University of Technology). Based on the non-Gaussian maximization principle, FastICA, which is a fast algorithm for ICA , uses fixed-point iteration theory to search for the non-Gaussian maximum of $\boldsymbol{W}^{\mathrm{T}}\boldsymbol{x}$. The algorithm adopts Newton's iterative algorithm to process a large number of sampled points of observation variable $\boldsymbol{x}$ as a batch and separates an independent component each time using maximum negative entropy as the target function. There are two methods for obtaining the independent component: the deflation method, which respectively updates each column of the separation matrix and extracts an independent component each time, and the symmetry method, which updates each column of the separation matrix, corresponding to all the independent components. Taking data from the Tuanjiegou gold deposit in Heilongjiang province and geochemical data of Gaoyao City, a copper mining district in Guangdong province, as examples, we applied FastICA to the mineral data to investigate mineral prediction. Compared with the predicted results of the original complete units, the analysis results are more in line with the actual results of the exploration, as confirmed later by further exploration results. The delineation of the prediction range is more clearly defined and more accurate. When used in the classification of remote sensing images, the classification results of FastICA are more accurate than those of direct classification. Moreover, when applied to the reduction of image noise, the performance of FastICA is superior to traditional reduction methods, such as Wiener filtering and wavelet noise filtering, among others.

Blind Source Separation: Theory and Applications, First Edition. Xianchuan Yu, Dan Hu and Jindong Xu.

We also applied the FastICA algorithm to preprocessing high-resolution remote sensing images with each band having a clear distinction after preprocessing.

6.1.1 Deflation Method

Through learning, this method finds the direction, which is a weight vector $\boldsymbol{w}_i$ of $\boldsymbol{W}$. Moreover, it creates a projection, $y_i = \boldsymbol{w}_i^{\mathrm{T}}\boldsymbol{x}$, which is the largest non-Gaussian. This method separates an independent component from the observation signals at every iteration. FastICA uses negative entropy as the objective function. As mentioned in Chapter 2, negative entropy can be approximated by kurtosis, which is particularly sensitive to specific values. So Hyvärinen and Karhunen (2001) proposed a more robust and faster method to approximate negative entropy:

$$J(y_i) \approx [E\{G(y_i)\} - E\{G(v)\}]^2 \tag{6.1}$$

where $\mathbf{y}_i$ is an output variable with zero mean and unit variance; $\boldsymbol{v}$ is a Gaussian random variable with zero mean and unit variance; and G is a nonlinear function of a non-square, which in practice is defined as

$$G_1(y) = \frac{1}{a_1}\log\cosh a_1 y \tag{6.2}$$

$$G_2(y) = -\exp(-y^2/2) \tag{6.3}$$

$$G_3(y) = \frac{1}{4}y^4 \tag{6.4}$$

where $1 \leq a_1 \leq 2$ and generally set to 1. Here, G_1 is suitable for the situation in which sub-Gaussian and super-Gaussian signals coexist, G_2 is suitable for the separation of super-Gaussian mixed signals, and G_3 is suitable for the separation of sub-Gaussian mixed signals, usually by maximizing the following function $J_G(\boldsymbol{W})$:

$$J_G(\boldsymbol{W}) = [E\{G(\boldsymbol{W}^{\mathrm{T}}\boldsymbol{x}) - E\{G(\boldsymbol{v})\}\}]^2 \tag{6.5}$$

The variance of independent components can be arbitrarily assigned, so we can specify $E\{(\boldsymbol{W}^{\mathrm{T}}\boldsymbol{x})^2\} = 1$. It is observed that the maximum value of $J_G(\boldsymbol{W})$ is usually obtained when $E\{G(\boldsymbol{W}^{\mathrm{T}}\boldsymbol{x})\}$ reaches its maximum. The problem is then transformed into that which, under the condition $E\{(\boldsymbol{W}^{\mathrm{T}}\boldsymbol{x})^2\} = 1$, seeks the maximum of $E\{G(\boldsymbol{W}^{\mathrm{T}}\boldsymbol{x})\}$. This can be solved by the Lagrange multiplier method, with Lagrange multiplier β, giving

$$\begin{aligned} J_G(\boldsymbol{W}) &= E\{G(\boldsymbol{W}^{\mathrm{T}}\boldsymbol{x})\} - \beta(\|\boldsymbol{W}\|^2 - 1) \\ &= E\{G\{\boldsymbol{W}^{\mathrm{T}}\boldsymbol{x}\}\} - \beta(\boldsymbol{W}^{\mathrm{T}}\boldsymbol{W} - 1) \end{aligned} \tag{6.6}$$

Let function g be the derivative of function G. Then by creating a solution for the derivative of the above equation $\boldsymbol{W}$ with the order of the derivative zero, we obtain the following equation:

$$E\{\boldsymbol{x}g(\boldsymbol{W}^{\mathrm{T}}\boldsymbol{x})\} + \beta \boldsymbol{W} = 0 \tag{6.7}$$

Newton's iterative method is used to solve this equation. If the order of the left-hand side of Equation 6.7 is F, and g' is the derivative of function g, then its gradient is given by

$$\frac{\partial F}{\partial \boldsymbol{W}} = E\{\boldsymbol{x}\boldsymbol{x}^{\mathrm{T}} g'(\boldsymbol{W}^{\mathrm{T}}\boldsymbol{x})\} + \beta \boldsymbol{I} \tag{6.8}$$

Since the data are standardized, it is possible to approximate the first term of the above equation as

$$E\{\boldsymbol{x}\boldsymbol{x}^{\mathrm{T}} g'(\boldsymbol{W}^{\mathrm{T}}\boldsymbol{x})\} \approx E\{\boldsymbol{x}\boldsymbol{x}^{\mathrm{T}}\}E\{g'(\boldsymbol{W}^{\mathrm{T}}\boldsymbol{x})\} = E\{g'(\boldsymbol{W}^{\mathrm{T}}\boldsymbol{x})\}\boldsymbol{I}$$

Thus, we can get the iterative equation of **W** as

$$\boldsymbol{W}(k+1) = \boldsymbol{W}(k) - [E\{\boldsymbol{x}g(\boldsymbol{W}(k)^{\mathrm{T}}\boldsymbol{x})\} + \beta \boldsymbol{W}]/[E\{g'(\boldsymbol{W}(k)^{\mathrm{T}}\boldsymbol{x})\} + \beta] \tag{6.9}$$

Equation 6.9 can be further simplified as

$$\boldsymbol{W}(k+1) = E\{\boldsymbol{x}g(\boldsymbol{W}(k)^{\mathrm{T}}\boldsymbol{x}) - E\{g'(\boldsymbol{W}(k)^{\mathrm{T}}\boldsymbol{x})\}\boldsymbol{W}(k)\} \tag{6.10}$$

where k is the number of iterations and g' is selected from the equations below:

$$\begin{aligned} g'(y) &= a_1(1 - \tanh^2(a_1 y)) \\ g'(y) &= (1 - y^2)\exp(-y^2/2) \\ g'(y) &= 3y^2 \end{aligned} \tag{6.11}$$

After the iteration, a normalization process of the weight vector is required:

$$\boldsymbol{W}(k+1) = \frac{\boldsymbol{W}(k+1)}{\|\boldsymbol{W}(k+1)\|} \tag{6.12}$$

6.1.2 Symmetry Method

In practical applications, the deflation method is relatively time-consuming. If we consider computing multiple independent components at the same time, the parallel mode of FastICA can be used. The symmetry method aims to satisfy constraint $E\{(\boldsymbol{w}_k^{\mathrm{T}}\boldsymbol{x})(\boldsymbol{w}_j^{\mathrm{T}}\boldsymbol{x})\} = \delta_{jk}$ and maximize $\sum_{i=1}^{M} J_G(\boldsymbol{w}_i), i = 1, \ldots, n$.

Running a total of M element algorithms, we can find the weight matrix $\boldsymbol{W} = [\boldsymbol{w}_1, \boldsymbol{w}_2, \ldots, \boldsymbol{w}_M]^{\mathrm{T}}$. To prevent different maximum entropies from converging to the same vector, vectors $\boldsymbol{w}_1, \boldsymbol{w}_2, \ldots, \boldsymbol{w}_M$ must be orthogonal. If the corresponding weight value of p independent components has been estimated, when evaluating the corresponding weight value of the $(p+1)$th independent component, we can use the following method for orthogonalization and normalization of the weight values:

$$\boldsymbol{w}_{p+1}(k+1) = \boldsymbol{w}_{p+1}(k) - \sum_{j=1}^{p} \boldsymbol{w}_{p+1}^{\mathrm{T}}(k)\boldsymbol{C}\boldsymbol{w}_j\boldsymbol{w}_j \tag{6.13}$$

$$\boldsymbol{w}_{p+1}(k+1) = \frac{\boldsymbol{w}_{p+1}(k+1)}{\sqrt{\boldsymbol{w}_{p+1}^{\mathrm{T}}(k+1)c\boldsymbol{w}_{p+1}(k+1)}}. \tag{6.14}$$

In an actual application, the following equation can be used to iterate:

$$\boldsymbol{W} = (\boldsymbol{W}\boldsymbol{C}\boldsymbol{W}^{\mathrm{T}})^{-1/2}\boldsymbol{W} \tag{6.15}$$

where $\boldsymbol{W} = [\boldsymbol{w}_1, \boldsymbol{w}_2, \ldots, \boldsymbol{w}_M]^{\mathrm{T}}$ and $(\boldsymbol{W}\boldsymbol{C}\boldsymbol{W}^{\mathrm{T}})^{-1/2}$ is obtained from the decomposition of eigenvalue $\boldsymbol{W}\boldsymbol{C}\boldsymbol{W}^{\mathrm{T}} = \boldsymbol{E}\boldsymbol{D}\boldsymbol{E}^{\mathrm{T}}$:

$$(\boldsymbol{W}\boldsymbol{C}\boldsymbol{W}^{\mathrm{T}})^{-1/2} = \boldsymbol{E}\boldsymbol{D}^{-1/2}\boldsymbol{E}^{\mathrm{T}} \tag{6.16}$$

Compared with other algorithms, FastICA has the following advantages:

1. The algorithm based on gradient descent has linear convergence, whereas the convergence rate of FastICA is cubic (at least quadratic), which means that FastICA converges very quickly.
2. FastICA does not need to select a step size parameter, which differs from gradient-based algorithms.
3. FastICA can calculate an independent non-Gaussian signal using an arbitrary nonlinear function g, while other algorithms need to evaluate the probability density function related to the selection of the nonlinear function.
4. FastICA inherits many advantages from neural network algorithms, including distributed parallel calculation, simple calculation, and smaller memory requirements.

The performance of FastICA is very good and suitable for all types of non-Gaussian data, and is thus widely used.

For whitened data, FastICA of a signal vector has the form:

$$\mathbf{w}(k) = E\{\boldsymbol{x}g(\boldsymbol{w}(k-1)^{\mathrm{T}}\boldsymbol{x})\} - E\{g'(\boldsymbol{w}(k-1)^{\mathrm{T}}\boldsymbol{x})\}\boldsymbol{w}(k-1) \tag{6.17}$$

After iteration, weight vector $\mathbf{w}$ is normalized to a unit vector. In practical application, the mathematical expectation is estimated from the average value of the input samples. The vector from the FastICA algorithm can be linked to a system utilized to estimate independent components. This system can estimate independent components by progressively eliminating correlation or eliminating correlation in parallel. FastICA based on a recursive fixed point, as proposed by Hyvärinen, is a fast and robust algorithm. By translating the independent analysis of n original signals to n optimization problems, it can separate mixed signals one by one. Before applying FastICA, it is necessary to ensure that the correlation matrix of $\boldsymbol{x}$ satisfies $E\{\boldsymbol{x}\boldsymbol{x}^{\mathrm{T}}\} = \boldsymbol{I}$, which is known as whitening.

6.2 FastICA Algorithm

6.2.1 Ordinary FastICA

FastICA realizes the learning process of $\boldsymbol{W}$ ($\boldsymbol{W} = \boldsymbol{B}^{\mathrm{T}}$) by maximizing negative entropy with its objective function

$$J(y) \approx \sum_{i=1}^{p} k_i\{E[G_i(y)] - E[G_i(\boldsymbol{v})]\}^2 \tag{6.18}$$

where k_i is a positive constant and $\boldsymbol{v}$ is a random variable of the Gaussian with zero mean and unit variance. Supposing that $\boldsymbol{y}$ has zero mean and unit variance, $G_i(\cdot)$, which is a non-quadratic function, is usually selected as follows:

1. If the original signals are super-Gaussian and sub-Gaussian:

$$G_1(\boldsymbol{u}) = \frac{1}{a_1}\log\cosh(a_1\boldsymbol{u}), \quad G'_1(\boldsymbol{u}) = \tanh(a_1\boldsymbol{u}), \quad 1 \le a_1 \le 2 \tag{6.19}$$

2. If all the original signals are super-Gaussian or Gaussian, demanding robustness:

$$G_2(\boldsymbol{u}) = -\frac{1}{a_2}\exp(-a_2\boldsymbol{u}^2/2), \quad G'_2(\boldsymbol{u}) = \boldsymbol{u}\exp(-a_2\boldsymbol{u}^2/2), \quad a_2 \approx 1 \tag{6.20}$$

3. If all the original signals are sub-Gaussian:

$$G_3(\boldsymbol{u}) = 0.25\boldsymbol{u}^4, \quad G'_3(\boldsymbol{u}) = \boldsymbol{u}^3 \tag{6.21}$$

4. If all the original signals have a skewed distribution:

$$G_4(\boldsymbol{u}) = \boldsymbol{u}^3/3, \quad G'_4(\boldsymbol{u}) = \boldsymbol{u}^2 \tag{6.22}$$

Specifically, when $p = 1$, the approximation of negative entropy is expressed as

$$J(\boldsymbol{y}) \propto \{E[G(\boldsymbol{y})] - E[G(\boldsymbol{v})]\}^2 \tag{6.23}$$

Since $\boldsymbol{y} = \boldsymbol{w}^{\mathrm{T}}\boldsymbol{X}$ ($\mathbf{y}$ is one of the independent components, $\boldsymbol{w}$ is a row in the separation matrix, and $\boldsymbol{X}$ is a matrix of mixed signals), the approximation function of negative entropy can be defined as

$$J_G(\boldsymbol{W}) \propto \{E[G(\boldsymbol{w}^{\mathrm{T}}\boldsymbol{X})] - E[G(\boldsymbol{v})]\}^2 \tag{6.24}$$

The problem is transformed to solve separation matrix $\boldsymbol{W}$ and maximize function $J_G(\boldsymbol{W})$ using the separate evaluated signal $\boldsymbol{y} = \boldsymbol{w}^{\mathrm{T}}\boldsymbol{X}$. Since $E\{(\boldsymbol{w}^{\mathrm{T}}\boldsymbol{x})^2\} = 1$, the objective function is defined as

$$\begin{cases} \max & J_G(\boldsymbol{w}) = E\{G(\boldsymbol{w}^{\mathrm{T}}\boldsymbol{X})\} \\ s.t. & \|\boldsymbol{w}\| = 1 \end{cases} \tag{6.25}$$

Under Kuhn–Tucker conditions, the problem is transformed into an unrestricted optimization problem with the following objective function:

$$F(\boldsymbol{w}) = E[G(\boldsymbol{w}^{\mathrm{T}}\boldsymbol{X})] + c(\|\boldsymbol{w}\|^2 - 1) \tag{6.26}$$

where c is a constant. The optimization of the objective function is solved by Newton's method, which is the iterative equation for FastICA.

$$\boldsymbol{w}^{+} = E\{\boldsymbol{X}g(\boldsymbol{w}^{\mathrm{T}}\boldsymbol{X})\} - E\{g'(\boldsymbol{w}^{\mathrm{T}}\boldsymbol{X})\}\boldsymbol{w} \tag{6.27}$$

$$\boldsymbol{w}^{*} = \frac{\boldsymbol{w}^{+}}{\|\boldsymbol{w}^{+}\|} \tag{6.28}$$

The steps in the FastICA iteration algorithm based on maximum negative entropy are given below.

Algorithm FastICA iteration algorithm based on maximum negative entropy

Input: Whitened matrix $\boldsymbol{X}$
Output: Independent component $\boldsymbol{Y}$
Steps:

1. $n = 0$ and initialize weight vector $\boldsymbol{w}(0)$;
2. $n = n + 1$;
3. Adjust $\boldsymbol{w}$ and $\boldsymbol{w}(n+1) = E\{\boldsymbol{X}g(\boldsymbol{w}^{\mathrm{T}}(n)\boldsymbol{X})\} - E\{g'(\boldsymbol{w}^{\mathrm{T}}(n)\boldsymbol{X})\}\boldsymbol{w}(n)$;
4. Perform a normalization process: $\boldsymbol{w}(n+1) = \frac{\boldsymbol{w}(n+1)}{\|\boldsymbol{w}(n+1)\|}$;
5. If there is no convergence, go back to step (3);
6. Once the algorithm has converged, an independent component is obtained, $y_1 = \hat{s}_1 = \boldsymbol{w}\boldsymbol{X}$.

The expected value in the equation for weight adjustment in step (4) must be solved; this is approximated by the mean of the samples in the actual calculation. Theoretically, it is hoped that all the samples are involved in the computation, although this leads to a vast amount of calculation. It is possible to randomly extract a moderate length data block to obtain the average value each time, where the locations of the extracted data blocks are as dispersed as possible and the lengths of the data blocks is determined by the convergence performance. The iterative steps of the fixed point algorithm given above for extracting an independent component were developed to extract multiple independent components, with the weight vectors $\boldsymbol{w}_1^{\mathrm{T}}, \boldsymbol{w}_2^{\mathrm{T}}, \cdots, \boldsymbol{w}_n^{\mathrm{T}}$, consisting of the row vectors of the ICA separation matrix $\boldsymbol{W}$, iteratively obtained one by one. To prevent the convergence direction of the new iterative vector $\boldsymbol{w}_{i+1}$ from being the same as the directions of the i solved weight vectors given above, it is vital that the row vector of the separation matrix, corresponding to the new independent component, and the other row vectors of the obtained separation matrix are orthogonal to each other. Therefore, when solving the new vector w_{i+1}, an orthogonal projection operation must be included in the loop to remove correlation. The corresponding equations are given below.

$$w_{i+1} = w_{i+1} - \sum_{j=1}^{i} w_{i+1}^{\mathrm{T}} w_j w_j \tag{6.29}$$

$$w_{i+1} = \frac{w_{i+1}}{\sqrt{w_{i+1}^{\mathrm{T}} w_{i+1}}} \tag{6.30}$$

There are several characteristics of FastICA based on the criterion of approximate negative entropy:

1. The fast convergence speed is an important feature of the algorithm, which converges after 5–10 iterations. The rate of convergence is usually quadratic, although it can be cubic if the independent components are symmetric.
2. Compared with the descent-based method, the algorithm does not need a parameter for the learning step μ as well as various other parameters.
3. The algorithm extracts only a single separate component each time, rather than all of the components. Therefore, if it extracts a component and there is enough prior knowledge, the next component can be quickly extracted with the amount of calculation reduced.
4. Any nonlinear function G can iteratively obtain an independent component based on the greatest non-Gaussian. Moreover, if the nonlinear function G is appropriately selected, it can optimize the performance of the algorithm.

6.2.2 *Optimization of FastICA*

To reduce the number of requirements of the Jacobian matrix in Newton's iterative process for nonlinear equations, a general improvement of the Newton iterative method is realized if the Jacobian matrices of all the iterative process are taken as $J(w_0)$. The iterative equation then becomes

$$w_{k+1} = w_k - F(w_k)/J(w_0) \tag{6.31}$$

Though this improvement can reduce the computational load, it also reduces the rate of convergence. As a result, when applied to ICA, it may lead to the algorithm ultimately not converging. Thus, use of the algorithm for ICA is actually not beneficial. On the other hand, the improved method given below can reduce the number of iterations by a large margin on condition that the number of requirements for the Jacobian matrix does not increase, thereby reducing the computational load and increasing the speed of the algorithm. Assuming that $\boldsymbol{w}_k$ has been obtained, we get $\boldsymbol{w}_{k+1}$ as follows:

$$\begin{cases} \boldsymbol{w}_k^{(0)} = \boldsymbol{w}_k \\ \boldsymbol{w}_k^{(i)} = \boldsymbol{w}_k^{(i-1)} - \frac{F(\boldsymbol{w}_k^{(i-1)})}{J(\boldsymbol{w}_k^{(0)})} \quad i = 1, 2, \dots, m \\ \boldsymbol{w}_{k+1} = \boldsymbol{w}_k^{(m)} \end{cases} \tag{6.32}$$

We can prove that the convergence order of Equation 6.32 is $m + 1$. Since it needs to compute a $J(\boldsymbol{w})$ every m iterations, the amount of computation decreases and the speed of iteration increases. If $m = 2$, Equation 6.32 can be simplified as

$$\boldsymbol{w}_{k+1} = \boldsymbol{w}_k - \frac{F(\boldsymbol{w}_k) + F(\boldsymbol{w}_k - F(\boldsymbol{w}_k - F(\boldsymbol{w}_k)/J(\boldsymbol{w}_k)))}{J(\boldsymbol{w}_k)} \tag{6.33}$$

Substituting $F(\boldsymbol{w}_k)$ and $J(\boldsymbol{w}_k)$ in Equation 6.33, the iteration of the improved FastICA (M-FastICA) can be obtained for $m = 2$. The algorithm for M-FastICA is given below.

Algorithm M-FastICA

Input: Whitened matrix $\boldsymbol{X}$
Output: Independent component $\boldsymbol{Y}$
Steps:

1. Select the initial random weight vector $\boldsymbol{w}_0$;
2. Utilize Equation 6.33 to update $\boldsymbol{w}_{k+1}$;
3. Normalize $\boldsymbol{w}_{k+1} : \boldsymbol{w}_{k+1} = \frac{\boldsymbol{w}_{k+1}}{\|\boldsymbol{w}_{k+1}\|}$;
4. If $|\boldsymbol{w}_{k+1} - \boldsymbol{w}_k| > \varepsilon$, the algorithm has not converged; go back to step (2). Otherwise, estimate an independent component and terminate the algorithm.

The extraction of multiple independent components can be completed by imitating the typical and fast deflation method of the ICA algorithm and repeatedly adopting the basic form of M-FastICA. After each component has been extracted, it needs to subtract the independent component from the observed signals. This is done repeatedly, until all the independent components have been extracted. The method for eliminating extracted independent components is given as:

$$\boldsymbol{w}_{k+1} = \boldsymbol{w}_{k+1} - \sum_{j=1}^{k} \boldsymbol{w}_{k+1}^{\mathrm{T}} \boldsymbol{w}_j \boldsymbol{w}_j, \quad \boldsymbol{w}_{k+1} = \frac{\boldsymbol{w}_{k+1}}{\sqrt{\boldsymbol{w}_{k+1}^{\mathrm{T}} \boldsymbol{w}_{k+1}}} \tag{6.34}$$

6.3 Application and Analysis

To further explain the basic theory of ICA, several simulated experiments are provided in this section. In the first subsection, through two preprocessing experiments for whitening random noise, we account for the relationship between irrelevance and statistical independence and the importance of the whitening process by analyzing the joint probability distribution of the signal. In the next two subsections, through separation experiments of mixed voice and mixed images, respectively, the fixed-point algorithm is shown to be a fast and effective method for ICA compared with the learning performance of the natural descent algorithm. When applied to the analysis of exploration data and classification of remote sensing images, FastICA is shown to be very effective in Sections 6.3.4 and 6.3.5. In Sections 6.3.6 and 6.3.7, experiments in which FastICA is applied to image noise reduction and face recognition show that the noise reduction performance is good, while the rate of face recognition is excellent. Finally, FastICA is employed in the extraction of image information used for classification and recognition. The results indicate that the images have a greater rate of recognition when using the information extracted by FastICA.

6.3.1 Whitening Preprocessing

Two uniformly distributed random signals s_1 and s_2 with zero mean and dimensions of 1×1000, were randomly generated in $[-1, 1]$. The joint probability distribution of s_1 and s_2 is shown in Figure 6.1a, where the horizontal axis denotes s_1 and the vertical axis s_2.

The joint probability distribution density of mixed variables x_1 and x_2 is shown in Figure 6.1b, where the horizontal axis denotes x_1 and the vertical axis x_2. The mixed matrix is $\boldsymbol{A} = \begin{bmatrix} 0.4 & 0.6 \\ 0.7 & 0.3 \end{bmatrix}$.

Variables z_1 and z_2 can be obtained by the method of whitening, the joint probability distribution density of which is shown in Figure 6.1c, where the horizontal axis denotes z_1 and the vertical axis z_2.

Whitened signals z_1 and z_2 are orthogonal and the difference between their joint probability distribution density and that of the original signals s_1 and s_2 is only a rotation angle. Thus, the whitening process greatly decreases the complexity of the ICA algorithm, and is widely used in most ICA algorithms. Meanwhile, the whitening process ensures that the unknown mixed matrix is orthogonal and greatly decreases the number of parameters that need to be estimated.

6.3.2 Separation of Blind Signals

Four randomly generated signals, each with 1000 data points, including one Gaussian signal, were used in a simulation experiment. A mixed signal was created by randomly producing a 4×4 linear mixed matrix from the four random signals. Then FastICA was adopted to separate the mixed signals with the results of this process illustrated in

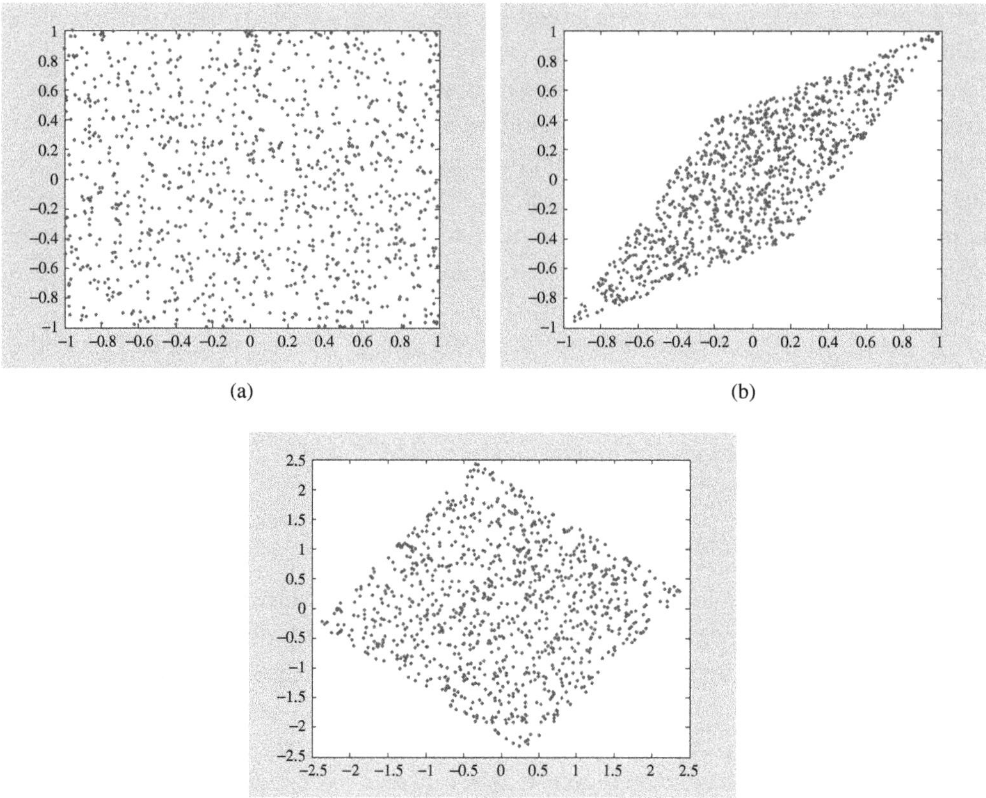

Figure 6.1 Application of the ICA algorithm for whitening preprocessing. (a) Joint probability distribution density of independent variables, (b) joint probability distribution density of mixed variables, and (c) joint probability distribution density of two whitened variables

Figure 6.2. To evaluate the performance of separation, we give the randomly generated mixed matrix.

$$\boldsymbol{A} = \begin{bmatrix} 0.5468 & 0.4465 & 0.6187 & 0.7265 \\ 0.4399 & 0.8652 & 0.6660 & 0.3144 \\ 0.6950 & 0.3945 & 0.8024 & 0.7543 \\ 0.3176 & 0.5544 & 0.0367 & 0.5633 \end{bmatrix}$$

Voice signals were used in a subsequent experiment with the results illustrated in Figure 6.3.

Two criteria, namely, convergence speed and separation accuracy, were used to evaluate the algorithms. Convergence speed is measured by computation time, while separation accuracy is measured as the distance between the product $\boldsymbol{C} = \boldsymbol{WA}$ ($\boldsymbol{C}$ is usually called the global matrix) of the separation matrix $\boldsymbol{W}$ and the actual mixed matrix

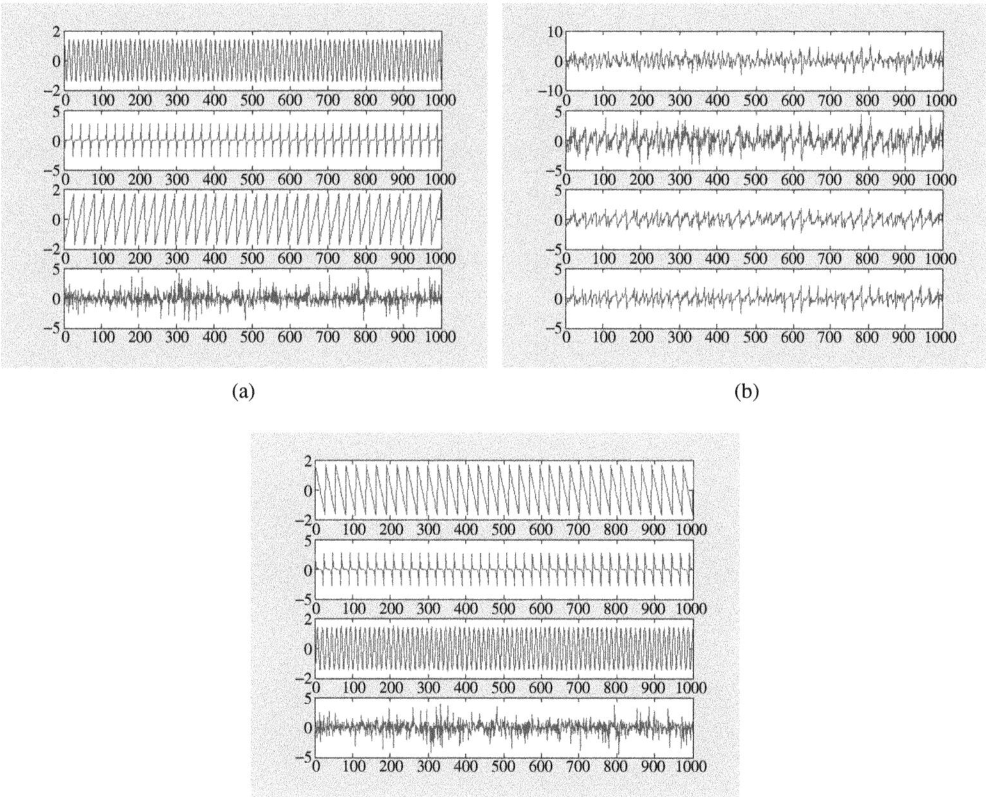

(a) (b) (c)

Figure 6.2 Separation results of FastICA for simulated mixed signals. (a) Four original independent signals, (b) four mixed signals, and (c) separation results of FastICA (deflation method)

$\boldsymbol{A}$, and unit matrix $\boldsymbol{I}$. In practice, the performance value (Equation 5.9) is adopted as the metric, where PI indicates the separation performance of the algorithm: the smaller it is, the better the separation performance. The ideal value is $PI = 0$, which denotes complete separation. The performance metrics for FastICA obtained in the above experiments are given in Tables 6.1 and 6.2.

6.3.3 Separation of Image Signals

We scanned two images (Figure 6.4a,b) into a one-dimensional vector by row or column (when observing and handling the results, we need to restore the two-dimensional image in the opposite order). Then, we randomly blended the two images to obtain mixed images (Figure 6.4c,d), which were subsequently processed by FastICA to restore the two separate images (Figure 6.4e,f).

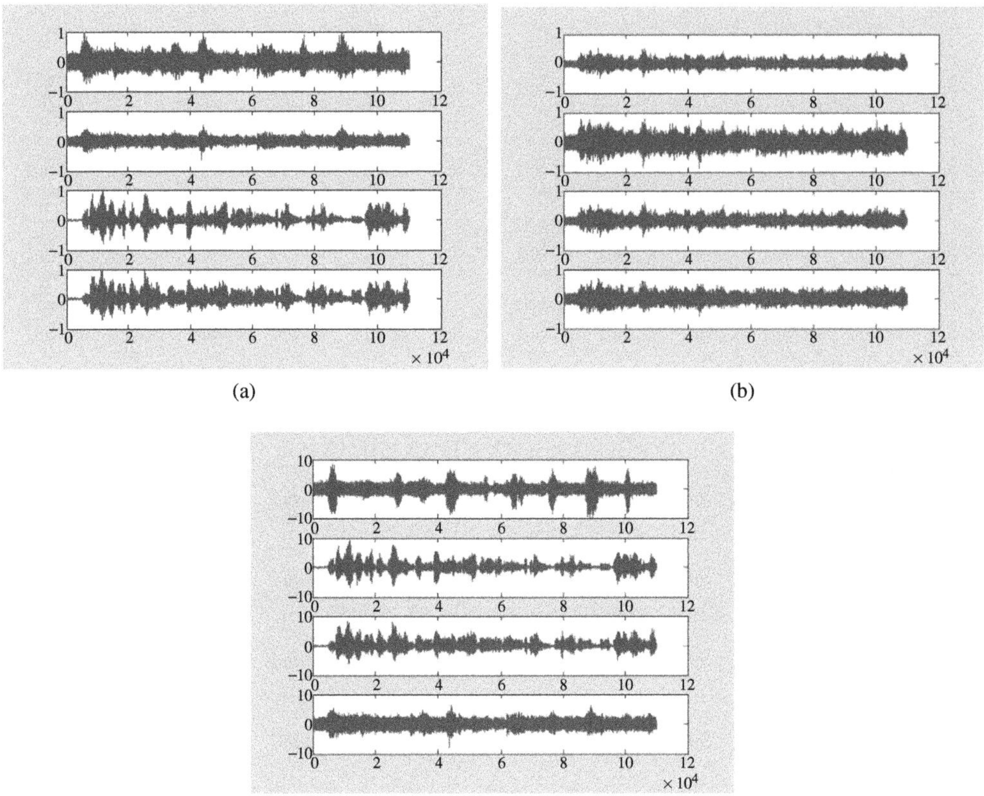

(a) (b) (c)

Figure 6.3 Separation results of FastICA for mixed voice signals. (a) Four original independent signals, (b) four mixed signals, and (c) separation results of FastICA (deflation method)

Table 6.1 Separation performance of simulated mixed signals using FastICA

Algorithm	Computational time (s)	Separation accuracy
FastICA	0.062	0.0013

Table 6.2 Separation performance of mixed voice signals using FastICA

Algorithm	Computational time (s)	Separation accuracy
FastICA	4.954	0.1205

Figure 6.4 ICA applied to image separation: (a) and (b) are the original images; (c) and (d) are mixed images; and (e) and (f) are the separated images

Figure 6.4 shows that FastICA accurately restores the original images. Owing to the uncertainty of BSS (blind source separation), the sequence between the separate and original images is different. Although image quality deteriorates slightly, the separation performance is good.

6.3.4 ICA Algorithm for Geochemical Exploration Data Analysis

The example of FastICA applied to geochemical exploration data analysis makes use of the 1/10 000 secondary halos data of the gold field in Tuanjiegou, Heilongjiang province, including 6779 sample points and 11 chemical elements (Au, As, Hg, Sb, Bi, Cu, Pb, Zn, Ag, Mo, W). In practice, we focus on gold.

The purpose of the experiment is to obtain relatively independent components and a specific combination of each using FastICA. Five predicted areas were known from previous work, and were subsequently proved by practical exploration. The areas marked by gray boxes in Figure 6.5 are the known predicted areas (Yu *et al.*, 2007a,b).

The data was first preprocessed using FastICA. To avoid the error caused by the vulnerability of the algorithm to the impact of outliers, the data was first modified

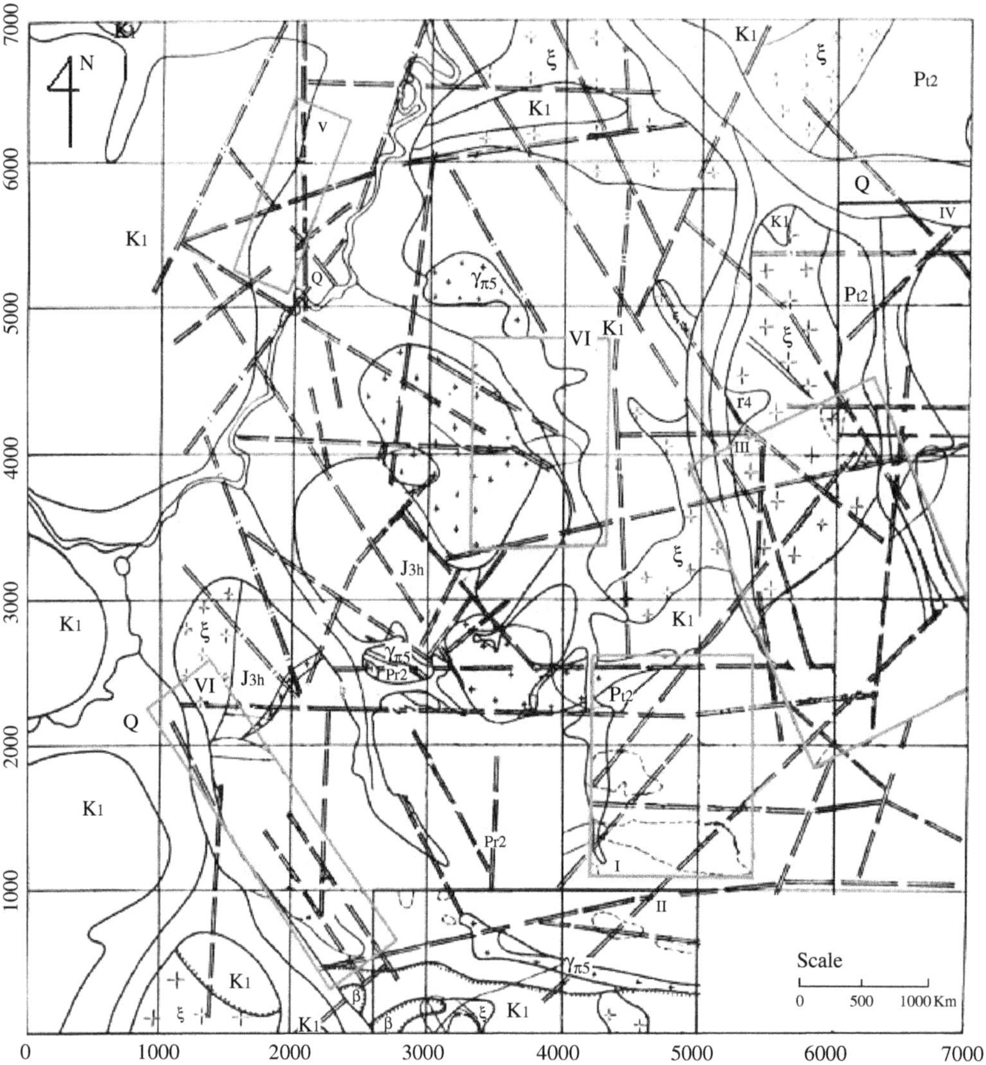

Figure 6.5 Five predicted gold areas in Tuanjiegou

according to the rules below.

$$x = \begin{Bmatrix} E + 4\sigma & if\ x > E + 4\sigma \\ E - 4\sigma & if\ x < E - 4\sigma \end{Bmatrix} \tag{6.35}$$

where E is the mean value and σ is the variance. After preprocessing the data, FastICA was again applied. The results of the final experiment were compared with the five predicted areas.

The waveforms of 11 elements resulting from applying FastICA twice are illustrated in Figure 6.6, where the x-axis denotes each sample point and the y-axis denotes the purity of each element. We find that the ICA results including the data preprocessing correspond closely to the predicted areas.

The regions with intensive gray sample points shown in Figures 6.7–6.9 are the mineral regions forecast using the ICA algorithm.

From the experimental results, we can see that not only can the ICA algorithm extract the principal characteristics of mass data, indicating the locations of gold deposits, but it can also better illustrate the actual distribution of deposits after robust preprocessing.

6.3.5 FastICA Algorithm Used for Remote Sensing Image Classification

The FastICA algorithm was applied to ETM^+ (enhanced thematic mapper) remote sensing images of the Zhaoqing district from December 2002 (Figure 6.10a) and TM (thematic mapper) remote sensing images of the Shunyi district in Beijing from April 1999 (Figure 6.10b). Both these images comprised six bands, namely, 1, 2, 3, 4, 5, and 7. Based on the characteristics of the images, different band combinations were selected to achieve the best performance with less computation (He and Yu, 2005b). To guarantee the comparability of the experimental results, combinations of different parameters were adopted when dealing with the same image. The FastICA algorithm with different parameter combinations was applied to the same images to find the most accurate set (Wang, Zeng and Xia, 2005). To test the universality and robustness of the FastICA algorithm, the experimental data included several datasets, from small to large (128×128 to 512×512). Owing to space constraints, the 512×512 results are not given. Blocking of the FastICA algorithm was attempted in the experiment, with the results shown in Figure 6.11. The independent component images obtained by the process of ICA comprise the R, G, and B components of the image.

From Figures 6.10 and 6.11, it is clear that FastICA can remove image correlation and restore relatively independent components, thereby improving image quality and contributing to image classification. Though FastICA is better than the PCA (principal component analysis) algorithm in removing correlation, some features are lost, especially nonlinear features and specific texture information. Figure 6.11 further indicates that the blocking results for each sub-region have a clearer hierarchy, and are significantly superior to the overall results in terms of maintaining the detail and class discrimination.

Owing to the rich vegetation information in the Shunyi district image, the combined effect of bands 5, 4, and 3 is preferable, while the non-linear function *pow* yields better results than the others (Figure 6.11a). Since it is necessary to distinguish different vegetation types (woodland, vegetable plots, rice fields, and so on) in the Zhaoqing district image, the independent component figure from the combination of bands 6, 4, and 3 is better than that from other band combinations according to the experimental results (Figure 6.10c). In general, choosing a nonlinear function that can be executed

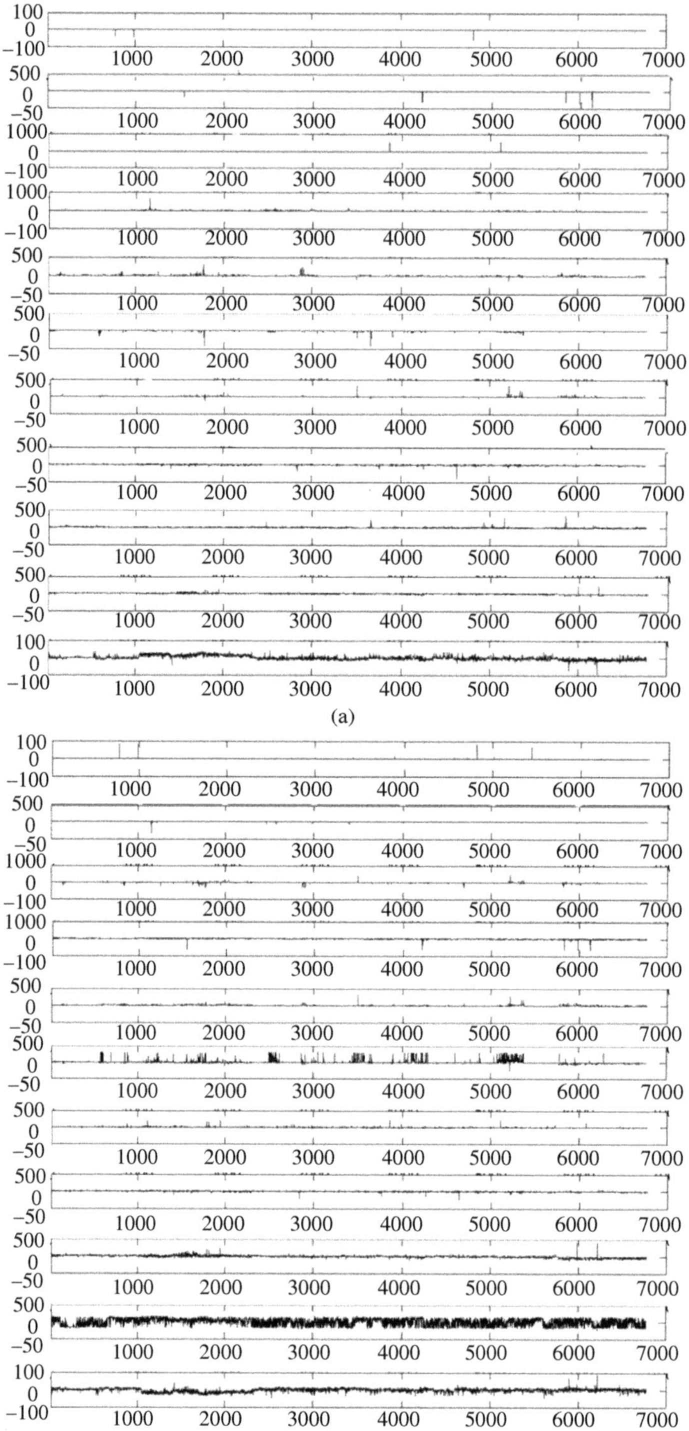

Figure 6.6 Experimental results processed twice by ICA. (a) Results without the robust preprocessing and (b) results with the robust preprocessing

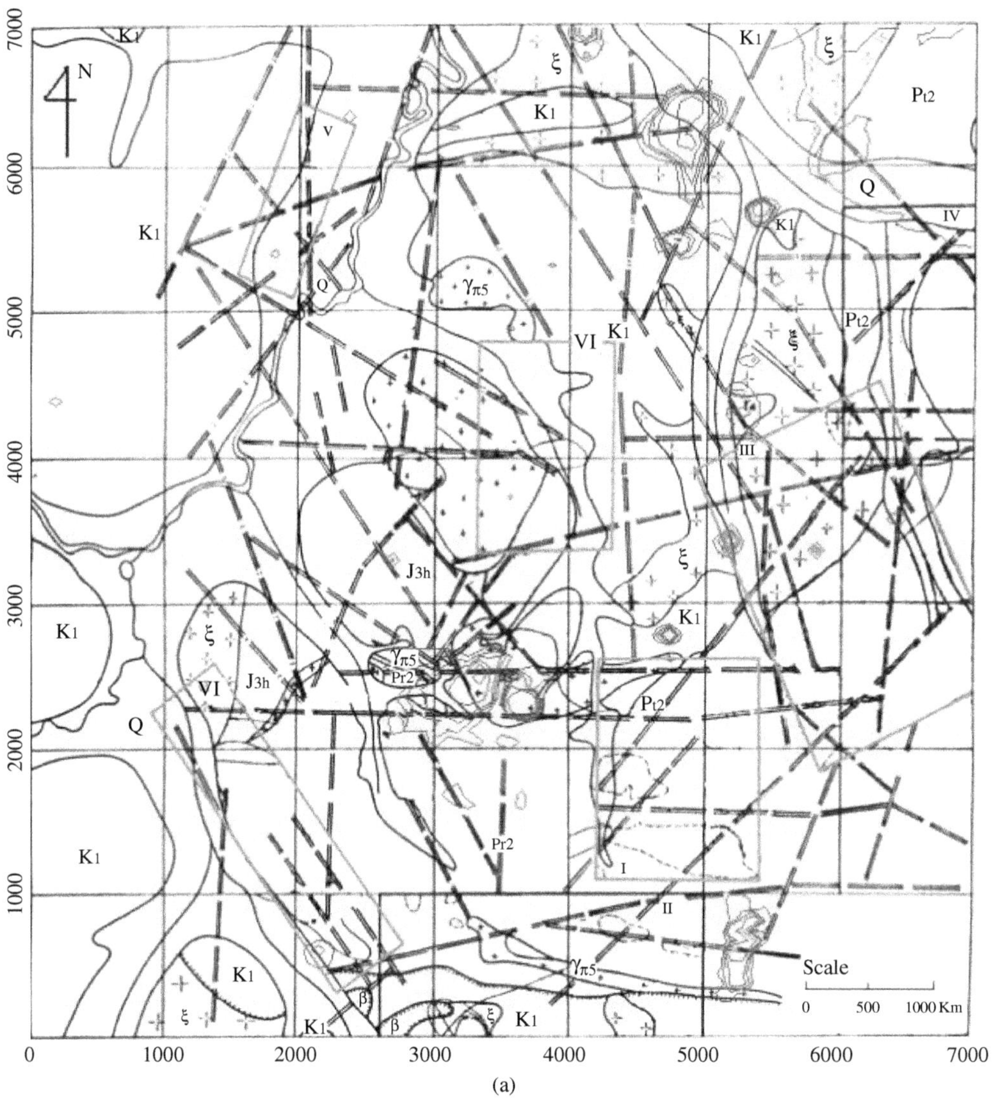

(a)

Figure 6.7 The anomaly map of Hg. (a) Results without robust preprocessing and (b) results with robust preprocessing

the fastest, results in less impressive results. The algorithm with the fastest execution speed uses the nonlinear function *skew*, but the performance thereof is inferior to that of the algorithm using *pow*.

Using the above independent component figures as input images for classification, we applied the maximum likelihood ratio to supervised classification. From the classification results in Figure 6.12, it can be seen that, compared with the classification results obtained directly from the original images, the classification results after

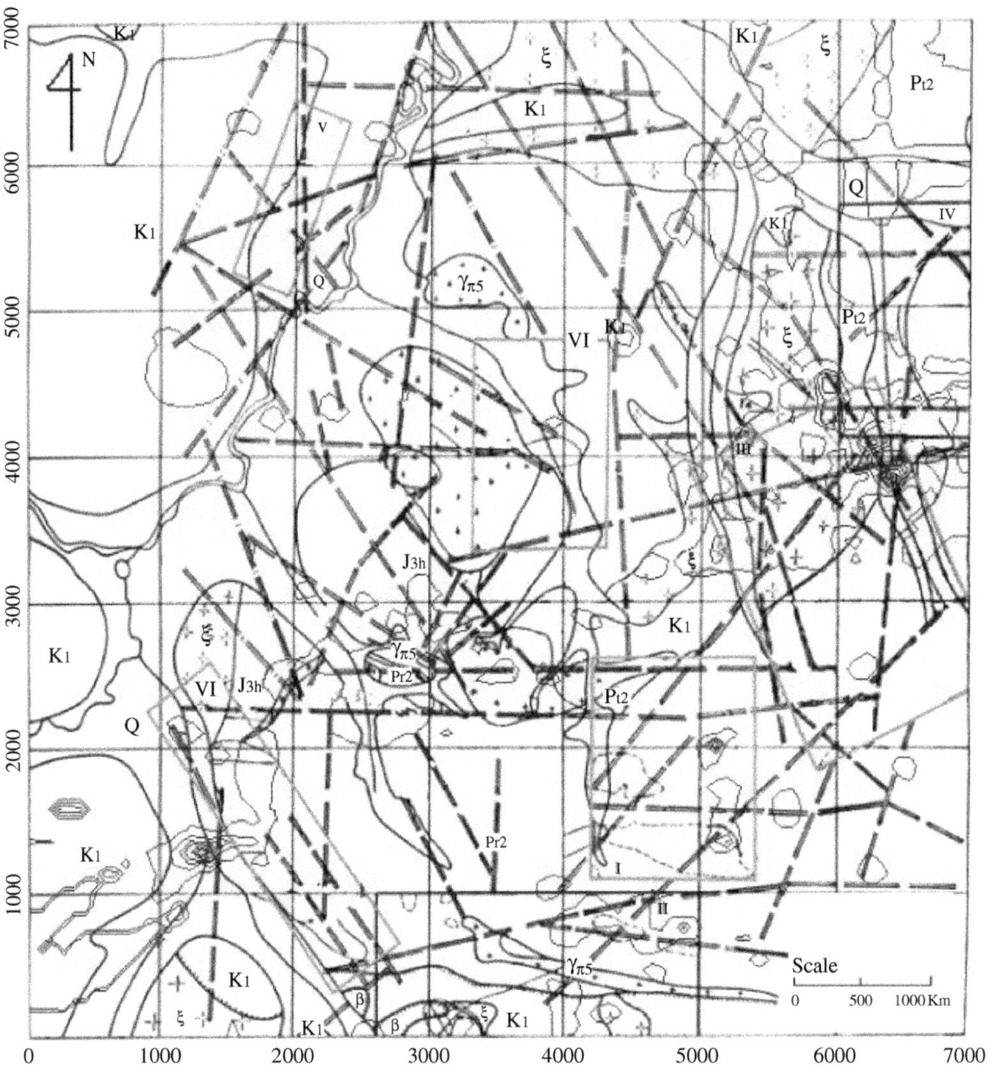

(b)

Figure 6.7 *(continued)*

processing by the FastICA algorithm are more defined and better clustered. However, FastICA processing leads to the loss of some detail, which results in the loss of small individual targets (Tables 6.3–6.5).

6.3.6 Application of the ICA Algorithm to Image Noise Reduction

In practical applications, because the measurement signal is often a linear superposition of a number of independent source signals and observation noise, it is difficult to

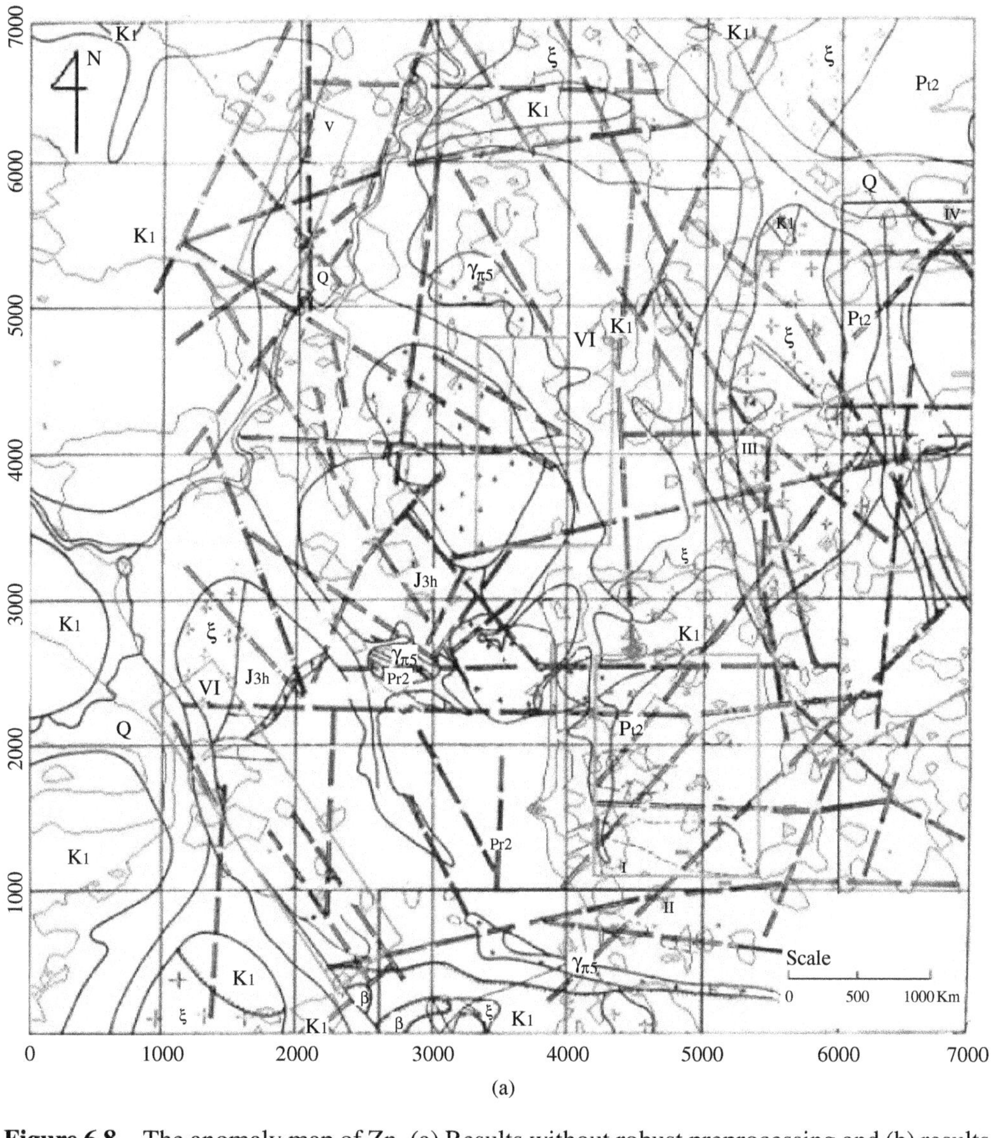

(a)

Figure 6.8 The anomaly map of Zn. (a) Results without robust preprocessing and (b) results with robust preprocessing

obtain ideal separation results when the existing ICA algorithm is directly applied to an observed signal with noise. Figure 6.13 shows a typical ICA model with noise.

In the following experiment, a series of 8×8 training sub-image blocks were randomly extracted from a group of natural scene images, and the mean value of each image sub-block was removed. Then, a 1×64 column vector was created by expanding the columns (from the first to the last) of the image sub-block to produce a column vector of data z without noise. In the experiments, we selected 13 images as training

(b)

Figure 6.8 *(continued)*

images, which formed a $64 \times 10\,000$ training sample dataset as the input sample T of M-FastICA. Details of the experimental process are given below.

1. Evaluate sparse transformation matrix $\boldsymbol{W}$. M-FastICA is used to process the training data and obtain 64×64-dimensional ICA basis vectors. Each column of $\boldsymbol{W}_1$ is a 1×64 vector, corresponding to an 8×8 block of the base image. Thus, we obtained 64 base images as shown as Figure 6.14. Orthogonalization of matrix $\boldsymbol{W}$ leads to the sparse transformation matrix $\boldsymbol{W}$:

$$\boldsymbol{W} = \boldsymbol{W}_1(\boldsymbol{W}_1^{\mathrm{T}}\boldsymbol{W}_1)^{-1/2} \tag{6.36}$$

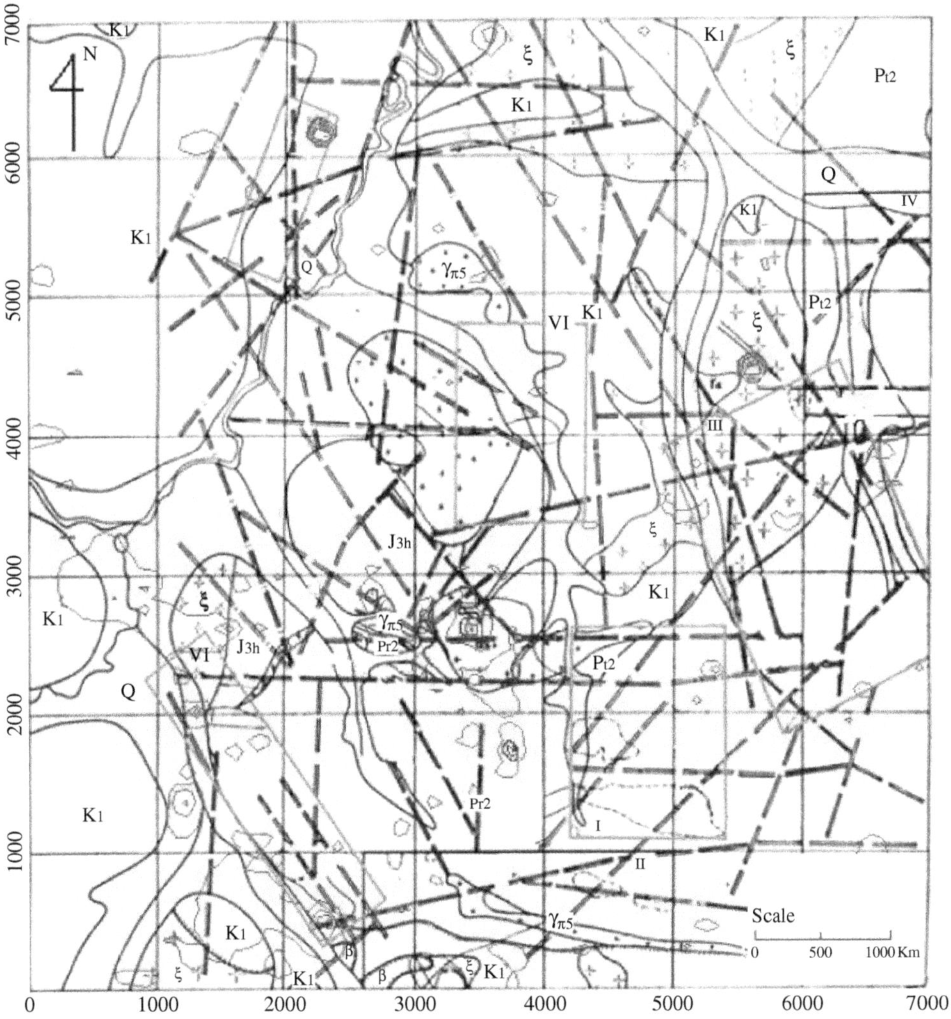

(a)

Figure 6.9 The anomaly map of Mo. (a) Results without robust preprocessing and (b) results with robust preprocessing

2. Using the sparse coding shrinkage method, estimate the contraction function $g_i(.)$ corresponding to each base vector. For each component of the training data $\boldsymbol{s}_i = \boldsymbol{W}_i^{\mathrm{T}}\boldsymbol{z}, \quad i = 1, 2, \ldots, n$, use Equation 6.37 to estimate the corresponding shrinkage function $g_i(.)$:

$$g(\boldsymbol{u}) = \frac{1}{1 + \sigma^2 a} \text{sign}(\boldsymbol{u}) \max(0, |\boldsymbol{u}| - b\sigma^2) \tag{6.37}$$

where σ^2 is the variance of the noise. If the variance of the noise is unknown, the product of the mean absolute deviation of y_i corresponding to each sparse set s_i

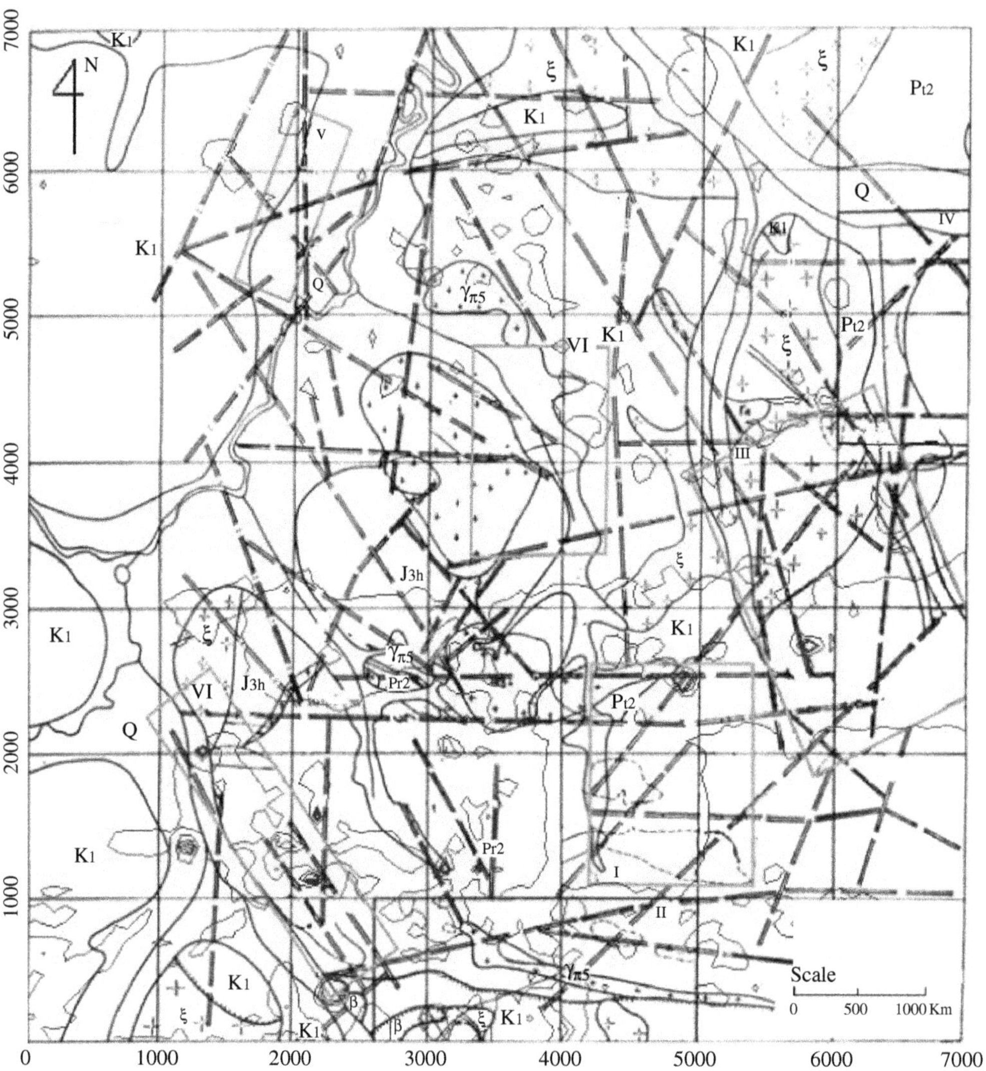

(b)

Figure 6.9 *(continued)*

and constant factor 0.6475 can be used to estimate it. Parameters a and b can be evaluated by the following equations:

$$b = \frac{2p_s(0)E\{\mathbf{s}^2\} - E\{|\mathbf{s}|\}}{E\{\mathbf{s}^2\} - E\{|\mathbf{s}|\}^2}$$

$$a = \frac{1}{E\{\mathbf{s}^2\}}[1 - E\{|\mathbf{s}|\}b] \tag{6.38}$$

where $p_s(.)$ denotes the value of the density function when $\mathbf{s}$ is 0.

(a) (b) (c)

Figure 6.10 Comparison of pre-and post-processing performance on the Zhaoqing district image. (a) Original ETM$^+$ image (512 × 512); (b) independent component figure of combined-bands 5, 4, and 3; and (c) independent component figure of combined-bands 6, 4, and 3

(a) (b) (c)

Figure 6.11 Independent component figures of the Shunyi district TM image. (a) Non-linear function *pow*, (b) non-linear function *tanh*, and (c) results from block process

Figure 6.12 Independent component figures from FastICA processing

Table 6.3 Covariance matrix of principal components

PC	1	2	3
1	1.000000	0.521387	0.579347
2	0.521387	1.000000	0.308342
3	0.579347	0.308342	1.000000

Table 6.4 Covariance matrix of independent components

IC	1	2	3
1	1.000000	−0.111410	−0.057916
2	−0.111410	1.000000	0.058096
3	−0.057916	0.058096	1.000000

Table 6.5 Covariance matrix of original images

Band	5	4	3
5	1.000000	0.710994	0.721963
4	0.710994	1.000000	0.584670
3	0.721963	0.584670	1.000000

3. For data with noise $\boldsymbol{x}(t), t = 1, 2, \ldots, T$, calculate its projection on the sparse base

$$\boldsymbol{y}(t) = \boldsymbol{W}\boldsymbol{x}(t) \tag{6.39}$$

4. For each element y_i of vector $\boldsymbol{y}$, take the corresponding contraction function $g_i(.)$ of each base vector as the soft threshold operator and remove the Gaussian noise with a smaller value in higher-order information. Then, calculate the average of the 64 corresponding shrinkage functions for each pixel.

$$\widehat{s}_i(t) = \text{mean}(g_i(y(t))) \tag{6.40}$$

5. Project the projection data after removing the noise using Equation 6.41 onto the image space and rearrange the image in the opposite order. Then, we can obtain the image using the sparse coding contraction method to remove noise.

$$\widehat{\boldsymbol{x}}(t) = \boldsymbol{W}^{\mathrm{T}}\widehat{\boldsymbol{s}}(t) \tag{6.41}$$

In our experiments, we used the *wiener2* function in MATLAB® for 5×5 areas as the Wiener denoising method. *Ddencmp* and *wdencmp* functions in MATLAB®

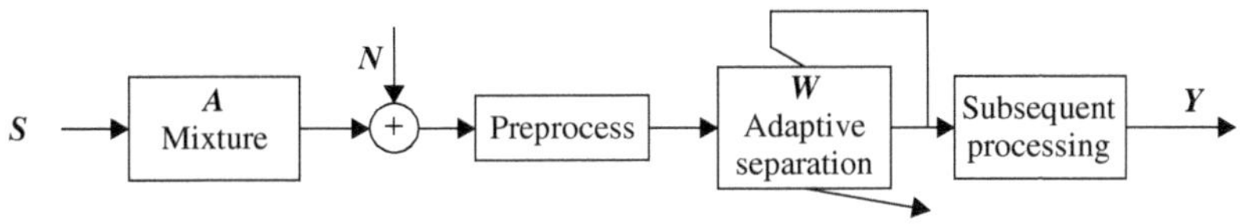

Figure 6.13 ICA model with noise

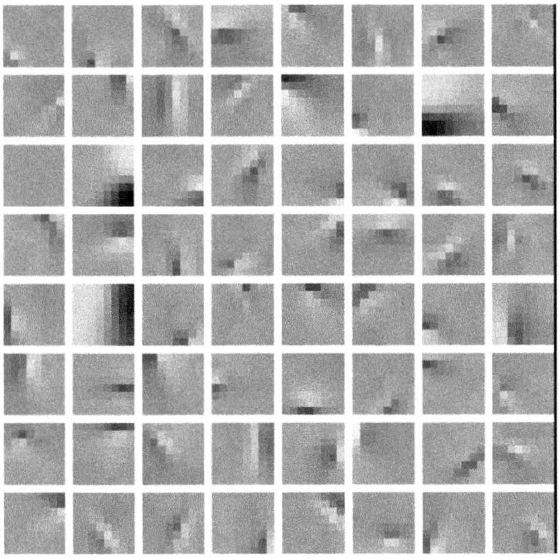

Figure 6.14 ICA base vectors

were used as wavelet transforms. To compare several denoising methods, the minimum square error (MSE) and signal-to-noise ratio (SNR) were introduced to evaluate performance.

$$MSE = \frac{\sum_{i=0}^{M-1}\sum_{j=0}^{N-1}(f_{ij} - f'_{ij})^2}{M * N} \tag{6.42}$$

$$SNR = -10\log_{10}\left(\frac{\sum_{i=0}^{M-1}\sum_{j=0}^{N-1}\left(f_{ij} - f'_{ij}\right)^2}{\sum_{i=0}^{M-1}\sum_{j=0}^{N-1}(f_{ij})^2}\right) \tag{6.43}$$

where f_{ij} represents the original image without noise, f'_{ij} represents the denoising image using different methods, and $M \times N$ denotes the image size. A comparison of various denoising methods is shown in Figure 6.15.

(a) (b) (c) (d) (e)

Figure 6.15 Comparison of various noise reduction methods. (a) Original Lena image, (b) mixed image with Gaussian noise (mean 0 and variance 0.01), (c) noise reduction image by the wavelet threshold, (d) noise reduction image by Wiener filtering, and (e) noise reduction image by ICA domain

It can clearly be seen from Figure 6.15 that the denoising method based on the ICA domain achieves the best performance. The sparse coding contraction method based on the ICA algorithm can successfully separate noise mixed into images according to the MSE and SNR criteria. Though Wiener filtering also yields a small MES, since its optimal criterion is based on MSE and it handles all errors with equal weight, the performance thereof based on visual cues is not satisfactory. Humans have greater sensitivity for error perception in bright areas and less sensitivity in dark areas and high gradient regions.

The performance comparison of several denoising methods is shown in Table 6.6, while a comparison of the iterations in FastICA and M-FastICA is given in Table 6.7. It can be seen from Table 6.7 that M-FastICA has a faster convergence speed than the ordinary FastICA algorithm. Moreover, from Table 6.6, we see that M-FastICA has good performance in processing noisy images. Since the initial value is randomly generated, each result must be different. So the number of iterations for M-FastICA exceeds that for FastICA in some experiments. Generally speaking, M-FastICA retains the good separation performance of FastICA and further accelerates the convergence rate of FastICA.

Table 6.6 Performance comparison of several denoising methods

Image	SNR (dB)	Minimum MSE
Noisy image	14.997	646.42
Wiener denoising image	22.1376	125.47
Wavelet denoising image	20.6286	169.78
M-FastICA denoising image	24.2865	123.785

Table 6.7 Iteration comparison between FastICA and M-FastICA

No.	FastICA	M-FastICA
1	489.5313	303.9219
2	366.4844	212.4063
3	304.2344	363.4531
4	483.1719	267.5781
5	433.1563	312.2656
6	530.8906	270.0469
7	418.1094	312.6250
8	276.7696	250.3281
9	331.1563	233.0625
10	354.5781	255.3906
Mean value	398.8082	278.1078

Although the Wiener and wavelet denoising methods can, to a certain extent, remove noise mixed into an image and be applied successfully, since they do not take advantage of the different statistical information characteristics, their denoising performance either does not remove the noise completely or damages the original image. The denoising method based on the ICA domain deals with the noise and image independently. Moreover, it obtains the projection of the image in the ICA domain by calculating the statistical information of the image; there are great differences between the image and noise with clear separation features. The image value is large in the ICA domain, while the noise value is small. Therefore, it is easy to remove noise using some threshold and to preserve the original image information. The shortcoming of the denoising method based on the ICA domain is that it requires a reference image, which is usually difficult to obtain. The experiments in this section are based on Gaussian noise, but the sparse coding method is also suitable for removing Poisson noise. It is believed that, with the development of the ICA algorithm, the denoising method based on the ICA domain can be further enhanced.

6.3.7 M-FastICA for Face Recognition

The ICA algorithm for feature extraction in face images is based on the following assumptions: each face image is composed of a linear mixture separation matrix $\boldsymbol{S}$ of some independent statistical base images, and the separation matrix $\boldsymbol{W}$ can be calculated by the ICA algorithm to restore $\boldsymbol{S}$. M face vectors of the training set are included in the matrix. Because the ICA algorithm is usually applied to rows, we consider the training image size to be $M \times N$, $\boldsymbol{X} = [\boldsymbol{x}_1, \boldsymbol{x}_2, \dots, \boldsymbol{x}_M]^{\mathrm{T}}, \boldsymbol{X} \in \boldsymbol{R}^{M \times N}$. Suppose that M face images are linearly mixed with M independent base images $\boldsymbol{S} = [\boldsymbol{s}_1, \boldsymbol{s}_2, \dots, \boldsymbol{s}_M], \boldsymbol{S} \in \boldsymbol{R}^{M \times N}$ (Figure 6.16).

In Figure 6.16, each row of $\boldsymbol{X}$ represents a face image, each row of $\boldsymbol{S}$ represents a base image, and $\boldsymbol{A} \in \boldsymbol{R}^{M \times M}$ is the mixed matrix.

The separation matrix $\boldsymbol{W}$ is obtained from the ICA algorithm so as to make the row vectors of output $\boldsymbol{Y} = \boldsymbol{W}\boldsymbol{X} = \boldsymbol{W}\boldsymbol{A}\boldsymbol{S}$ independent of each other.

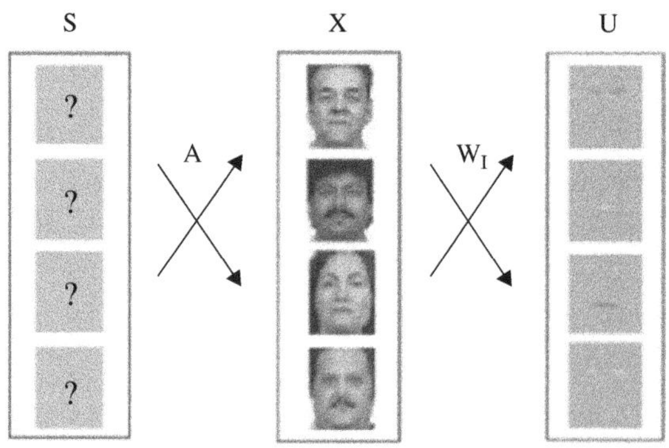

Figure 6.16 Face recognition model based on ICA algorithm

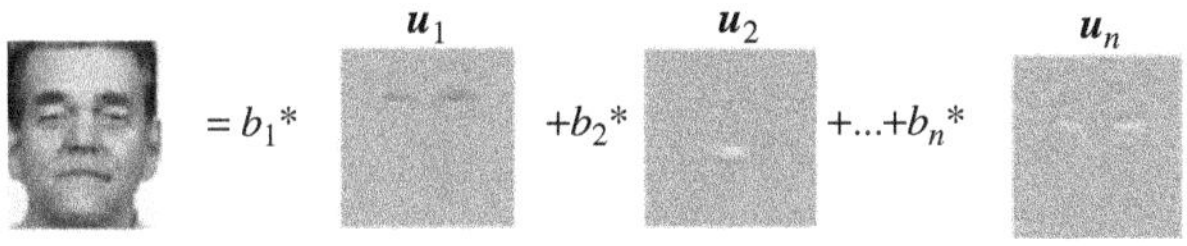

Figure 6.17 Face image linearly combined with base images

$Y = [y_1, y_2, \ldots, y_M]^T, Y \in R^{M \times N}$ is the evaluation of the independent base images. After Y has been obtained, it is possible to construct a feature subspace from the row vectors of Y. The face images to be identified are projected onto the subspace; in other words, that which is expressed by the linear combination of base images (Figure 6.17). Function f is set to the image to be identified and we can get:

$$f = a_1 y_1 + a_2 y_2 + \ldots + a_M y_M \tag{6.44}$$

where $a_1, a_2, \ldots, a_M$ are the projected parameters of the face image in the subspace, which can be solved by Equation 6.45.

$$[a_1, a_2, \ldots, a_M] = f^T \times pinv(Y) \tag{6.45}$$

where $pinv(\cdot)$ is the matrix pseudo-inverse process, $pinv(\boldsymbol{Y}) = \boldsymbol{Y}(\boldsymbol{Y}\boldsymbol{Y}^T)^{-1}$.

P different images of the same person in the training set are projected onto this subspace, and the mean projected coefficients thereof are taken as the projected coefficients of this person. This realizes extraction of face features, reducing N-dimensional face images to M-dimensional ones. Then, the nearest Euclidean distance and similarity of known patterns are adopted to classify the face data with reduced dimensions. In other words, the projected coefficients of the face image to be identified are compared with all the mean projected coefficients; the closest pattern is taken as the recognized face.

In our experiments, we used the Yale face library (http://cvc.yale.edu/projects/yalefaces/yalefaces.html), which includes 11 gray images of 50 persons with variations in posture, expression, wearing of glasses, facial hair, and so on. Some of these are shown in Figure 6.18, with the size of each image being 100×100. To remove the influence of specific data, each image was normalized, by subtracting the mean value of the image and dividing by the standard deviation. The steps in M-FastICA for face recognition are given below:

1. Preprocessing is carried out, which is the process of normalizing all images in the face library. Because images in the Yale face library include the whole face exactly, face extraction is omitted.
2. Select the N face images of M persons as the training set, and take the remaining images as test samples.
3. The above M images of persons are decomposed by the M-FastICA algorithm to generate M base images for ICA, thereby constructing the subspace.

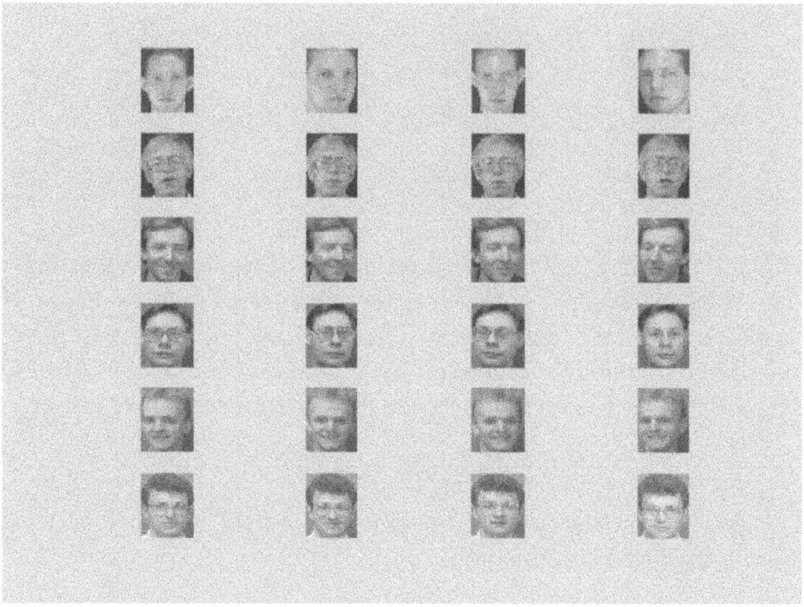

Figure 6.18 Some of the face images in the Yale face library

4. Project all the face images onto the subspace, and substitute the projected coefficients for the original images, thus forming new sample sets.
5. For an image in the test sample sets, calculate the distance between its projected coefficients and the projected coefficients of the images in the training sets; consider it to be the training image pattern with the minimum distance.

It is vital for the normalization process that the image brightness differs with respect to the radial.

The eigenface images depicted in Figure 6.19 were obtained from a PCA process on 24 face images composed of the first 4 images of each person. Independent base images from the M-FastICA process are shown in Figure 6.20. In the eigenface space, each orthogonal basis presents the face shape, depicts the distribution information, and preserves the topological relation between the face parts. However, in the independent base images, the face shape and topological relations between the face parts are mainly reflected, while the other parts stress the local outline and gray information. In addition, information for PCA decreases from front to back, whereas information for ICA does not.

In our experiments, we took the first two images of the ORL face library as the training set and the last eight images as the test set. Then, successively increased to select each of the first three images as the training set, and the remaining images are used as test sets. Then the first four images are taken as the training set and the

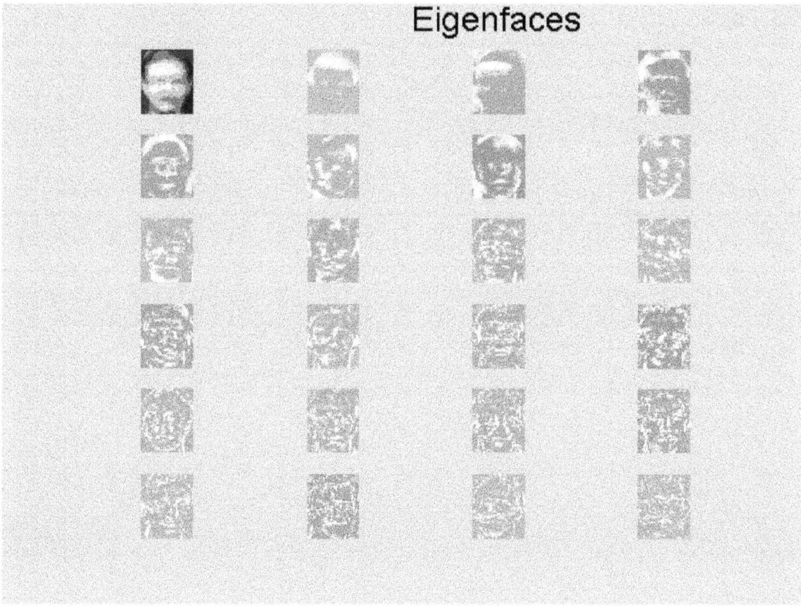

Figure 6.19 Eigenfaces from the training images by PCA

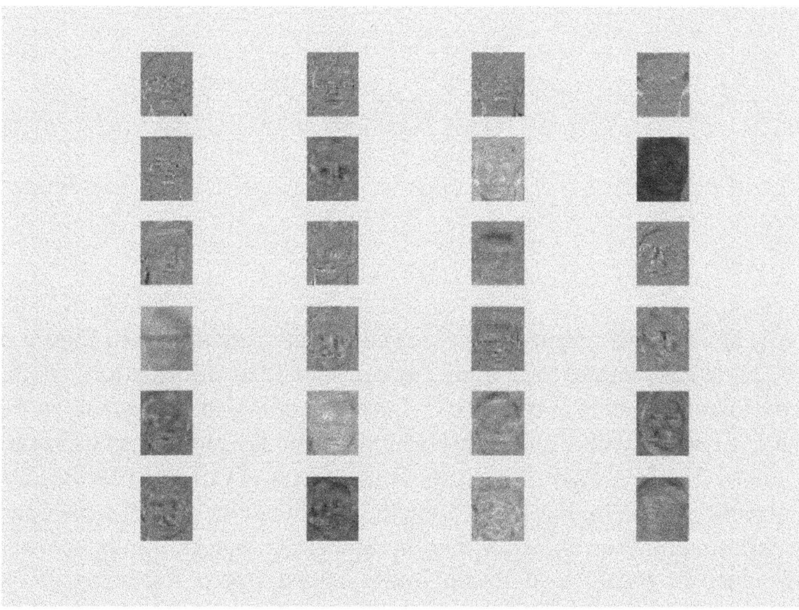

Figure 6.20 ICA base images from the training images by M-FastICA

Table 6.8 Comparison of the recognition ratio among PCA, FastICA, and M-FastICA with different training sets

Algorithm	6×2	6×3	6×4	6×5	6×6	6×7
PCA (%)	68.75	73.81	77.78	83.33	87.5	88.89
FastICA (%)	75	83.33	86.11	83.33	91.67	94.44
M-FastICA (%)	75	83.33	86.11	86.67	91.67	94.44

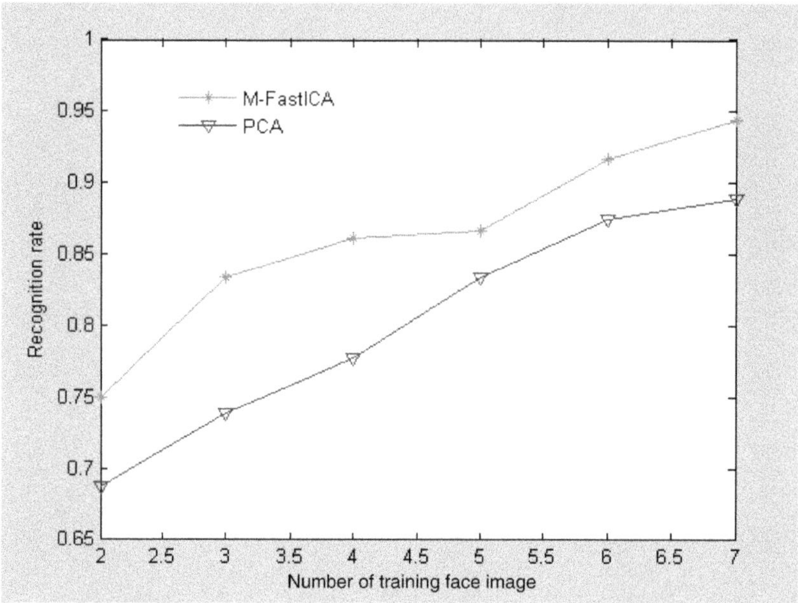

Figure 6.21 Comparative figures for different numbers of training face images. (*) M-FastICA and (∇) PCA

remaining images as tests sets. The recognition rates for PCA and FastICA are shown in Table 6.8. Comparative figures for recognition rates are shown in Figure 6.21.

The ICA algorithm can extract higher-order statistical information. So, according to the experimental results (Figure 6.21), there is some improvement in the recognition rate compared with the PCA algorithm. Although the recognition ratio of the FastICA and M-FastICA algorithms is the same, fewer iterations are needed by the M-FastICA algorithm than by FastICA (Table 6.9), which can speed up the convergence rate. The experimental results show that the M-FastICA algorithm is able to improve the performance of FastICA while retaining satisfactory performance in extracting features.

Table 6.9 Comparison of mean iterations for PCA and M-FastICA

Number algorithm	6×2	6×3	6×4	6×5	6×6	6×7
FastICA (%)	25.3333	34.1667	59	51.9667	54.0833	56.2857
M-FastICA (%)	23.6667	33.7778	48.7083	41.9667	45.5556	44.4762

6.3.8 FastICA for Extracting Image Information

In this experiment, a high resolution QuickBird satellite image, captured on 26 April 2005 and including panchromatic and four multi-spectral bands, of an area in Beijing was selected for preprocessing using FastICA (He, Zhang, and Yu, 2006). To guarantee convergence, the number of iterations for each independent component was only 50. A printscreen illustration of this image is shown in Figure 6.22.

Three categories of objective images, including roads, vegetation, and houses, were randomly cut out of the image in Figure 6.22 as experimental data, which comprised a total of 120 images of size 30×30 pixels. Meanwhile grayscale processing was performed on all the experimental data. Some of the training set images are shown in Figure 6.23.

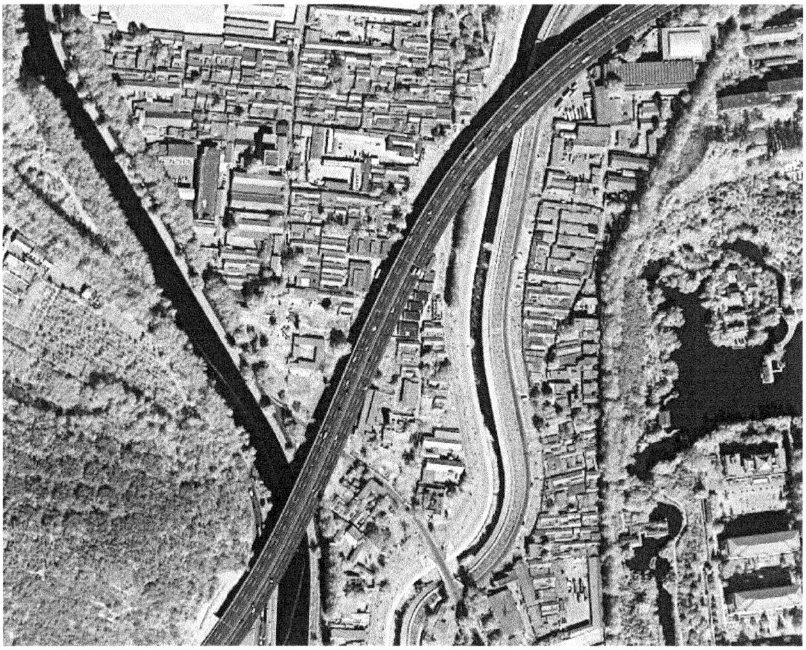

Figure 6.22 High resolution QuickBird satellite image of an area in Beijing

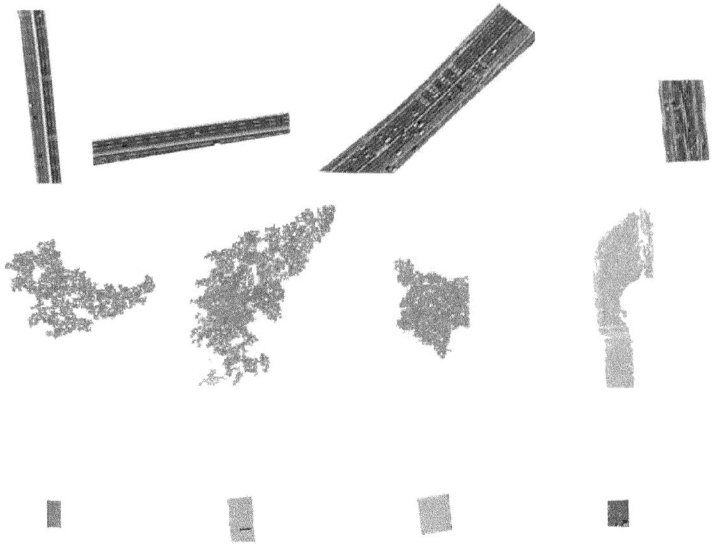

Figure 6.23 Some of the training sample set data

Table 6.10 Comparison of recognition rates in the case of small samples

Classification	FastICA (%)	PCA (%)
Roads	78	72
Vegetation	77	73
House	79	72

Thirty images were selected from each target image as the training set, and another sixty images were selected as the test set. Both FastICA and PCA were applied separately, and traditional ICA and PCA algorithms with Euclidean distance were used for recognition. It can be seen from the experimental results (Table 6.10) that the recognition rate of FastICA is greater than that of PCA.

6.4 Conclusion

The FastICA algorithm is a learning fixed point algorithm, with a convergence rate superior to that of the descent-based methods and type of convergence fast cubic. FastICA is able to achieve good separation accuracy in a short time. As a result of ICA that extracts higher-order information, the recognition rate of the FastICA algorithm is superior to that of PCA. Thus, it can reflect the actual distribution of elements when applied to mineral prediction and obtains a more defined outline than direct classification. The denoising method based on the ICA domain considers the noise and image data to be mutually independent. A projection of the image in the ICA domain is obtained by calculating the statistical information of the image data, which differs greatly from that of noise. The separation characteristics are so good that the noise values are smaller and the image values are greater in the ICA domain. Thus, it is easy to remove the noise and preserve the original information based on some threshold. There are fewer iterations in M-FastICA than in FastICA; in other words, M-FastICA is able to improve the performance of FastICA while maintaining good performance in extracting features.

References

He, H. and Yu, X.C. (2005) A comparison of PCA /ICA for data preprocessing in remote sensing imagery classification. *Proceedings of the SPIE MIPPR*, **6044**, 60–65.

He, H., Zhang, T., and Yu, X.C. (2006) Application of fast independent component analysis on extracting the information of remote sensing imagery. Proceedings of the 5th International Conference on Machine Learning and Cybernetics, pp, 1066–1071.

Hyvärinen, A. (1998) New approximations of differential entropy for independent component analysis and projection pursuit. *Advances in Neural Information Processing System*, **10**, 273–279.

Hyvärinen, A. (1999) Fast and robust fixed-point algorithms for independent component analysis. *IEEE Transactions on Neural Networks*, **10**(3), 626–634.

Hyvärinen, A., Hoyer, P., and Oja, E. (1998) Sparse code shrinkage for image denoising. *Proceedings of Neural Networks IEEE World Congress on Computational Intelligence*, **2**, 859–864.

Hyvärinen, A. and Karhunen, J. (eds) (2001) *Independent Component Analysis*, John Wiley & Sons, Inc..

Hyvärinen, A. and Oja, E. (1997) A fast fixed-point algorithm for independent component analysis. *Neural Computation*, **9**(7), 1483–1492.

Hyvärinen, A. and Oja, E. (2000) Independent component analysis: algorithms and applications. *Neural Networks*, **13**(4–5), 411–430.

Wang, X.M., Zeng, S.G. and Xia, D.S. (2005) Remote image classification based on loose modified fast independent component analysis algorithm. *Computer Engineering and Application*, **41**(7), 566–765.

Yu, X.C., Liu, S.H., Ren, J.M. *et al.* (2007) Robust fast independent component analysis applied to mineral resources prediction. Proceedings of the Geomathematics and GIS Analysis of Resources, Environment and Hazards, pp. 94–97.

Yu, X.C., Zhang, T., Zhang, L.B. *et al.* (2007) Independent component analysis and its applications to manufacturing problems in (eds D. Laha and P. Mandal) *Handbook of Computational Intelligence in Manufacturing and Production Management*, Information Science Reference, New York, Hershey, PA.

7

Maximum Likelihood Independent Component Analysis and Its Application

7.1 Overview

7.1.1 Likelihood Estimation

A popular approach for estimating the independent component analysis (ICA) model is maximum likelihood (ML) estimation, which is a statistical method. ML estimation is the process of finding parametric values for the model parameters that make the observations the most probable (for the given model). This chapter explains how to use ICA ML estimation, and analyzes the relationship between the method and the maximization information flow (infomax) principle in neural networks.

The goal of ML estimation is to find variables $\boldsymbol{s}$, which through a linear mapping can produce the observed data, $\boldsymbol{x} = \boldsymbol{A}\boldsymbol{s}$. In the field of blind signal processing, the only known data are the observation data $\boldsymbol{x}$, so ML estimation is a more natural selection. The source signal probability density function (PDF) is $p_s(\boldsymbol{s})$.

Suppose $\widehat{p}_x(\boldsymbol{x})$ is the probability density $p_x(\boldsymbol{x})$ estimation of the observation vector $\boldsymbol{x}$. According to the linear transformation relationship above, the estimated PDF $\widehat{p}_x(\boldsymbol{x})$ and the source signal PDF satisfy

$$\widehat{p}_x(\boldsymbol{x}) = \frac{p_s(\boldsymbol{A}^{-1}\boldsymbol{x})}{|\det \boldsymbol{A}|} \tag{7.1}$$

Given the model, the likelihood function of the observed data $\boldsymbol{x}$ is a function of the model parameters $\boldsymbol{A}$, defined as

$$L(\boldsymbol{A}) = E\{\log \widehat{p}_x(\boldsymbol{x})\} = \int p_x(\boldsymbol{x}) \log p_s(\boldsymbol{A}^{-1}\boldsymbol{x})\mathrm{d}\boldsymbol{x} - \log|\det \boldsymbol{A}| \tag{7.2}$$

Blind Source Separation: Theory and Applications, First Edition. Xianchuan Yu, Dan Hu and Jindong Xu.

If the model parameters for the separation matrix are $W = A^{-1}$, the logarithmic likelihood function is given by

$$L(W) \approx \frac{1}{T}\sum_{t=1}^{T}\{\log p_s(Wx(t))\} + \log|\det W| \tag{7.3}$$

where T is the number of samples of the independent distribution from the observed data. Maximizing the likelihood function is achieved with the best parameter estimation.

7.1.2 Probability Density Estimation

In Section 7.1.1, we expressed likelihood as a function of the parameters of the model, which are the elements of the mixing matrix. For simplicity, we used the elements of the inverse W of the mixing matrix. This is allowed since the mixing matrix can be directly computed from its inverse.

However, there is another estimate needed in the ICA model, namely, the densities of the independent components. Estimates of the probability density are usually nonparametric, which makes the problem much more complicated owing to an increase in the number of parameters. Thus, we should avoid nonparametric estimation of densities. There are two ways to achieve this. First, in some cases we may know the densities of the independent components in advance, and then we can simply use these prior densities in the likelihood. If reasonably small errors in the specification of these prior densities have little influence on the estimator, this procedure gives reasonable results. In fact, it is shown below that this is in fact the case.

A second method is to approximate the densities of the independent components by a family of densities that are specified by a limited number of parameters. If the number of parameters in the density family is very large we do not gain much from this approach, since the goal we are aiming for is to reduce the number of parameters. However, for any density p_i, if it is possible to use a very simple family of densities to estimate the ICA model, we get a simple solution. Fortunately, this idea is feasible. We can use an extremely simple parametrization of the p_i consisting of a single binary parameter; in other words, the choice of one of two densities.

The effectiveness of these methods is explained below.

Theorem 7.1 Let the assumed independent component density be $\widetilde{p}_i$, then

$$g_i(s_i) = \frac{\partial}{\partial s_i}\log\widetilde{p}_i(s_i) = \frac{\widetilde{p}'_i(s_i)}{\widetilde{p}_i(s_i)} \tag{7.4}$$

The independent component estimation $y_i = b_i^{\mathrm{T}}x$ constraints are determined for each related output with unit variance. If the density of the hypothesis $\widetilde{p}_i$ for all i satisfies the requirements, then

$$E\{s_i g_i(s_i) - g_i'(s_i)\} > 0 \tag{7.5}$$

The ML estimator is locally consistent.

Theorem 7.1 shows rigorously that small mis-specifications in the densities p_i do not affect the local consistency of the ML estimator, since sufficiently small changes do not change the sign in Equation 7.5.

Moreover, the theorem shows how to construct families consisting of only two densities, so that the condition in Equation 7.5 is true for one of these densities. For example, consider the following log-densities:

$$\log \widetilde{p}_i^{+}(s) = \alpha_1 - 2 \log \cosh(s) \tag{7.6}$$

$$\log \widetilde{p}_i^{-}(s) = \alpha_2 - [s^2/2 - \log \cosh(s)] \tag{7.7}$$

where α_1 and α_2 are positive parameters that are fixed so as to make these two functions logarithms of the probability densities. Actually, these constants can be ignored in the following. Factor 2 in Equation 7.6 is not important, but it is normally used; also, factor 1/2 in Equation 7.7 could be changed.

Simple computations show that the value of the nonpolynomial moment in Equation 7.5 for $\widetilde{p}_i^{+}$ is:

$$2E\{-\tanh(s_i)s_i + (1 - \tanh^2(s_i))\} \tag{7.8}$$

and for $\widetilde{p}_i^{-}$ is:

$$E\{\tanh(s_i)s_i - (1 - \tanh(s_i)^2)\} \tag{7.9}$$

From the above, we see that the signs of these expressions are always opposite. Thus, for practically any distribution of the s_i, one of these functions satisfies the condition.

Thus we can merely calculate the nonpolynomial moments of the two prior distributions in Equations 7.6 and 7.7, and choose the one that satisfies the stability condition in Equation 7.5. This can be done online during maximization of the likelihood. This always provides a consistent estimator, and solves the problem of semiparametric estimation.

7.2 Algorithms for Maximum Likelihood Estimation

7.2.1 Gradient Algorithms

The infomax BSS algorithm proposed by Bell and Sejnowski integrated the work done by Linsker (1988) in 1988 on the information transmission maximum criterion of neural network unsupervised learning. Jutten and Herault (1991) proposed the independent component extraction nonlinear independent law, and provided the framework for a BSS algorithm based on maximum information transmission theory (Linsker, 1992).

The block diagram in Figure 7.1 shows that the infomax algorithm is a kind of information theoretic approach based on the forward feedback self-organizing neural network algorithm, where $\boldsymbol{x}$ is a multiple observation signal vector, consisting of n independent sources that are linearly mixed. The output of network $\boldsymbol{u} = \boldsymbol{W}\boldsymbol{x}$ is an approximation of the real source $\boldsymbol{s}$. $g(\cdot) = (g_1(\cdot), g_2(\cdot), \cdots, g_n(\cdot))^{\mathrm{T}}$ is a reversible

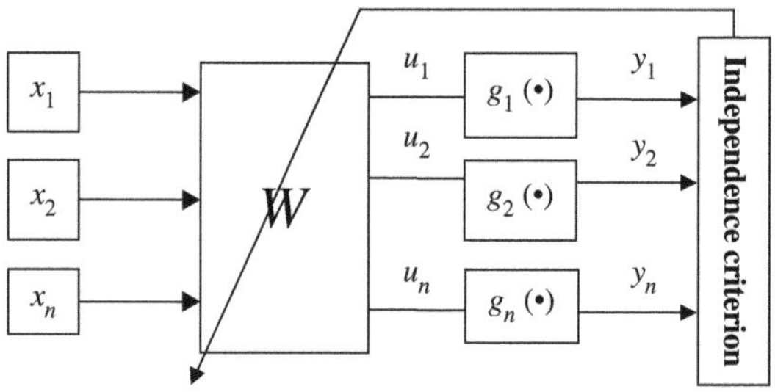

Figure 7.1 Block diagram of infomax algorithm

monotonic nonlinear function, and the nonlinear output is $\boldsymbol{y} = (y_i)$. The independence criterion is the maximum information transmission principle, adjusted to the separation matrix $\boldsymbol{W}$ (neural network connection weight matrix) to find the optimization of $\boldsymbol{W}$ so as to create maximum mutual information between the output of network $\boldsymbol{y}$ and the input $\boldsymbol{x}$. From information theory

$$I(\boldsymbol{x}|\boldsymbol{y}) = H(\boldsymbol{y}) - H(\boldsymbol{y}|\boldsymbol{x}) \tag{7.10}$$

where $H(\boldsymbol{y})$ is the joint entropy of the network and $H(\boldsymbol{y}|\boldsymbol{x})$ is the conditional entropy. If there is any noise $\boldsymbol{N}$ in the system, that is, $\boldsymbol{y} = g(\boldsymbol{u}) + \boldsymbol{N} = g(\boldsymbol{Wx}) + \boldsymbol{N}$ and $H(\boldsymbol{y}|\boldsymbol{x}) = H(\boldsymbol{N})$, Equation 7.10 can be written as

$$I(\boldsymbol{x}|\boldsymbol{y}) = H(\boldsymbol{y}) - H(\boldsymbol{N}) \tag{7.11}$$

The mutual information $I(\boldsymbol{x}|\boldsymbol{y})$ between $\boldsymbol{y}$ and $\boldsymbol{x}$ is equivalent to the joint entropy $H(\boldsymbol{y})$ (noise $\boldsymbol{N}$ is irrelevant in the system). From information theory, joint entropy and mutual information are considered as the target function.

$$H(\boldsymbol{y}) = H(y_1) + H(y_2) + \cdots + H(y_n) - I(y_1, y_2, \ldots, y_n) \tag{7.12}$$

$H(y_i)$ is the nonlinear output edge entropy, while $I(y_1, y_2, \ldots, y_n)$ is the nonlinear output mutual information, which is always negative. If only the nonlinear y_i are independent, $I(y_1, y_2, \ldots, y_n) = 0$. A monotonic reversible nonlinear mapping has no effect on mutual information and, therefore, when $I(\boldsymbol{y})$ takes the minimum value 0, $I(\boldsymbol{u})$ is set to the minimum value 0, each component is independent, and the problem of ICA is solved. Then,

$$H(\boldsymbol{y}) = H(y_1) + H(y_2) + \cdots + H(y_n) = -E\{\ln(p_y(\boldsymbol{y}))\} \tag{7.13}$$

where $p_y(\boldsymbol{y})$ is the PDF of $\boldsymbol{y}$, while the maximum $H(\boldsymbol{y})$ contains maximal edge entropy and minimal mutual information. Entropy is used as the target function because entropy is an unordered measurement of random variables measuring the amount of information. If the statistical independence of each component of $\boldsymbol{y}$ is higher, the entropy of $\boldsymbol{y}$ is larger and there is more information. It can be proved that if the nonlinear function $g_i(\cdot)$ is a source of the cumulative distribution function (cdf), which is the integration of the PDF, the edge entropy is at its maximum. Since $\boldsymbol{y} = g(\boldsymbol{u}) = g(\boldsymbol{Wx})$,

$$p_y(\mathbf{y}) = \frac{p_x(\mathbf{x})}{|J|} \tag{7.14}$$

where $|J|$ is the absolute value of the determinant of separation matrix $\boldsymbol{W}$. Then,

$$J = \det\begin{pmatrix} \frac{\partial y_1}{\partial x_1} & \cdots & \frac{\partial y_1}{\partial x_n} \\ \vdots & \ddots & \vdots \\ \frac{\partial y_n}{\partial x_1} & \cdots & \frac{\partial y_n}{\partial x_n} \end{pmatrix} = (\det \boldsymbol{W})\prod_{i=1}^{n}\frac{\partial y_i}{\partial u_i} \tag{7.15}$$

and

$$H(\boldsymbol{y}) = E\{\ln|J|\} - E\{\ln p_x(\boldsymbol{x})\}. \tag{7.16}$$

From Equation 7.16, if $\boldsymbol{x}$ is known, the size of $H(\boldsymbol{y})$ changes with $E\{\ln|J|\}$, that is, it changes with the weight matrix $\boldsymbol{W}$. Using Equations 7.15 and 7.16 and obtaining the gradient, we have

$$\Delta \boldsymbol{W} \propto \frac{\partial H}{\partial \boldsymbol{W}} = \frac{\partial}{\partial \boldsymbol{W}}\ln|J| = \frac{\partial}{\partial \boldsymbol{W}}\ln|\boldsymbol{W}| + \frac{\partial}{\partial \boldsymbol{W}}\ln\prod_{i=1}^{n}\frac{\partial y_i}{\partial u_i} \tag{7.17}$$

For the first element in Equation 7.17, the weight matrix $\boldsymbol{W}$ is expanded by row i, $\boldsymbol{W} = \sum_i \boldsymbol{w}_i A_{ii}$, where the A_{ii} are the corresponding algebraic elements, because

$$\frac{\partial}{\partial \boldsymbol{w}_i}\ln|\boldsymbol{W}| = \frac{A_{ii}}{\det \boldsymbol{W}} \tag{7.18}$$

Then,

$$\frac{\partial}{\partial \boldsymbol{W}}\ln|\boldsymbol{W}| = \frac{(\boldsymbol{W}^{\bullet})^{\mathrm{T}}}{\det \boldsymbol{W}} = (\boldsymbol{W}^{\mathrm{T}})^{-1} \tag{7.19}$$

where $\boldsymbol{W}^{\bullet}$ is the adjacency matrix of $\boldsymbol{W}$.

For the second element in Equation 7.17, we can expand the logarithmic element,

$$\varphi(\boldsymbol{u}) = -\frac{\partial p(\boldsymbol{u})/\partial \boldsymbol{u}}{p(\boldsymbol{u})} = \left[-\frac{\partial p(\boldsymbol{u}_1)/\partial \boldsymbol{u}_1}{p(\boldsymbol{u}_1)}, \ldots, -\frac{\partial p(\boldsymbol{u}_n)/\partial \boldsymbol{u}_n}{p(\boldsymbol{u}_n)}\right]^{\mathrm{T}} \tag{7.20}$$

Obviously, the second element is related to $g(\cdot)$, and the adjustment formula for the separation matrix is given as

$$\Delta \boldsymbol{W} \propto (\boldsymbol{W}^{\mathrm{T}})^{-1} - \varphi(\boldsymbol{u})\boldsymbol{x}^{\mathrm{T}} \tag{7.21}$$

Because we do not know the source probability distribution, we have to make use of an estimation thereof. This is, in fact, very efficient. $g_i(\cdot)$, which is a monotonically ascending function, such as the *sigmoid* or *tanh* function, selects a value between 0 and 1. In experiments using Bell's blind voice separation, the voice signal is usually super-Gaussian, so we use the sigmoid nonlinear function:

$$y = g(\boldsymbol{u}) = \frac{1}{1 + \mathrm{e}^{-u}} \tag{7.22}$$

Then, the formula for adjusting the parameters is given as

$$\Delta \boldsymbol{W} \propto (\boldsymbol{W}^{\mathrm{T}})^{-1} + (1 - 2\boldsymbol{y})\boldsymbol{x}^{\mathrm{T}} \tag{7.23}$$

$$\boldsymbol{W}(n+1) = \boldsymbol{W}(n) + \mu \Delta \boldsymbol{W}(n) \tag{7.24}$$

where n is the number of iterations and μ is the learning rate (a positive number less than 1). The infomax algorithm executes an iterative solution based on Equations 7.23 and 7.24, until the algorithm converges. Although Equation 7.23 can achieve iterative optimization of $\boldsymbol{W}$, it uses an inverse matrix, the computation of which is both time consuming and requires a large amount of calculation. So, using the natural gradient to simplify Equation 7.23, and multiplying by $\boldsymbol{W}^{\mathrm{T}}\boldsymbol{W}$ on the right-hand side, we get

$$\Delta \boldsymbol{W} \propto ((\boldsymbol{W}^{\mathrm{T}})^{-1} + (1 - 2\boldsymbol{y})\boldsymbol{x}^{\mathrm{T}})\boldsymbol{W}^{\mathrm{T}}\boldsymbol{W} = [I + (1 - 2\boldsymbol{y})\boldsymbol{u}^{\mathrm{T}}]\boldsymbol{W} \tag{7.25}$$

Here $\boldsymbol{W}^{\mathrm{T}}\boldsymbol{W}$ is a positive definite matrix, which does not affect stochastic gradient descent algorithm convergence. Moreover, Equation 7.25 avoids the matrix inversion process, thereby significantly reducing the amount of calculation and increasing convergence speed. Obviously, estimating the probability distribution of the source in this way is rough, but the result of BSS is very satisfactory. Bell and sejnowski (1995) also tried different forms of monotonic smooth nonlinear functions and obtained good results, except for the relation between convergence rate and nonlinear function. However, because the differential sigmoid function has a super-Gaussian distribution (since the kurtosis is positive), the infomax algorithm can only be applied to BSS of super-Gaussian signal sources.

As mentioned above, in statistical analysis of signals, we divide the signals into three types, namely, super-Gaussian, Gaussian, and sub-Gaussian, according to whether their kurtosis is positive. Since the original infomax algorithm could only be applied to the super-Gaussian probability model of Equation 7.22, it could only be used to separate super-Gaussian blind sources. However, as an actual observed signal is always a mixture of super-Gaussian and sub-Gaussian signals, using the same type of nonlinear model probability density estimation is often ineffective. Lee *et al* (1999) proposed a dual probability model based on the traditional infomax algorithm which, in a BSS process, according to changes in the statistical characteristics of the separation results, dynamically switches the probability model. Eventually unity of the probability model and the real source signal probability distribution is achieved, and the best blind separation results are obtained. This is the basic idea of the extended infomax algorithm.

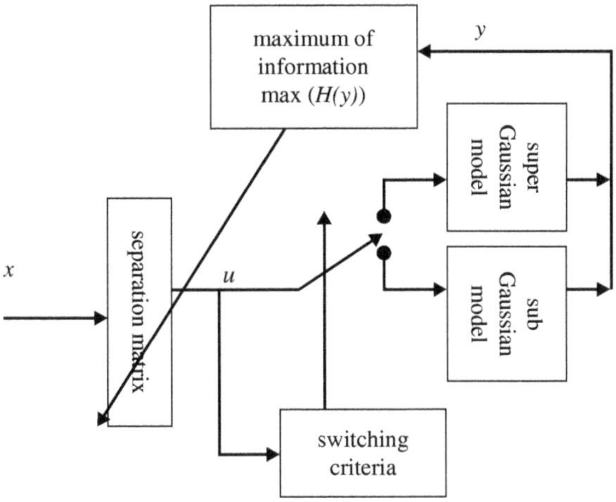

Figure 7.2 Block diagram of the extended infomax algorithm

Figure 7.2 illustrates the theoretical framework of the extended infomax algorithm, with the sub-Gaussian probability density model given as

$$p_1(\boldsymbol{u}) = \frac{1}{2}[N(\mu, \sigma^2) + N(-\mu, \sigma^2)] = \frac{1}{2}\left[\frac{1}{\sqrt{2\pi}\sigma} \exp \frac{(\boldsymbol{u} - \mu)^2}{2\sigma^2} + \frac{1}{\sqrt{2\pi}\sigma} \exp \frac{(\boldsymbol{u} + \mu)^2}{2\sigma^2}\right] \tag{7.26}$$

Assuming $a = \mu/\sigma^2$, the next formula is obtained from Equation 7.20 and the definition of the hyperbolic tangent function:

$$\varphi(\boldsymbol{u}) = -\frac{\partial p(\boldsymbol{u})/\partial \boldsymbol{u}}{p(\boldsymbol{u})} = \frac{\boldsymbol{u}}{\sigma^2} - a\left(\frac{\exp(a\boldsymbol{u}) - \exp(-a\boldsymbol{u})}{\exp(a\boldsymbol{u}) + \exp(-a\boldsymbol{u})}\right) = \frac{\boldsymbol{u}}{\sigma^2} - \frac{\mu}{\sigma^2}\tanh\left(\frac{\mu}{\sigma^2}\boldsymbol{u}\right) \tag{7.27}$$

Assuming $\mu = 1$ and $\sigma^2 = 1$, we can simplify Equation 7.27 as

$$\varphi(\boldsymbol{u}) = \boldsymbol{u} - \tanh(\boldsymbol{u}) \tag{7.28}$$

Thus, the adjustment formula for the separation matrix is

$$\Delta \boldsymbol{W} \propto [I + \tanh(\boldsymbol{u})\boldsymbol{u}^{\mathrm{T}} - \boldsymbol{u}\boldsymbol{u}^{\mathrm{T}}]\boldsymbol{W} \tag{7.29}$$

In Figure 7.2, the unimodal super-Gaussian probability density model is

$$p_2(\boldsymbol{u}) = p_G(\boldsymbol{u})\mathrm{sech}^2(\boldsymbol{u}) \tag{7.30}$$

where $p_G(\boldsymbol{u}) = N(0, 1)$ is the zero mean, unit variance of the Gaussian density function, $\mathrm{sech}(\boldsymbol{u}) = 2(\mathrm{e}^{u} + \mathrm{e}^{-u})^{-1}$. Also, the nonlinear function $\varphi(\boldsymbol{u})$ is calculated as

$$\varphi(\boldsymbol{u}) = -\frac{\partial p(\boldsymbol{u})/\partial \boldsymbol{u}}{p(\boldsymbol{u})} = \boldsymbol{u} + \tanh(\boldsymbol{u}) \tag{7.31}$$

Then, the adjustment formula for the separation matrix in the algorithm is given as

$$\Delta \boldsymbol{W} \propto [I - \tanh(\boldsymbol{u})\boldsymbol{u}^{\mathrm{T}} - \boldsymbol{u}\boldsymbol{u}^{\mathrm{T}}]\boldsymbol{W} \tag{7.32}$$

By introducing an n-dimensional diagonal matrix $\mathbf{K}$, called the probability model switching matrix, with the diagonal element k_i, the kurtosis symbol, we can determine the value according to normalized kurtosis symbol changes. From the definition of kurtosis,

$$k_i = \mathrm{sign}[\mathrm{kurt}(u_i)] = \mathrm{sign}\left(\frac{E\left(u_i^4\right)}{[E(u_i^2)]^2} - 3\right) \quad 1 \leq i \leq n \tag{7.33}$$

For a Gaussian signal, $k_i = 0$, for a super-Gaussian signal, $k_i > 0$, and for a sub-Gaussian signal, $k_i < 0$. The adjustment formula for the separation matrix in the extended infomax algorithm is given as

$$\Delta \boldsymbol{W} \propto [I - \mathbf{K}\tanh(\boldsymbol{u})\boldsymbol{u}^{\mathrm{T}} - \boldsymbol{u}\boldsymbol{u}^{\mathrm{T}}]\boldsymbol{W} \begin{cases} k_i = 1 \ : \ \text{super-Gauss} \\ k_i = -1 \ : \ \text{sub-Gauss} \end{cases} \tag{7.34}$$

The steps in the extended infomax algorithm are given below.

Algorithm Extended infomax algorithm

Input: Whitened matrix $\boldsymbol{X}$
Output: Independent component $\boldsymbol{Y}$
Procedure:

1. Subtract the mean to whiten the signal;
2. Initialize the resolved mixed matrix $\boldsymbol{W}$ (using a unit random matrix);
3. Calculate $\boldsymbol{u}$ and $\Delta \boldsymbol{W}$;
4. Adjust to $\boldsymbol{W}$;
5. Repeat steps (3) and (4) until $\boldsymbol{W}$ converges;
6. Get the independent component $\boldsymbol{y} = \boldsymbol{W}\boldsymbol{x}$.

7.2.2 A Fast Fixed-Point Algorithm

Likelihood can also be maximized using a fixed-point algorithm. The fixed-point algorithm defined by FastICA is a very fast and reliable maximization method that was introduced in Chapter 6 to maximize non-Gaussianity for use in ICA estimation (Hyvärinen and Oja, 1997; Hyvärinen, 1998; Hyvärinen, Hoyer, and Oja, 1998; Hyvärinen, 1999a, 1999b; Hyvärinen and Oja, 2000; Hyvärinen, Karhunen, and Oja, 2001). In fact, the FastICA algorithm can be directly applied to likelihood maximization.

As presented in Chapter 6, the FastICA algorithm is defined as

$$w \leftarrow w - [E\{zg(w^{\mathrm{T}}z)\} + \beta w]/[E\{g'(w^{\mathrm{T}}z)\} + \beta] \tag{7.35}$$

where $\beta = -E\{y_i g(y_i)\}$. If Equation 7.35 is given in matrix form, we have

$$w \leftarrow w + \mathrm{diag}(\alpha_i)[\mathrm{diag}(\beta_i) + E\{g(y)y^{\mathrm{T}}\}]w \tag{7.36}$$

where α_i is defined as $-1/(E\{g'(w^{\mathrm{T}}x)\} + \beta_i)$, and $y = Wx$. After each iteration, W should be mapped to a whitened matrix. This is achieved by classical methods for the square root of a matrix:

$$W \leftarrow (WCW^{\mathrm{T}})^{\frac{1}{2}}W \tag{7.37}$$

where $C = E\{xx^{\mathrm{T}}\}$ is the correlation matrix. Compared with the ML degree of the natural gradient method, FastICA can be considered as an optimized gradient algorithm. In FastICA, the speed of convergence is determined by matrices diag(α_i) and diag(β_i). These two matrices provide the available optimal step length.

The procedure for the ML fast fixed point algorithm is given below.

Algorithm FastICA algorithm

Input: Whitened matrix X
Output: Independent component Y
Procedure:

1. Subtract the mean to whiten the signal;
2. Initialize the resolved mixed matrix (using a unit random matrix);
3. Calculate:

$$y = Wx;$$

$$\beta_i = -E\{y_i g(y_i)\}, i = 1, \ldots, n;$$

$$\alpha_i = -1/(\beta_i + E\{g'(y_i)\}), i = 1, \ldots, n;$$

4. Adjust W:

$$W \leftarrow W + \mathrm{diag}(\alpha_i)[\mathrm{diag}(\beta_i) + E\{g(y)y^{\mathrm{T}}\}]W;$$

$$W \leftarrow (WCW^{\mathrm{T}})^{\frac{1}{2}}W;$$

5. Repeat steps (3) and (4) until W converges;
6. Get the independent component $y = Wx$.

Another advantage of FastICA is that it can estimate both sub- and super-Gaussian independent components without any additional steps. We can fix the nonlinearity to be equal to tanh nonlinearity for all the independent components. The reason for this is obvious: matrix diag(α_i) contains estimates based on the nature (sub- or super-Gaussian) of the independent components. These estimates are used in the gradient algorithm presented in the previous subsection. On the other hand, matrix diag(β_i) is considered as a scaling of the nonlinearities, since it can be reformulated.

Note that in FastICA, the outputs y_i are decorrelated and normalized to unit variance after every step. No such operations are needed in the gradient algorithm. FastICA is not stable if these additional operations are omitted. Thus the optimization space is slightly reduced.

7.3 Application and Analysis

7.3.1 Blind Signal Separation

We carried out a blind signal separation experiment to investigate the learning performance of the algorithm. We adopted four randomly generated signals for use in the simulation experiment, with one of these being a Gaussian signal. Each signal has 1000 data points. We obtained the original independent signals shown in

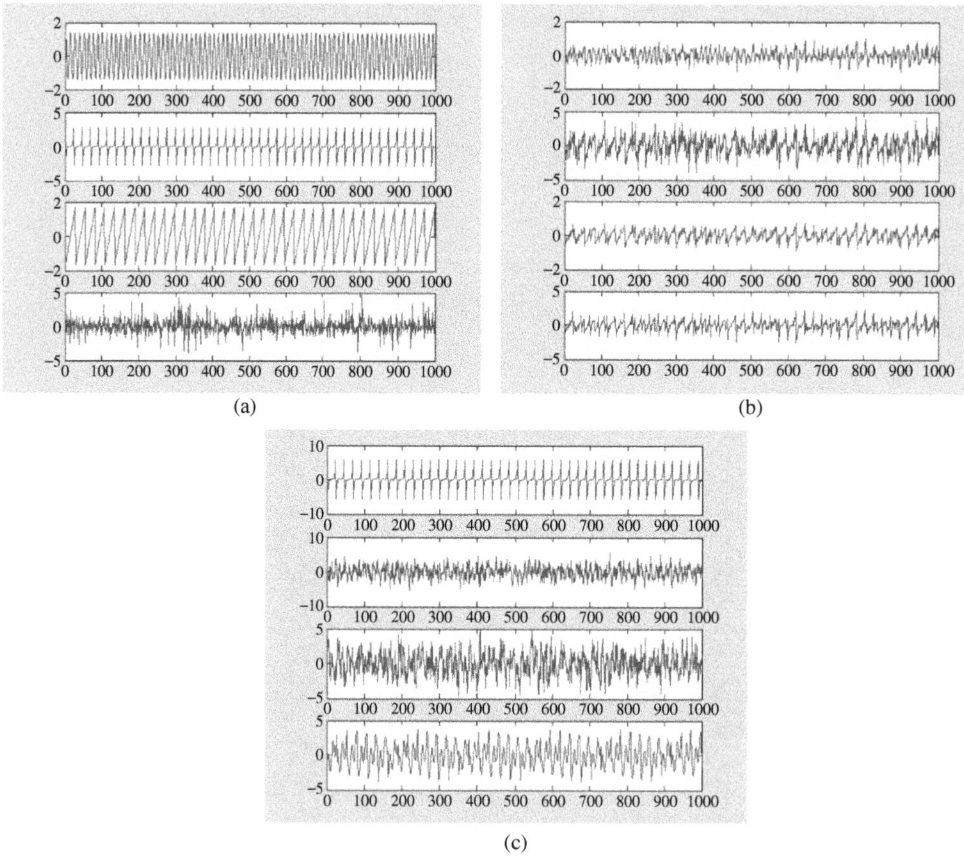

Figure 7.3 Mixed signal separation results using maximum likelihood estimation ICA (a) The original four independent signals (b) Blended to produce four mixed signals (c) Maximum likelihood estimation ICA separation results

Figure 7.3a. In the experiment, the four random signals are linearly mixed with a randomly generated 4×4 matrix to produce a mixed signal. We can get the mixed signals shown in Figure 7.3b and the results shown in Figure 7.3c. Then the ML estimation algorithm is used to separate the mixed signal. To evaluate separation performance, the randomly generated mixed matrix is given below.

$$A = \begin{bmatrix} 0.5468 & 0.4465 & 0.6187 & 0.7265 \\ 0.4399 & 0.8652 & 0.6660 & 0.3144 \\ 0.6950 & 0.3945 & 0.8024 & 0.7543 \\ 0.3176 & 0.5544 & 0.0367 & 0.5633 \end{bmatrix}$$

We also carried out a voice signal separation experiment as follows. The independent speech signals, mixed signals and results are shown in Figure 7.4a–c.

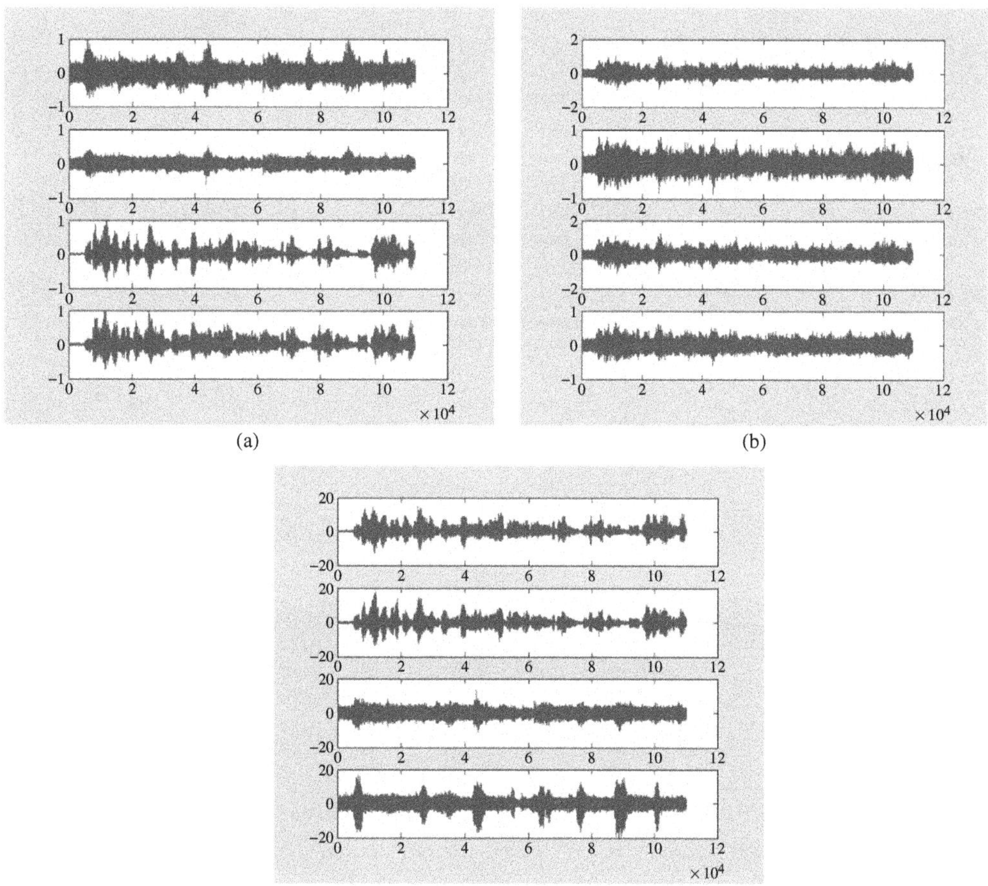

Figure 7.4 Hybrid speech signal separation results using maximum likelihood estimation ICA (a) Four independent speech signals (sources) (b) Blended to produce four mixed signals (c) Maximum likelihood estimation ICA separation results

Table 7.1 Performance of mixed analog signal separation using the maximum likelihood estimation ICA algorithm

Algorithm	Time consumed (s)	Separation accuracy
Maximum likelihood estimation ICA	0.969	0.3782

Table 7.2 Performance of mixed speech signal separation using the maximum likelihood estimation ICA algorithm

Algorithm	Time consumed (s)	Separation accuracy
Maximum likelihood estimation ICA	134.67	0.30143

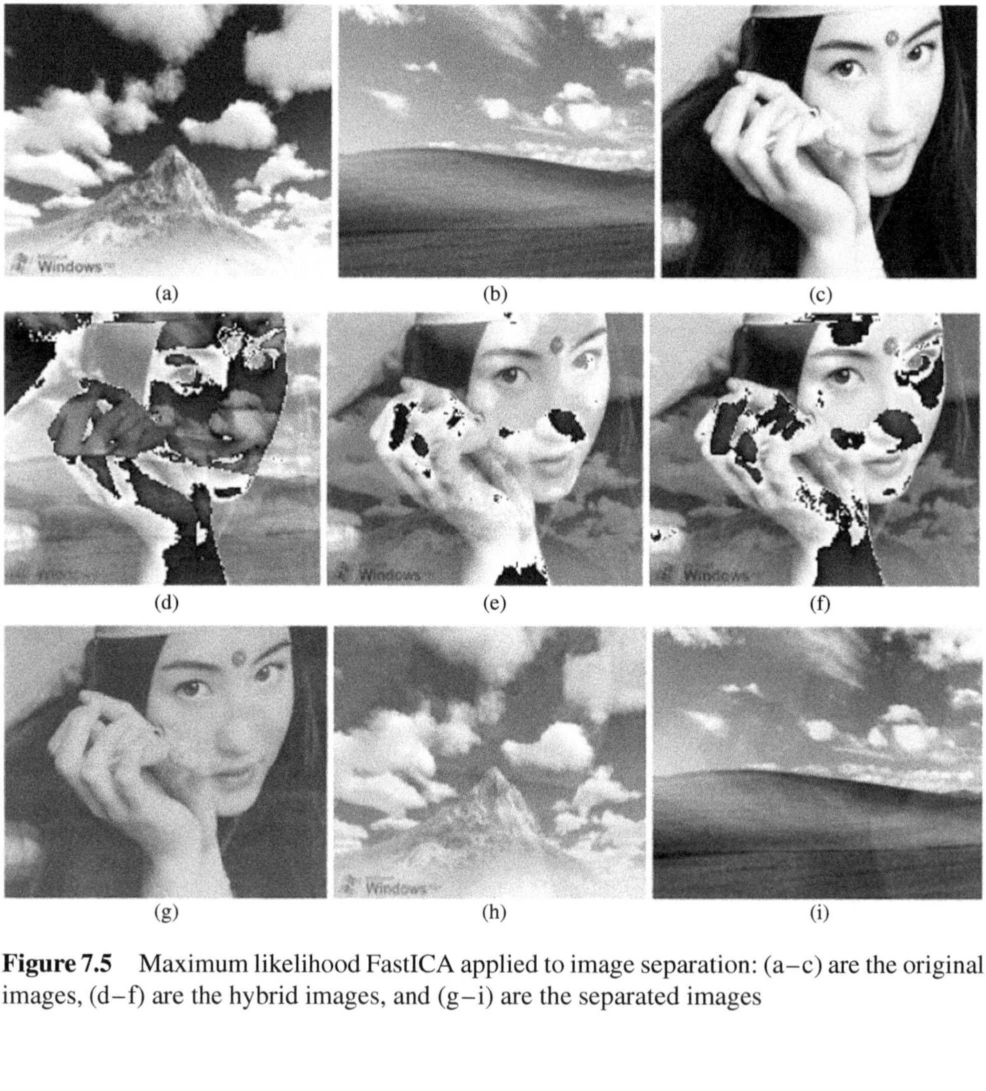

Figure 7.5 Maximum likelihood FastICA applied to image separation: (a–c) are the original images, (d–f) are the hybrid images, and (g–i) are the separated images

The performance of the separation algorithm was evaluated based on convergence speed and precision of the separation which are shown in Tables 7.1 and 7.2.

7.3.2 Image Signal Separation

Having scanned each of the three images given in Figure 7.5a–c by row or column, we saved them as one-dimensional vectors (in processing the result, we can restore the image by reverse processing). Then, we randomly mixed the three images yielding the three mixed images shown in Figure 7.5d–f. By applying the ML FastICA algorithm to these mixed images, we obtained the three separate images shown in Figure 7.5g–i.

From Figure 7.5, we can see that the images recovered by the ML FastICA algorithm are good, but the sequence is different from the original one owing to uncertainty of the BSS techniques.

7.4 Chapter Summary

ML estimation, perhaps the most commonly used statistical estimation technique, can also be used to estimate the ICA model. It is closely related to the infomax principle used in neural networks. If the densities of the independent components are known in advance, a very simple gradient algorithm can be derived. To speed up convergence, the natural gradient version and especially the FastICA fixed-point algorithm can be used. If ML ICA is applied to image signal separation, the results of separation are very similar to the original images, but with a change in the image sequence.

References

Bell, A.J., and Sejnowski, T.J. (1995). An information maximization approach to blind separation and blind deconvolution. *Neural Computation*, **7**(6), 1129–1159.

Hyvärinen, A. (1998). New approximations of differential entropy for independent component analysis and projection pursuit, in *Advances Neural Information Processing System*, MIT Press, pp. 273–279.

Hyvärinen, A. (1999a). Fast and robust fixed-point algorithms for independent component analysis. *IEEE Transactions on Neural Networks*, **10**(3), 626–634.

Hyvärinen, A. (1999b). Survey on independent component analysis. *Neural Computing Surveys*, **2**, 94–128.

Hyvärinen A., Hoyer P., and Oja E. (1998). Sparse code shrinkage for image denoising. Neural Networks Proceedings on IEEE World Congress on Computational Intelligence, pp. 859–864.

Hyvärinen, A., Karhunen, J., and Oja, E. (2001). *Independent Component Analysis*, John Wiley & Sons, Inc..

Hyvärinen, A., and Oja, E. (1997). A fast fixed-point algorithm for independent component analysis. *Neural Computation*, **9**(7), 1483–1492.

Hyvärinen, A., and Oja, E. (1998). Independent component analysis by general nonlinear Hebbian-like learning rules. *Signal Processing*, **64**(3), 301–313.

Hyvärinen, A., and Oja, E. (2000). Independent component analysis: algorithms and applications. *Neural Networks*, **13**(4), 411–430.

Jutten, C., and Herault, J. (1991). Blind separation of sources, part I: an adaptive algorithm based on neuromimetic architecture. *Signal Processing*, **24**(1), 1–10.

Lee, T.W., Girloami, M., and Sejnowski, T.J. (1999). Independent component analysis using extended infomax algorithm for mixed subGausssian and superGaussian sources. *Neural Computation*, **1**(2), 417–441.

Linsker, R. (1988). Self-organization in a perceptual network. *Computer*, **21**(3), 105–117.

Linsker, R. (1992). Local synaptic learning rules suffice to maximize mutual information in a linear network. *Neural Computation*, **4**(5), 691–702.

8

Overcomplete Independent Component Analysis Algorithms and Applications

When the number of sources is greater than the number of sensors, we refer to the blind source separation (BSS) problem as being overcomplete (Lee *et al.*, 1999). Lewicki and Sejnowski (2000) suggest that, if some information is known in advance about the source distribution, independent components (ICs) can be extracted to some extent by the overcomplete independent component analysis (ICA) model. Theis *et al.* (2002, 2004) proposed a geometric algorithm with two steps, namely, matrix recovery and source signal recovery. They improved the geometric algorithm for mixing matrix estimation when the number of source signals is equal to the number of mixed signals, and then applied the maximum likelihood method for source signal recovery. To improve the convergence properties of the algorithm, Li, Cichocki, and Amari (2003) proposed a sparse matrix factorization method for BSS. Davies and Mitianoudis (2004) proposed a simple hybrid model to solve the problem in the modified discrete cosine transform (MDCT) processing of an audio signal. Thus, in this chapter, we introduce the classic overcomplete ICA algorithm, the algebraic overcomplete ICA (AICA) algorithm, and the geometric overcomplete ICA (Geo-ICA) algorithm.

8.1 Overcomplete ICA Algorithms

8.1.1 Classic Overcomplete ICA Algorithm

Lewicki and Olshausen (1999) and Lewicki and Sejnowski (2000) proposed an algorithm based on statistical models to solve the overcomplete problem. This algorithm consists of a linear programming method and the natural gradient method. Given

---

Blind Source Separation: Theory and Applications, First Edition. Xianchuan Yu, Dan Hu and Jindong Xu.

observation data $\boldsymbol{X} = [\boldsymbol{X}_1, \boldsymbol{X}_2, \dots, \boldsymbol{X}_n]^{\mathrm{T}}$ mixed with additive noise, the model can be expressed as

$$\boldsymbol{X} = \boldsymbol{AS} + \varepsilon \tag{8.1}$$

where A is an $n \times m(m > n)$ matrix. We usually assume that ε is the Gaussian additive noise, so

$$\log P(X|\boldsymbol{A}, \boldsymbol{S}) \propto -\frac{1}{2\sigma^2}(X - \boldsymbol{AS})^2 \tag{8.2}$$

where σ^2 denotes the variance of the noise. Estimation of the source signal, based on a probabilistic model, attempts to find S which maximizes the posterior probability.

$$\hat{s} = \arg\max_S P(\boldsymbol{S}|\boldsymbol{X}, \boldsymbol{A}) = \arg\max_S P(\boldsymbol{X}|\boldsymbol{A}, \boldsymbol{S})P(\boldsymbol{S}) \tag{8.3}$$

where $\hat{s}$ is the estimation of S.

Given a matrix of the base vectors A and the observed data X, we can use the gradient descent method to optimize the logarithmic posterior probability distribution. In the case of an overcomplete model, the mixing matrix is not a square matrix, nor is it an inversion.

An alternative method, based on the assumption that $\varepsilon = 0$, is to solve the problem as a linear programming problem (Chen, Donoho, and Saunders, 1998):

$$\min \boldsymbol{c}^{\mathrm{T}}|\boldsymbol{S}| \ s.t. \ \boldsymbol{AS} = \boldsymbol{X} \tag{8.4}$$

Let $\boldsymbol{c}^{\mathrm{T}} = [1, \dots, 1]$, so $\boldsymbol{c}^{\mathrm{T}}|s| = \sum_k |s_k|$. By separating the positive and negative coefficients, Equation 8.4 can be transformed into a standard linear programming problem (with only positive coefficients). By executing the following transformations, $s \leftarrow [u; v]$, $\boldsymbol{c} \leftarrow [1; 1]$, and $\boldsymbol{A} \leftarrow [\boldsymbol{A}; -\boldsymbol{A}]$, Equation 8.4 becomes

$$\min [1; 1]^{\mathrm{T}}[u; v] s.t. [\boldsymbol{A}; -\boldsymbol{A}][u; v] = x, \ u, v \geq 0 \tag{8.5}$$

Equation 8.5 can be solved by a standard linear programming method, with solutions for u and v, while s is calculated by the following relationship.

$$s \leftarrow [u; v] \ s = u - v \tag{8.6}$$

To obtain a learning algorithm for A, we first need to determine an appropriate objective function; a natural goal is to maximize the probability of the observed data. For $\boldsymbol{X} = [\boldsymbol{X}_1, \boldsymbol{X}_2, \dots, \boldsymbol{X}_n]^{\mathrm{T}}$,

$$P(\boldsymbol{X}|\boldsymbol{A}) = \prod_{k=1}^{n} P(\boldsymbol{X}_k|\boldsymbol{A}) \tag{8.7}$$

$P(\boldsymbol{X}_k|\boldsymbol{A})$ can be obtained by marginalization of the internal state,

$$P(\boldsymbol{X}_k|\boldsymbol{A}) = \int \mathrm{d}sP(s)P(\boldsymbol{X}_k|\boldsymbol{A}, \boldsymbol{s}), \ k = 1, \dots, n. \tag{8.8}$$

Lewicki and Sejnowski (2000) used a multivariate Gaussian distribution on s to solve the above equation approximatively, and obtain the learning rule. This is given as

$$\Delta A \propto AA^{\mathrm{T}} \frac{\partial}{\partial A} \log P(x|A) \approx -A(z\hat{s}^{\mathrm{T}} + I) \tag{8.9}$$

where $z_k = \partial \log P(s_k)/\partial s_k$. Note that the learning rule in Equation 8.9 and standard ICA based on the natural gradient method are very similar in form.

8.1.2 *Algebraic Overcomplete ICA Algorithm*

See Waheed and Salam, 2002, 2003a, 2003b.

The idea behind the AICA and Geo-ICA algorithms is basically the same. The first step involves learning the projection of the matrix of base vectors A on the unit circle, while the second step restores the source vector.

The AICA algorithm calculates the distance between vectors using the dot product. The dot product between vectors α and β can be defined as

$$\alpha \bullet \beta = \alpha^{\mathrm{T}} \beta = |\alpha||\beta| \cos \theta_{\alpha\beta} \tag{8.10}$$

where $\theta_{\alpha\beta}$ represents the angle between vectors α and β. If α and β represent the projection of the two different vectors on the unit circle, the distance between them can be characterized completely by the direction angle $\theta_{\alpha\beta}$. Another benefit of using the dot product is that this type of operation is commutative, so $\alpha \bullet \beta = \beta \bullet \alpha$. As a result of the above characteristics, the AICA algorithm does not require the symbol transformation and replacement operation as in the Geo-ICA algorithm when calculating the distance of the base vector. Thus, the AICA algorithm only needs to calculate n dot product operations instead of $2^n - 1$ Euclidean distance calculations as is the case in the Geo-ICA algorithm.

The first step in the algorithm is to calculate the matrix of the base vectors, the implementation of which is described below.

Algorithm AICA algorithm: calculating the matrix of the base vectors

Input: Sample set $\boldsymbol{X} = [\boldsymbol{X}_1, \boldsymbol{X}_2, \ldots, \boldsymbol{X}_N]$
Output: Matrix of base vectors $\boldsymbol{A}_{n\times m}$
Steps:

1. Select m vectors in an n-dimensional unit circle to make up the weight matrix $\boldsymbol{A}_0 = [a_1, a_2, \ldots, a_m]$. When the matrix is initialized, an initial point can be selected randomly on the unit circle or by following a particular distribution.
2. Select an n-dimensional sample $\boldsymbol{X}_i$ from the observation vector group, making $\boldsymbol{X}_i \neq 0, i = 1, 2, \ldots, n$.
3. Selected sample $\boldsymbol{X}_i$ is projected onto the unit circle, such that $\boldsymbol{Y}_i = \frac{\boldsymbol{X}_i}{||\boldsymbol{X}_i||}$, $i = 1, 2, \ldots, n$.

4. Calculate the dot product between the column vectors of Y_i and A_i.
5. $a_j^* = \underset{a_j}{\operatorname{argmax}} |Y_i^{\mathrm{T}} \cdot a_j|, i = 1, 2, \ldots, n; j = 1, 2, \ldots, m$.
6. Apply iterative updating according to the nearest principle.

$$a_j(k+1)^* = \begin{cases} \psi\left(a_j(k)^* + \eta(t)\mathrm{sgn}(Y_i - a_j(k)^*)\right), Y_i^{\mathrm{T}} \cdot a_j^* > 0 \\ \psi(a_j(k)^* + \eta(t)\mathrm{sgn}(-Y_i - a_j(k)^*)), Y_i^{\mathrm{T}} \cdot a_j^* < 0 \end{cases}$$

   where $\eta(t)$ is the learning rate dependent on the number of iterations, similar to that in the Geo-ICA algorithm; $\psi(\cdot)$ is the nonlinear equation used to project a column vector in the weight matrix onto the unit circle.
7. Update the weight matrix.

$$\Delta A_i = A_i - A_{i-1}, i = 1, 2, \ldots, n$$

8. If $|\Delta A_i| < \varepsilon$ ($\varepsilon > 0$), the algorithm converges to its end state, finally yielding A_i, which is the requested matrix of the base vectors.
9. If the algorithm does not satisfy the convergence conditions, repeat steps (2)–(7) until convergence is achieved.

The second step in the algorithm is source vector recovery. The process can use the shortest path algorithm, which is also used in the Geo-ICA algorithm, or the linear programming algorithm that is used in the overcomplete ICA algorithm.

8.1.3 Geometric Overcomplete ICA Algorithm

See Theis *et al.*, 2002, 2004.

The idea behind this algorithm is explained below. To obtain A and S in an overcomplete own learning process is quite difficult; however, one could first find A', which is the A projection on the unit circle. Then S can be approximately obtained through this. The Geo-ICA algorithm includes two steps: matrix recovery and source signal recovery.

The matrix recovery process is described below.

1. Choose $2m$ unit roots $\omega_1, \omega'_1, \ldots, \omega_m, \omega'_m$ on the unit circle $S^{n-1} \subset R^n$, making the ω_i and ω'_i symbols opposite and paired, that is, $\omega_i = -\omega'_i$ ($i = 1, \ldots, m$). These ω_i are usually called neurons.
2. Fix the learning rate $\eta : N \to R$, making $\eta(t) > 0$, $\sum_{m\in N} \eta(m) = \infty$, and $\sum_{m\in N} \eta(m)^2 < \infty$.
3. Iterate by executing the following steps until the convergence conditions are satisfied. Select a sample $X(t) \in R^n$ of random variable X; if $X(t) = 0$, reselect a nonzero sample. By assuming that the probability distribution function of X is a continuous function, it could happen that the sample value is zero. $X(t)$ is projected

onto the unit circle, then $Y(t) = \frac{X(t)}{||X(t)||}$. Let ω_i or ω'_i be the nearest point from $Y(t)$ calculated by the Euclidean distance, and execute the following iteration

$$\omega_i(t+1) = \pi(\omega_i(t) + \eta(t)\pi(Y(t) - \omega_i(t))) \tag{8.11}$$

where $\pi : R^n \backslash 0 \to S^{n-1}$ represents the projection from R^n to the $n-1$-dimensional unit circle S^{n-1}, as well as

$$\omega'_i(t+1) = -\omega_i(t+1) \tag{8.12}$$

4. All neurons are updated by these iterations.

This algorithm can be considered the absolute winner-takes-all learning algorithm. It is similar to a special case of Kohonen's competitive learning algorithm for self-organizing maps, which is a zero-neighborhood algorithm. The difference is that the step in the sample direction in the former algorithm does not depend on the distance, and the learning process is conducted on the unit circle S^{n-1}, rather than on R^{n-1}.

After completion of the above iterative algorithm, we get the group $\omega_1, \omega'_1, \ldots, \omega_m, \omega'_m, \omega_1, \ldots, \omega_m$ composed of matrix A, which is the next operation.

The source signal recovery process uses the shortest path algorithm. For simplicity, we assume $n = 2$. The goal of the algorithm is $\text{argmin}_{x_\lambda = As}|S|_1$, where $|S|_1$ denotes the L1 paradigm of S, and mixing matrix A can be obtained by the first step in the Geo-ICA algorithm. Let $a_k (k = 1, \ldots, m)$ represent a column vector after matrix A is normalized. Theis *et al.* (2002) presented the following conclusions. For any given sample x_i, identify the two column vectors a_j and a_k of matrix A that are the closest to x_i; then $S_i \in R^m$ can be expressed as

$$(S_i)_l = \begin{cases} \left(\left(a_j|a_k\right)^{-1} x_i\right)_j, l = j \\ ((a_j|a_k)^{-1}x_i)_k, l = k \\ 0, \text{otherwise} \end{cases} \tag{8.13}$$

It is not difficult to prove that A and S_i obtained by the above operation satisfy $AS_i = X_i$.

8.2 Applications and Analysis

Most of the classic face recognition algorithms are subspace analysis methods. For example, the principal component analysis (PCA) and linear discriminant analysis (LDA) algorithms obtain a facial image by depicting the input data from a high-dimensional space to a low-dimensional subspace. Subspace refinement typically requires a massive facial database. However, in many practical application fields the available data for refining are limited, or only a small number of the resulting refined images are available. To improve recognition rate we must obtain as much information as possible from a limited number of input data. From the foregoing, compared with the number of obtained observation signals, overcomplete ICA can obtain more source signals. This section focuses on overcomplete ICA in facial feature extraction.

8.2.1 Overcomplete ICA Facial Feature Representation

A face image can be represented as a linear superposition of a group of independent base images. If x_i represents a face image, a training set comprising n training images is given as $\boldsymbol{X} = [\boldsymbol{X}_1, \boldsymbol{X}_2, \ldots, \boldsymbol{X}_n]^{\mathrm{T}}$. Next, each image is whitened. Training images can be considered to be linear combinations of m independent components $\boldsymbol{S} = (\boldsymbol{S}_1, \boldsymbol{S}_2, \ldots, \boldsymbol{S}_m)$. If $m = n$, the problem becomes an ICA feature extraction problem (Hyvärinen, 2001). However, the more general case is $m > n$, that is, the overcomplete case. $\boldsymbol{X} = \boldsymbol{AS}$ represents the relationship between the training samples and ICs; in other words, each image $\boldsymbol{X}_i$ can be represented as a linear combination of $\boldsymbol{S}_1, \boldsymbol{S}_2, \cdots, \boldsymbol{S}_m$, with its coefficients being the ith row vector of weight matrix $\boldsymbol{A}$, as shown in Figure 8.1 (Jiang and Zhu, 2005).

Finally, with the coordinate vectors, $(a_{i1}, a_{i2}, \ldots, a_{im})$ as the overcomplete ICA face representation, such an image can be expressed using the base image coordinates. Overcomplete ICA provides a better face representation method, since overcomplete basis vectors offer more facial features to represent all the face data.

8.2.2 Experiments and Conclusions

We used some of the face images in the FERET database for our experiments (Cheng *et al.*, 2003). First, we randomly selected 20 images from the test database, and then extracted the required data from these 20 images. The test database contained a total of 420 images of 70 individuals. Six 112×92 images, each of which differed with respect to illumination and expression, were included for each individual.

Using the model described in Equation 8.1, $\boldsymbol{X} = \boldsymbol{AS} + \varepsilon$, where $\boldsymbol{X}$ is a $20 \times 10\,304$ matrix with the first row comprising a facial image for processing. $\boldsymbol{A}$ is a 20×30 matrix. The source signal matrix $\boldsymbol{S}$ is obtained as a 30×10304 matrix using the overcomplete ICA algorithm described earlier. Each row of $\boldsymbol{S}$ is a source image. Figure 8.2 illustrates the 30 source images.

In our experiments, we compared overcomplete ICA, PCA, and standard ICA. Identification of facial features and a comparison of the recognition rate using PCA (Turk and Pentland, 2001) and standard ICA (Bartlett, Movellan, and Sejnowski,

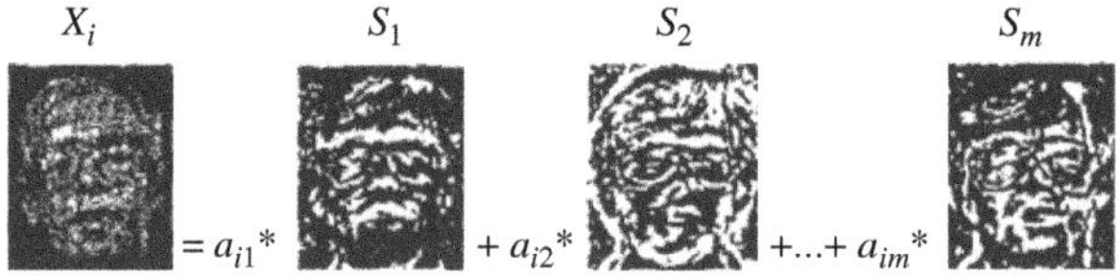

Figure 8.1 Representation of an overcomplete ICA human face

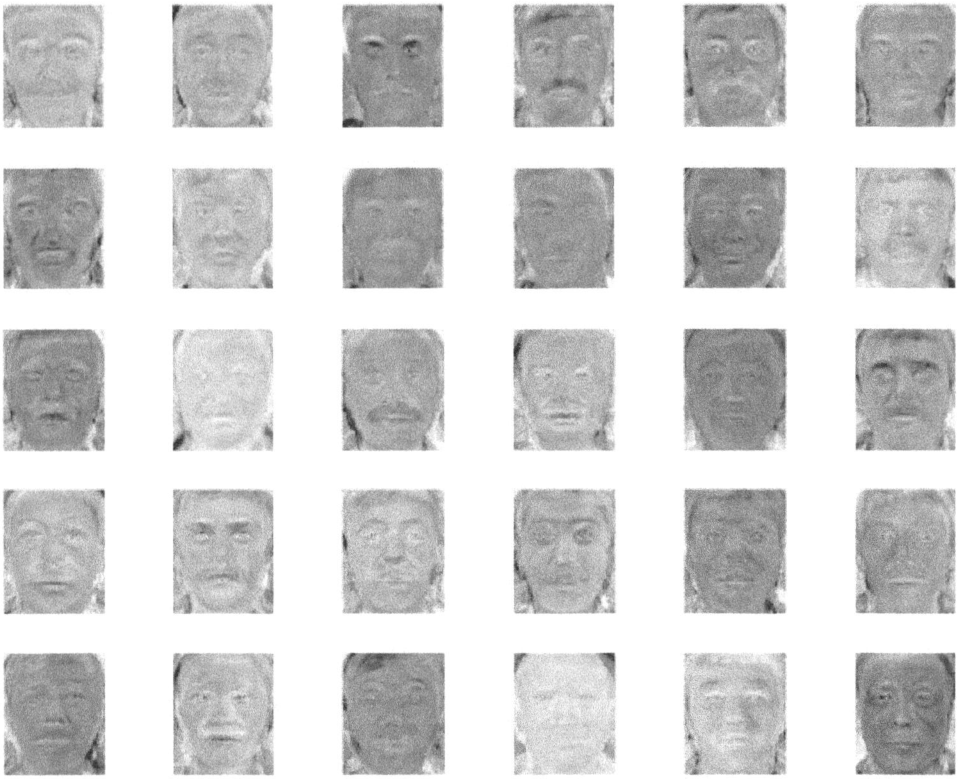

Figure 8.2 Source images

2002) was carried out using the same database and test set. The facial images extracted by overcomplete ICA, PCA, and standard ICA are referred to as the overcomplete IC face, the eigen face, and the IC face. These facial images were projected as a basic feature set $\boldsymbol{B}$, while each image from the test database was also projected onto the basic feature set $\boldsymbol{B}$. Coefficient f is a feature vector used to identify the face

$$f = \boldsymbol{B}a \tag{8.14}$$

The similarity measure was calculated as the distance between the feature vectors f, thereby evaluating the effect of the face recognition experiment: a smaller distance between the two vectors denotes greater similarity. The distance formula is calculated as

$$d_{ij} = \frac{f_i \cdot f_j}{||f_i|| \cdot ||f_j||} \tag{8.15}$$

The first and second rows in Figure 8.3 (Jiang and Zhu, 2005) show the results of PCA feature selection from the face samples, while the third and fourth rows show the

Figure 8.3 Eigen faces and IC faces

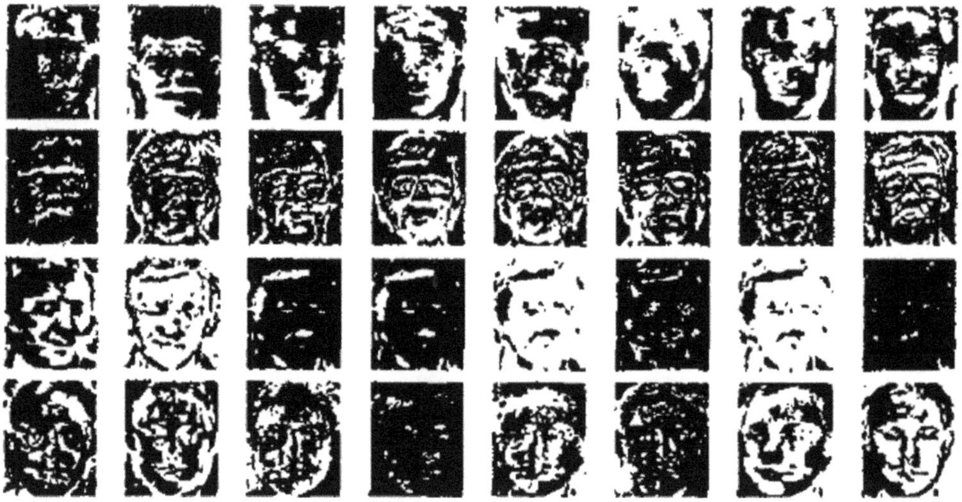

Figure 8.4 Overcomplete IC faces

results of ICA feature selection. In other words, a sample training face that participated in the learning is mapped to the subspace spanned by the IC and principle component using the ICA and PCA methods, respectively. Figure 8.4 shows the overcomplete IC extracted by overcomplete ICA.

In effect, the eigen faces extracted by PCA retain the basic shape of the face; whereas the IC faces extracted by ICA depict the contour lines of the face better, and reflect local statistical characteristics of the faces and image intensity. The overcomplete IC faces extracted by overcomplete ICA are able to characterize the facial contours better, and fully reflect the "sparse" characteristics. The overcomplete ICA algorithm

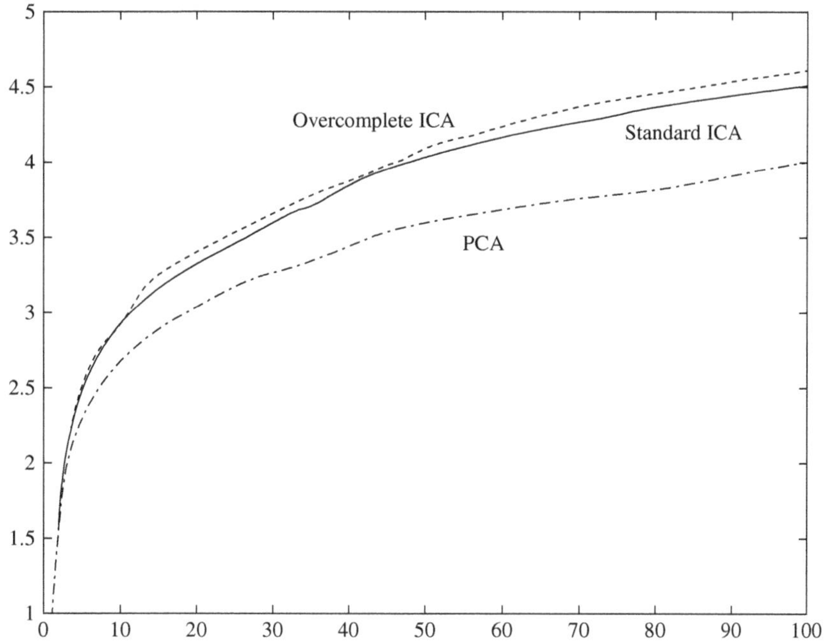

Figure 8.5 Comparison of experimental results using overcomplete ICA, basic ICA, and PCA

introduces some noise in the calculation process, which has some effect on the overcomplete eigen faces.

Figure 8.5 shows the comparison results of the face recognition experiment using the three methods: overcomplete ICA, basic ICA, and PCA. According to Figure 8.5, overcomplete ICA and basic ICA achieve better results than the PCA algorithm, while compared with basic ICA, overcomplete ICA offers certain improvements. Based on these results, use of overcomplete ICA for face recognition is beneficial, and if the feature extraction algorithm is used to further improve the performance, the advantage is even greater.

8.3 Chapter Summary

When the number of ICs is greater than the number of observed mixed signals, we refer to the BSS problem as being overcomplete. Similar to the case of noise ICA, overcomplete ICA has to solve two different problems, that is, how to estimate the mixing matrix and the ICs. This is completely different to ordinary ICA, because in general ICA these two problems can be addressed simultaneously.

If the basis vectors are overcomplete, it is more difficult to describe their likelihood, because the problem belongs to the class of problems with missing data. One approach for solving this problem is to use more prior knowledge about the density and time structure of the source, while another is to take advantage of the sparse representation of the data matrix. By first estimating the mixing matrix, and then the source, if the source is sufficiently sparse, it can be separated directly. The AICA and Geo-ICA algorithms described in this chapter adopt the latter approach.

References

Bartlett, M.S., Movellan, J.B., and Sejnowski, T.J. (2002) Face recognition by independent component analysis. *IEEE Transactions on Neural Networks*, **13**(6), 1450–1464.

Chen, S.S., Donoho, D.L., and Saunders, M.A. (1998) Atomic decomposition by basis pursuit. *Scientific Computing*, **20**(1), 33–61.

Cheng, J., Lu, H.Q., Chen, Y.W., and Zeng, X.Y. (2003) Face recognition using overcomplete independent component analysis. *Knowledge-Based Intelligent Information and Engineering Systems*, **27**(3), 1443–1448.

Davies, M., and Mitianoudis, N. (2004) Simple mixture model for sparse overcomplete ICA. *IEE Proceedings-Vision Image and Signal Processing*, **151**(1), 35–43.

Hyvärinen, A. (2001) Blind source separation by nonstationarity of variance: a cumulate based approach. *IEEE Transactions on Neural Networks*, **12**(6), 1471–1474.

Jiang, Y.W., and Zhu, S.M. (2005) Face feature extraction based on overcomplete ICA feature extraction. *Computer Science*, **32**(7), 162–165, (in Chinese).

Lee, T., Lewicki, M.S., Girolami, M., and Sejnowski, T.J. (1999) Blind source separation of more sources than mixtures using overcomplete representations. *IEEE Transactions on Signal Processing*, **6**(4), 87–90.

Lewicki, M.S., and Olshausen, B.A. (1999) Probabilistic framework for the adaptation and comparison of image codes. *Journal of the Optical Society of America A: Optics Image Science and Vision*, **16**(7), 1587–1601.

Lewicki, M.S., and Sejnowski, T.J. (2000) Learning overcomplete representations. *Neural Computation*, **12**(2), 337–365.

Li, Y., Cichocki, A., and Amari, S.I. (2003) Sparse component analysis for blind source separation with less sensors than sources. Proceedings of ICA-2003, pp. 89–94.

Theis, F.J., Lang, E.W., and Puntonet, C.G. (2004) A geometric algorithm for overcomplete linear ICA. *Neural Computation*, **56**(1), 1–18.

Theis, F.J., Lang, E.W., Westenhuber, T., and Puntonet, C.G. (2002) Overcomplete ICA with a geometric algorithm, in *Artificial Neural Networks – ICANN 2002*, Vol. **2415**, Springer, Berlin, pp. 1049–1054.

Turk, M., and Pentland, A. (1991) Eigenfaces for recognition. *Journal of Cognitive Neuroscience*, **3**(1), 77–86.

Waheed, K., and Salam, F.M. (2002) State-space blind source recovery for mixtures of multiple source distributions. *IEEE International Symposium on Circuits and Systems*, **1**, 197–200.

Waheed, K., and Salem, F.M. (2003) Algebraic independent component analysis: an approach for separation of overcomplete speech mixtures. *Proceedings of the International Joint Conference on Neural Networks*, **1**, 775–780.

Waheed, K., and Salam, F.M. (2003b) Algebraic overcomplete independent component analysis. Circuits, Proceedings of ICA-2003, pp. 1–6.

9

Kernel Independent Component Analysis

KICA (kernel independent component analysis) includes KCCA (kernel canonical correlation analysis) and KGV (kernel generalized variance) (Fyfe and Lai, 2000; Bach and Jordan, 2002) and its contrast function is based mainly on correlativity in RKHS (reproducing kernel Hilbert spaces). Most ICA algorithms are based on the expectation form of a single fixed nonlinear function to define the contrast function (Hyvärinen, 1999); however, KICA is based on the entire possible nonlinear function space for solving the problem (Harmeling *et al.*, 2001, 2000a,b, 2003). Use of the function space makes KICA suitable for different distribution sources and robust to varying sources.

9.1 KICA Algorithm

9.1.1 Object Function of KICA

In RKHS, KICA uses a contrast function to perform ICA. The contrast function is constructed according to the regular correlation rule. Since kernel techniques are used (in eigenspaces, the kernel can be used to calculate inner products), KICA is suitable for many source signal distributions.

The contrast function for KICA is defined based on the correlation of random variables. For simplicity, we consider two random variables, x_1 and x_2, and let F and $F-$ denote the vector function space and correlation, respectively. ρ_F denotes the highest correlation coefficients between $f_1(x_1)$ and $f_2(x_2)$, where $f_1, f_2 \in F$ (Bach and Jordan, 2001, 2002), that is,

$$\rho_F = \max_{f_1, f_2 \in F} \operatorname{corr}\{f_1(x_1), f_2(x_2)\} = \max_{f_1, f_2 \in F} \frac{\operatorname{cov}\{f_1(x_1), f_2(x_2)\}}{(\operatorname{var}\{f_1(x_1)\})^{1/2} (\operatorname{var}\{f_2(x_2)\})^{1/2}} \tag{9.1}$$

Blind Source Separation: Theory and Applications, First Edition. Xianchuan Yu, Dan Hu and Jindong Xu.

If x_1 and x_2 are independent, their correlation is zero, and if there are sufficient transformations in F space, the inverse proposition also exists. To calculate the correlation more easily, we can use the RKHS idea of making F the RKHS in $\mathbf{R}$. $K(x, y)$ denotes the associated kernel. The characteristic pattern $\Phi(x) = K(\cdot, x)$ denotes a mapping from the input space to RKHS, where $K(\cdot, x)$ is the functional of variable x in F space.

Owing to the reproductive property of kernel Hilbert spaces

$$f(x) = (\Phi(x), f), \forall f \in F, \forall x \in \mathbf{R} \tag{9.2}$$

there exist

$$\mathrm{corr}\{f_1(x_1), f_2(x_2)\} = \mathrm{corr}\{(\Phi(x_1), f_1), (\Phi(x_2), f_2)\} \tag{9.3}$$

Therefore, the existence of Equation 9.3 allows ρ_F in Equation 9.1 to be calculated in functional space. Equation 9.1 is the regular correlated definition, which manifests that KICA is, for ICA, just about using "kernel" regular correlated analysis to compute a flexible contrast function.

If correlation is used as the contrast function for ICA, we need to compute the regular correlation in the characteristic pattern. We have defined the F – correlation based on population expectation; however, in general, only limited samples are available instead of the whole population. Therefore, we need an empirical estimation of the F – correlation to construct the contrast functions. By minimizing these contrast functions, we can produce some flexible and stable ICA algorithms, such as KCCA and KGV.

9.1.2 KCCA Algorithm

The steps in the KCCA algorithm are given below.

Input: Samples $\boldsymbol{X} = (\boldsymbol{x}_1, \boldsymbol{x}_2, \cdots, \boldsymbol{x}_N)$, where $\boldsymbol{x}_i \in \mathbf{R}^n, i = 1, \ldots, N$, and kernel function $K(\boldsymbol{x}, \boldsymbol{y}) = G_\sigma(\boldsymbol{x} - \boldsymbol{y}) = \exp\left(\frac{1}{2\sigma^2}||\boldsymbol{x} - \boldsymbol{y}||^2\right)$.

Output: Separation matrix $\boldsymbol{W}_{m\times n}$.

Steps:

1. Whiten data and ascertain the number of sources $m(1 < m < n)$.
2. Initialize separation matrix $\boldsymbol{W} \in \mathbf{R}^{m\times n}$.
3. Minimize the objective function $C(\mathbf{W})$:
   a. Estimate sources $\hat{\boldsymbol{s}}_1, \ldots, \hat{\boldsymbol{s}}_N \in \mathbf{R}^n$, $\hat{\boldsymbol{s}}_i = \boldsymbol{W}\boldsymbol{x}_i$, and calculate their central Gram matrix $K_1, \ldots, K_m$.
   b. $\lambda_F(K_1, \ldots, K_m)$ is the minimum eigenvalue of the generalized eigenvector equation $\boldsymbol{K}\alpha = \lambda\boldsymbol{D}\alpha$, where $\boldsymbol{K}$ is an $mN \times mN$ dimensional matrix consisting of $K_1, \ldots, K_m$ and $\boldsymbol{D}$ is an $mN \times mN$ dimensional block diagonal matrix.

c. Calculate objective function $C(\boldsymbol{W}) = -\frac{1}{2}\log \lambda_F(K_1, \ldots, K_m)$.
d. Use the steepest descent method to minimize $C(\boldsymbol{W})$.

9.1.3 KGV Algorithm

KGV is basically the same as KCCA, with the only difference being the definition of $C(\boldsymbol{W})$:

$$C(\mathbf{W}) = -\frac{1}{2}\log \widehat{\delta}_F(K_1, \ldots \ldots K_m) \tag{9.4}$$

where

$$\widehat{\delta}_F(K_1, \ldots \ldots K_m) = \det \boldsymbol{K}/\det \boldsymbol{D} \tag{9.5}$$

9.2 Application and Analysis

Compared with FastICA, KICA is more robust. When applied in image processing, KICA retains more texture information. In this section, we introduce various applications of KICA (related to remote sensing image preprocessing).

We selected a Quickbird remote sensing image (2005.05.31, the Summer Palace, Beijing, 763×613 pixels) with 1 m resolution as the experimental data (see Figure 9.1a). This image has three bands and depicts mainly houses, roads, water, and a greenbelt. We used FastICA and KGV to preprocess the data and carry out some comparisons. Figure 9.1b,c show the processed results by FastICA and KGV, respectively. From the experimental results, we can see that both FastICA and KGV obtained all the independent components. However, FastICA loses some texture information, whereas KGV retains abundant texture information.

Taking the original image and preprocessed result as the input image for our KeenEye 2.0 system, we used the nearest neighbor method (unsupervised) for classification. The classified results are shown in Figure 9.2. We can see that the classified results after preprocessing by the two ICA methods have a clearer outline than that without preprocessing. Comparing FastICA and KGV, we again found that KGV retains abundant texture information, while FastICA can only retain some of the information.

Classification accuracy and Kappa coefficients (Cohen, 1960) are given in Table 9.1. Both the average classification accuracy and the Kappa coefficient of KGV are higher than those of FastICA.

This method is also efficient for low resolution remote sensing images, and we carried out various tests on TM remote sensing images (He and Yu, 2005) . The experimental results are shown in Figures 9.3 and 9.4, while Table 9.2 gives the Kappa coefficients for evaluating classification accuracy. The KGV algorithm requires fewer iterations than FastICA (Figure 9.5). In the experiment, we also found that KGV was

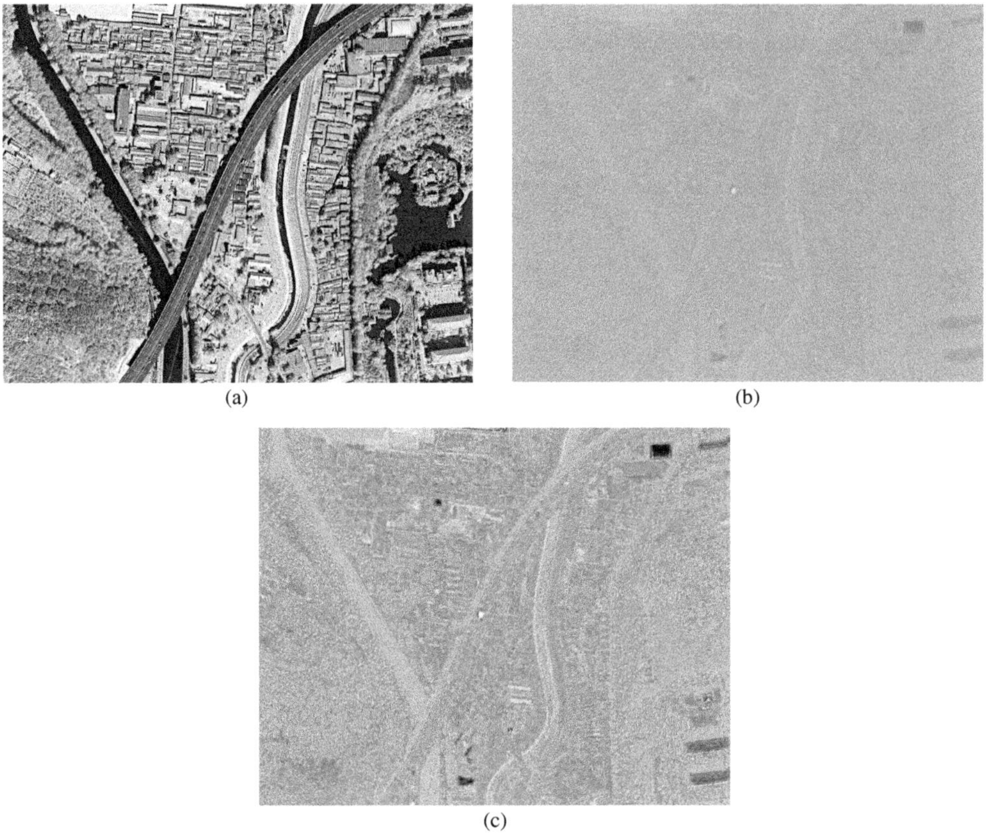

Figure 9.1 Contrast experiment with FastICA and KGV on a Quickbird remote sensing image (763 × 613 pixels) (a) Original image (b) Processed result by FastICA (c) Processed result by KGV

Table 9.1 Comparison of classification results (%)

Algorithm	Classification accuracy	Kappa coefficient
FastICA	64	41
KGV	73	61

Table 9.2 Kappa coefficients (%)

Algorithm	Maximum	Minimum	Average
FastICA	73.52	43.21	57.94
KGV	86.32	60.25	76.68

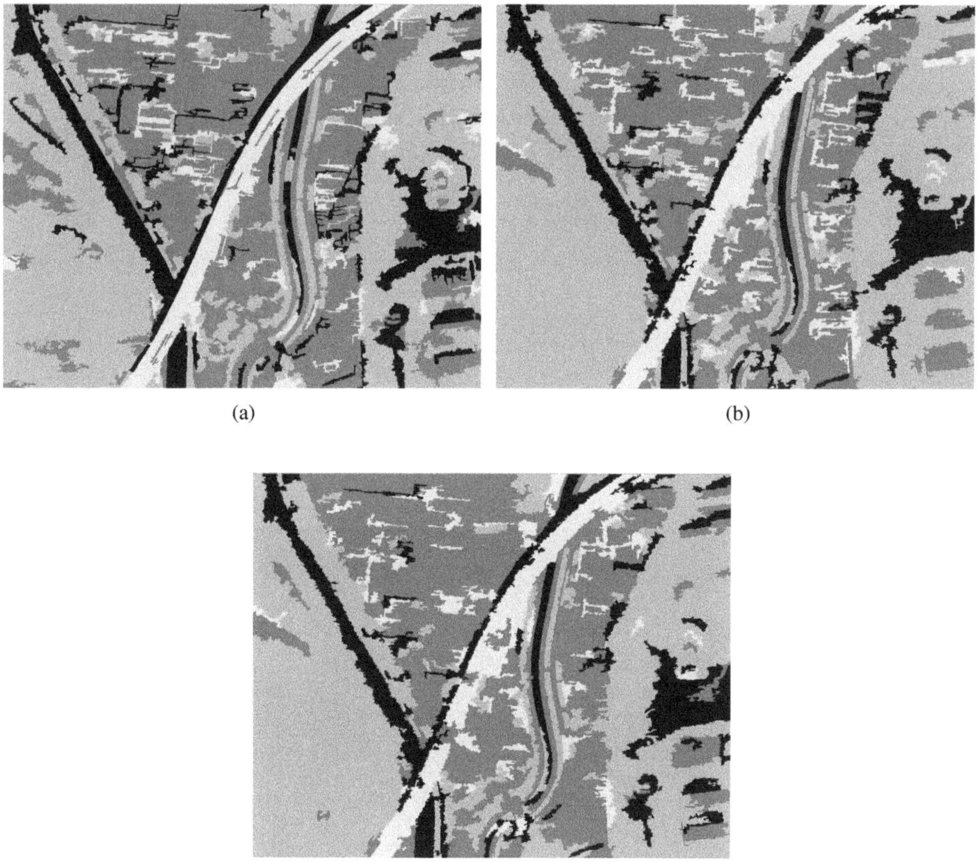

(a) (b) (c)

Figure 9.2 Classification results (763 × 613 pixel) (a) Directly classified result by nearest neighbor method (b) Classified result after FastICA preprocessing (c) Classified result after KGV preprocessing

not very efficient for large data. The main reason for this may be that the kernel function is too complex and fragile. Thus, we need a more efficient KICA algorithm to process large data.

9.3 Concluding Remarks

According to the independence criterion and by some optimization algorithms, ICA is defined as decomposing multi-channel observed signals into several independent components, which are approximate estimations of the source signals. Multispectral or hyperspectral remote sensing images have high dimensional data, which is not

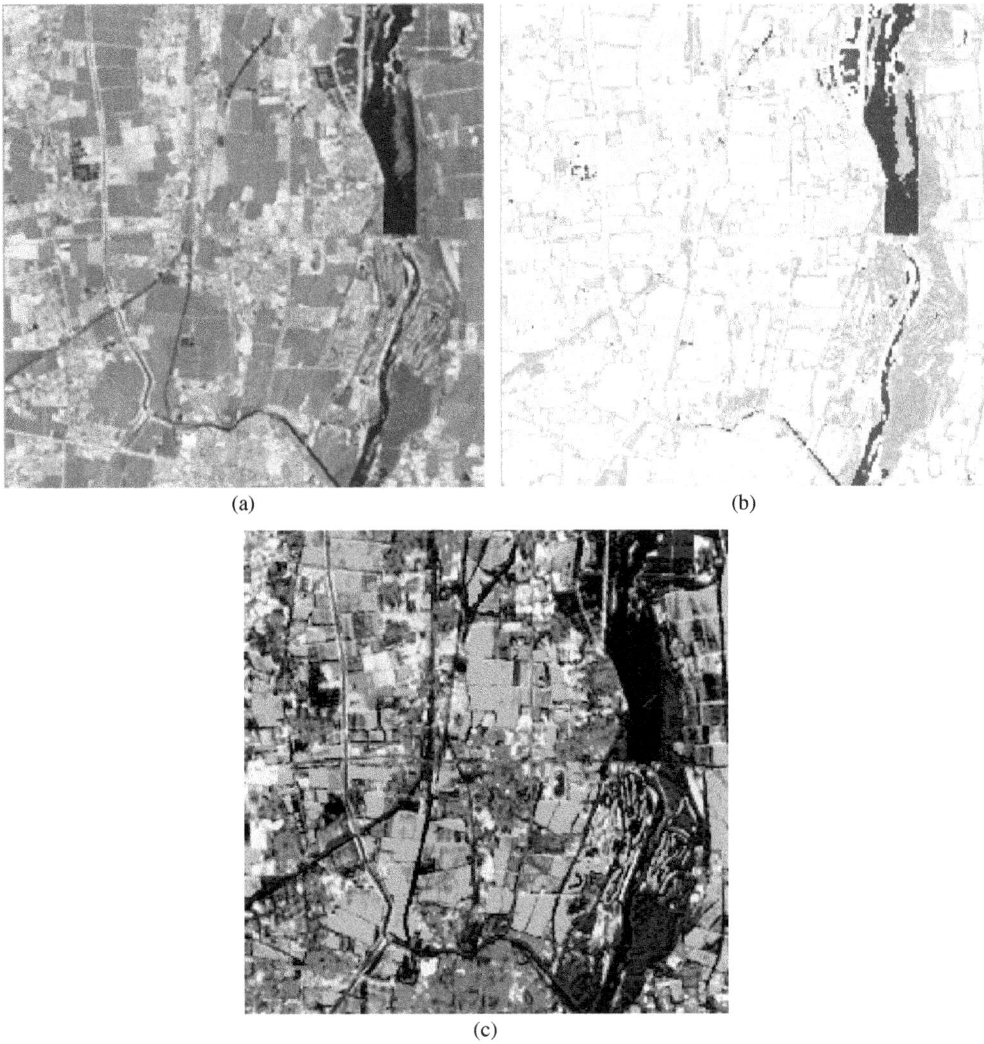

Figure 9.3 Comparison experiment using a TM remote sensing image (256 × 256 pixels): (a) original TM image; (b) processed result by FastICA; and (c) processed result by KGV

suitable for classification. ICA, like PCA, reduces the dimensions of the data and uses a lower dimensional subspace to represent each band in the image.

In multiband remote sensing images, some correlation between the bands could exist, making these images unsuitable for classification. ICA, which utilizes higher-order statistical information and obtains independent components to as large a degree as possible, effectively removes all types of correlations (include nonlinear) and improves classification accuracy. Nevertheless, remote sensing images are typically extremely large with nonlinear noise and some singular value points. As

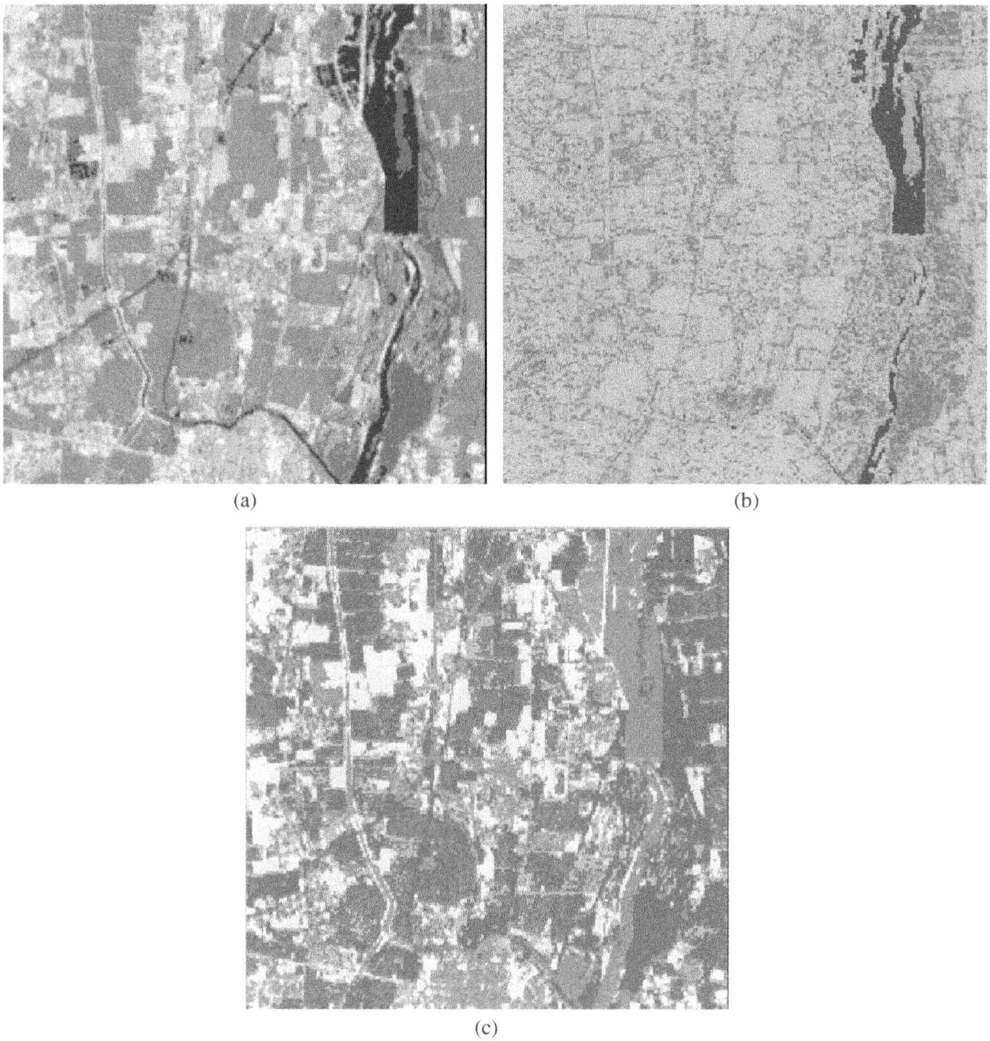

Figure 9.4 Classification results: (a) directly classified by nearest neighbor method; (b) classified after FastICA preprocessing; and (c) classified after KGV preprocessing

a result, FastICA with a single fixed contrast function cannot achieve convergence and is highly sensitive to the singular values.

From the experimental results, we can see that there is a loss of texture information in the classified results after FastICA preprocessing, whereas KICA retains abundant texture information. The KGV algorithm is more flexible and robust. Compared with FastICA, KGV has higher classification accuracy and a faster convergence rate. However, KICA needs to search for a contrast function in the entire RKHS, and thus has large computational complexity with a high RAM requirement. Thus, if

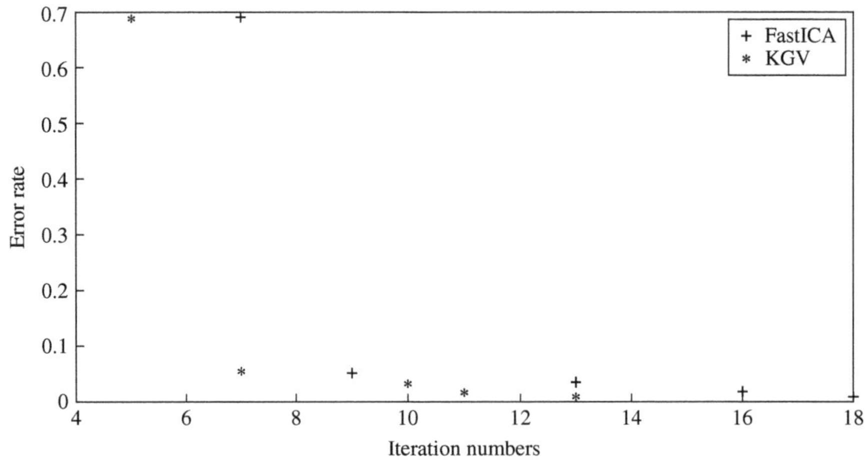

Figure 9.5 Number of iterations by error rate (horizontal axis denotes number of iterations, while vertical axis denotes error rate × 100)

the size of the image is greater than 512×512, we should consider a more rapid calculation method.

References

Bach, F.R. and Jordan, M.I. (2001) Kernel Independent Component Analysis. Technical Report No. CSD-01-1166. Computer Science Division University of California, Berkeley.

Bach, F.R. and Jordan, M.I. (2002) Kernel independent component analysis. *Journal of Machine Learning Research*, **3**(1), 1–48.

Cohen, J. (1960) A coefficient of agreement for nominal scales. *Educational and Psychological Measurement*, **20**(1), 37–46.

Fyfe, C. and Lai, P.L. (2000) ICA using kernel canonical correlation analysis. Proceedings of the International Workshop on Independent Component Analysis and Blind Signal Separation, Helsinki, Finland, pp. 279–284.

Harmeling, S., Ziehe, A., Kawanabe, M. and Müller, K.-R. (2002a) Kernel feature spaces and nonlinear blind source separation, *Advances in Neural Information Processing Systems*, MIT Press, pp. 761–768.

Harmeling, S., Ziehe, A. M. Kawanabe, and K.-R. Müller (2002b) Kernel-Based Nonlinear Blind Source Separation. Technical Report 1/2002, BLISS Project.

Harmeling, S., Ziehe, A., Kawanabe, M. and Müller, K.-R. (2003) Kernel-based nonlinear blind source separation. *Neural Computation*, **15**(5), 1089–1124.

Harmeling, S., Ziehe, A., Kawanabe, M. *et al.* (2001) Nonlinear blind source separation using kernel feature spaces. Proceedings of the International Workshop on Independent Component Analysis and Blind Signal Separation (ICA2001), TALK.

He, H. and Yu, X.C. (2005) A comparison between FastICA and KERNELICA in remote sensing image classification, *International Geoscience and Remote Sensing Proceedings*, IEEE Press, Seoul, pp. 1658–1661.

Hyvärinen, A. (1999) Fast and robust fixed-point algorithms for independent component analysis. *IEEE Transactions on Neural Networks*, **10**(3), 626–634.

10

Natural Gradient Flexible ICA Algorithm and Its Application

Natural gradient flexible ICA proposed by Choi *et al.* (Choi (1999); Choi *et al.* (2000)) is an adaptive non-linear ICA learning algorithm. The algorithm implements a parametrized generalized Gaussian density model on a natural Riemannian gradient structure to ascertain the distribution of the source, and is able to process real-time mixed data containing super-Gaussian and sub-Gaussian source signals. The nonlinear function for flexible ICA is controlled by a Gaussian index, which is an estimate of the kurtosis of the decomposition filter output. In the following sections we discuss the details of the natural gradient flexible ICA algorithm and several of its selected nonlinear functions.

10.1 Natural Gradient Flexible ICA Algorithm

10.1.1 Basic Algorithm for Flexible ICA

In the ICA algorithm, the relationship between the separated signal and the source signal can be represented by the linear model

$$\boldsymbol{y}(t) = \boldsymbol{W}\boldsymbol{x}(t) \tag{10.1}$$

where $\boldsymbol{W}_{m\times n}$ is the separation matrix, $\boldsymbol{x}(t)$ is the mixed signal, and $\boldsymbol{y}(t)$ is the separated signal. The objective function can be obtained through the maximum likelihood or minimization of mutual information estimation methods (Amari *et al.*, 1996; Cardoso, 1997).

$$L(\boldsymbol{W}) = -\log|\det\boldsymbol{W}| - \sum_{i=1}^{m}\log p_i(y_i) \tag{10.2}$$

where $p_i(\cdot)$ represents the probability density function. Let

$$\varphi_i(y_i) = -\frac{\mathrm{d}\log p_i(y_i)}{\mathrm{d}y_i} \tag{10.3}$$

Blind Source Separation: Theory and Applications, First Edition. Xianchuan Yu, Dan Hu and Jindong Xu.

Then, the gradient of the objective function is

$$\nabla L(\boldsymbol{W}) = \frac{\partial L(\boldsymbol{W})}{\partial \boldsymbol{W}} = -\boldsymbol{W}^{-\mathrm{T}} + \varphi(\boldsymbol{y})\boldsymbol{x}^{\mathrm{T}} \tag{10.4}$$

where the ith element of $\varphi(\boldsymbol{y})$ is $\varphi_i(y_i)$.

The natural gradient flexible ICA algorithm for $\boldsymbol{W}$ is given in the following form:

$$\begin{aligned}\boldsymbol{W}(t+1) &= \boldsymbol{W}(t) - \eta_t \tilde{\nabla} L(\boldsymbol{W}) \\ &= \boldsymbol{W}(t) - \eta_t \frac{\partial L(\boldsymbol{W})}{\partial \boldsymbol{W}} \boldsymbol{W}^{\mathrm{T}}(t)\boldsymbol{W}(t) \\ &= \boldsymbol{W}(t) - \eta_t \{I - \varphi(\boldsymbol{y}(t))\boldsymbol{y}^{\mathrm{T}}(t)\}\boldsymbol{W}(t)\end{aligned} \tag{10.5}$$

where $\tilde{\nabla} L(\boldsymbol{W})$ is the natural gradient.

10.1.2 Related Improvement of the Flexible ICA Basic Algorithm

Amari (1999) proposed a natural gradient with orthogonal constraints. Assuming that whitening of the hybrid vector $\boldsymbol{x}(t)$ has been carried out and that the source signal is normalized,

$$E\{\boldsymbol{x}(t)\boldsymbol{x}^{\mathrm{T}}(t)\} = \boldsymbol{I}_n \tag{10.6}$$

$$E\{\boldsymbol{s}(t)\boldsymbol{s}^{\mathrm{T}}(t)\} = \boldsymbol{I}_m \tag{10.7}$$

The following formula can be obtained by Equations 10.6 and 10.7:

$$\boldsymbol{A}\boldsymbol{A}^{\mathrm{T}} = \boldsymbol{I}_n \tag{10.8}$$

where the nth row vector of $\mathbf{A}$ is a group of orthogonal m-dimensional vectors. The m-dimensional subspace in $\mathbb{R}^n$ is called the Stiefel manifold. The natural gradient of a Stiefel manifold is calculated according to the following formula (Amari, 1999):

$$\tilde{\nabla} L(\boldsymbol{W}) = \nabla L(\boldsymbol{W}) - \boldsymbol{W}\{\nabla L(\boldsymbol{W})\}^{\mathrm{T}}\boldsymbol{W} \tag{10.9}$$

Using Equation 10.9, we obtain the following equation:

$$\tilde{\nabla} L(\boldsymbol{W}) = \varphi(\boldsymbol{y})\boldsymbol{x}^{\mathrm{T}} - \boldsymbol{y}\varphi^{\mathrm{T}}(\boldsymbol{y})\boldsymbol{W} \tag{10.10}$$

The learning algorithm for $\boldsymbol{W}$ is given as

$$\boldsymbol{W}(t+1) = \boldsymbol{W}(t) - \eta_t \tilde{\nabla} L(\boldsymbol{W}) = \boldsymbol{W}(t) - \eta_t\{\varphi(\boldsymbol{y}(t))\boldsymbol{x}^{\mathrm{T}}(t) - \boldsymbol{y}(t)\varphi^{\mathrm{T}}(\boldsymbol{y}(t))\boldsymbol{W}(t)\} \tag{10.11}$$

It is noteworthy that when $m = n$, in the iterative process, $\boldsymbol{W}$ is orthogonal, and therefore Equation 10.11 becomes

$$\boldsymbol{W}(t+1) = \boldsymbol{W}(t) - \eta_t\{\varphi(\boldsymbol{y}(t))\boldsymbol{y}^{\mathrm{T}}(t) - \boldsymbol{y}(t)\varphi^{\mathrm{T}}(\boldsymbol{y}(t))\}\boldsymbol{W}(t) \tag{10.12}$$

For the actual processing, since whitening and separating can be done simultaneously, Cardoso and Laheld (1996) proposed an equivariant adaptive learning separation algorithm, with its learning rules as follows:

$$\Delta \boldsymbol{W}(t) = \eta_t\{\boldsymbol{I} - \boldsymbol{y}(t)\boldsymbol{y}^{\mathrm{T}}(t) - \varphi(\boldsymbol{y}(t))\boldsymbol{y}^{\mathrm{T}}(t) + \boldsymbol{y}(t)\varphi^{\mathrm{T}}(\boldsymbol{y}(t))\}\boldsymbol{W}(t) \tag{10.13}$$

where $\boldsymbol{W}(t)$ is the separation matrix, η_t is the learning rate, $\boldsymbol{I}$ is a unit matrix, $\boldsymbol{y}(t)$ is the separated signal, and $\varphi(\cdot)$ is a nonlinear function.

According to the parametrized and generalized Gaussian model, the nonlinear function in Equation 10.11 is defined as

$$\varphi_i(y_i) = -\frac{\mathrm{d}\log p_i(y_i)}{\mathrm{d}y_i} = |y_i|^{\alpha_i - 1}\mathrm{sgn}(y_i) \tag{10.14}$$

where $\mathrm{sgn}(y_i)$ is the sign function of y_i.

10.1.3 Determination of α_i

If $\alpha_i = 1$, $\varphi_i(y_i)$ in Equation 10.14 becomes a symbolic function. This nonlinear symbolic function is applicable to the separation of voice signals, because a natural language model usually obeys a Laplace distribution.

If $\alpha_i = 4$, $\varphi_i(y_i)$ in Equation 10.14 becomes a cubic function, applicable to the separation of a source signal that obeys the sub-Gaussian distribution.

To select a suitable Gaussian index α_i, we estimate the kurtosis of the output signal y_i and select a corresponding α_i according to the relationship in Figure 10.1. The kurtosis k_i of y_i is estimated by the following iterative algorithm:

$$k(t+1) = \frac{M_{4i}(t+1)}{M_{2i}^2(t+1)} - 3 \tag{10.15}$$

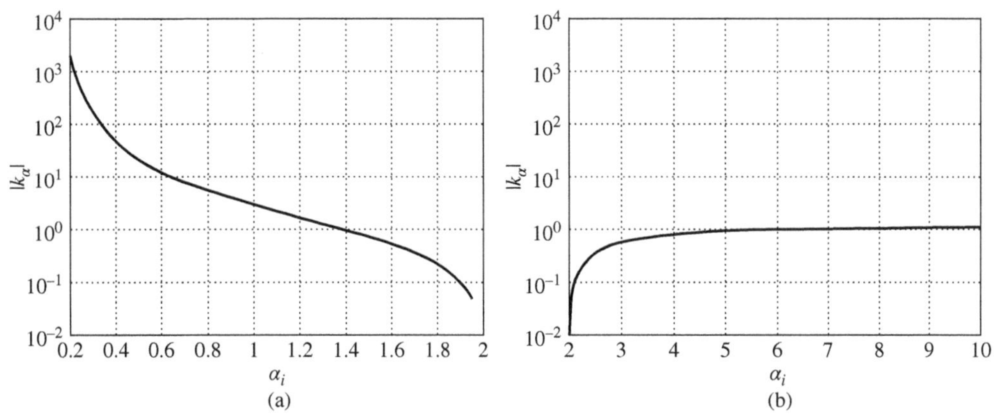

Figure 10.1 Relationship showing that the absolute value of kurtosis varies with a change in the Gaussian index, where is the axis and the coordinate (a) Super-Gaussian distribution (b) Sub-Gaussian distribution

where

$$M_{4i}(t+1) = (1-\delta)M_{4i}(t) + \delta|y_i(t)|^4 \tag{10.16}$$

$$M_{2i}^2(t+1) = (1-\delta)M_{2i}(t) + \delta|y_i(t)|^2 \tag{10.17}$$

$$M_n = \int_{-\infty}^{\infty} y^n p(y;\alpha)\mathrm{d}y \tag{10.18}$$

where δ is a small constant, for example, $\delta = 0.01$.

Generally, the estimated output kurtosis of the decomposition filter and the kurtosis of the source signal are not exactly the same. However, it is possible to distinguish whether the estimated source obeys a super-Gaussian or sub-Gaussian distribution. In addition, Amari and Cardoso (1997) and Cardoso (1997) pointed out that despite the assumed density model and the true density model not matching exactly, the performance of source separation is not affected. Therefore, we can establish a practical model, in which several different forms of the nonlinear function based on this model are used.

Changes in the Gaussian index affect the kurtosis of a sub-Gaussian source only slightly (see Figure 10.1b) and, therefore, if the estimated kurtosis of y_i is negative seasonal, $\alpha_i = 4$. The corresponding cubic nonlinear function of a sub-Gaussian source is also associated with the minimization method for kurtosis (Cardoso and Laheld, 1996). For a super-Gaussian source, the kurtosis Gaussian index varies greatly (see Figure 10.1a), and therefore, in different situations, α_i takes more than one different value. Generally, a Gaussian index that takes two to three different values is sufficient to handle a variety of super-Gaussian sources. If α_i can take different values, several typical nonlinear functions are possible, as shown in Figure 10.2.

10.2 Application and Analysis

At present, much research is being carried out on despeckling synthetic aperture radar (SAR) images, from traditional multi-view processing to spatial filtering and the introduction of the wavelet transform. Researchers have, to a certain extent, succeeded in suppressing speckle noise, but noise reduction capabilities, edge features, details, goals, and algorithm adaptations have yet to be further improved. Next we introduce some traditional image denoising algorithms (including median filtering and mean filtering) and ICA denoising algorithms (FastICA denoising, ERICA (equivariant robust independent component analysis) denoising, and speckle noise suppression based on the natural gradient flexible ICA). SAR images are used to compare and analyze the performance of these denoising algorithms.

10.2.1 Experimental Data

The experimental data used was a European Space Agency SAR Envisat-ASAR C-band image of the northern part of Beijing (with VV polarization and HV polarization) captured on 26 January 2004. The 512×512 image, with the number of independent views greater than 3.00, shows intercepting roads and a small lake area (see Figure 10.3).

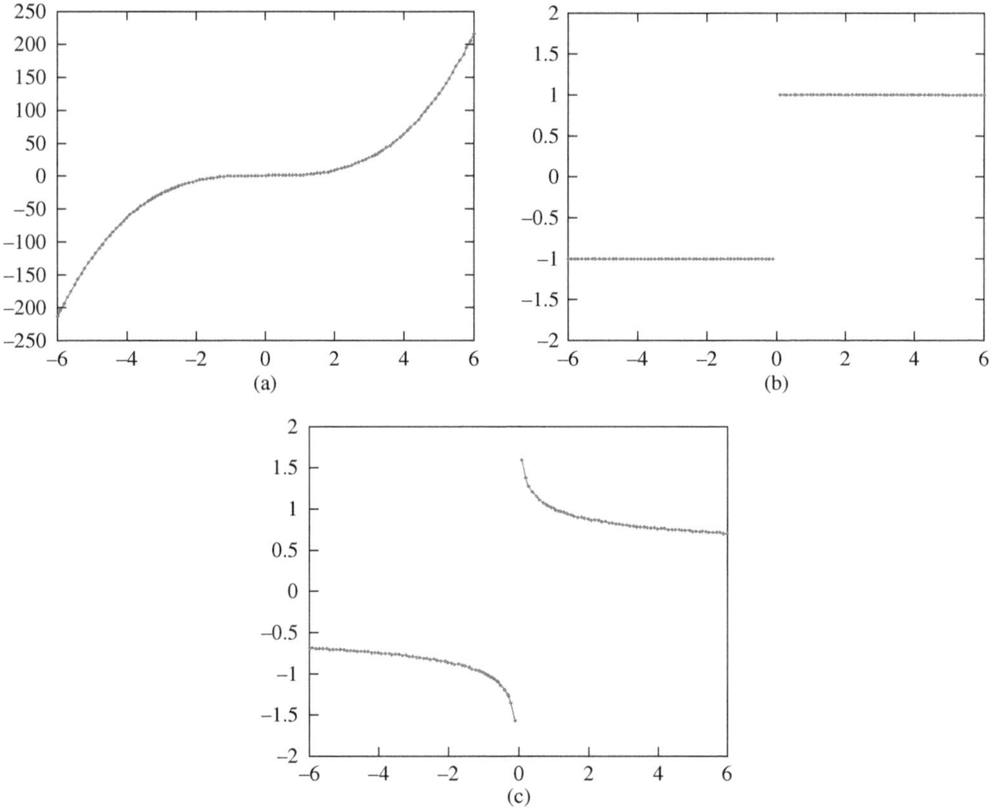

Figure 10.2 Simulations of nonlinear function $\varphi_i(y_i)$: (a) $\alpha_i = 4$, (b) $\alpha_i = 1$, and (c) $\alpha_i = 0.8$

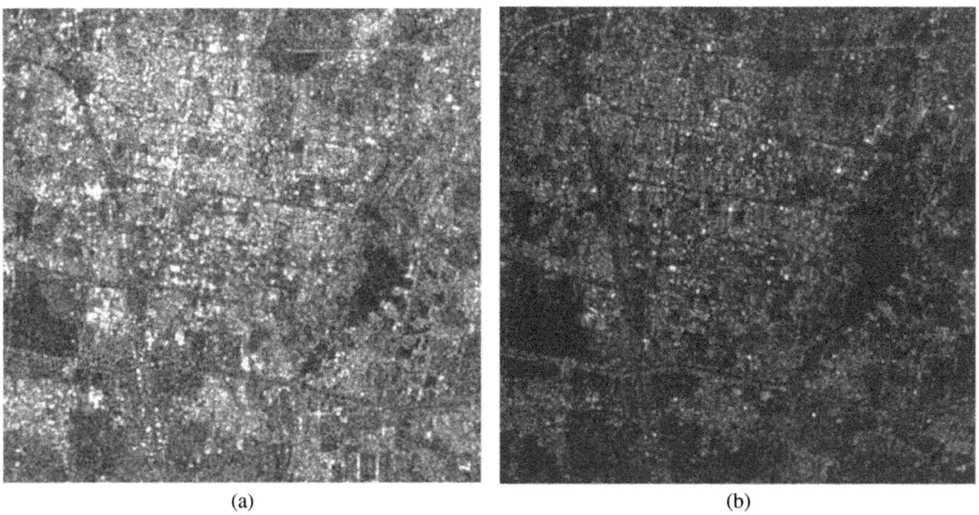

Figure 10.3 Original SAR image with different polarization (a) Original image with VV polarization (b) Original image with HV polarization

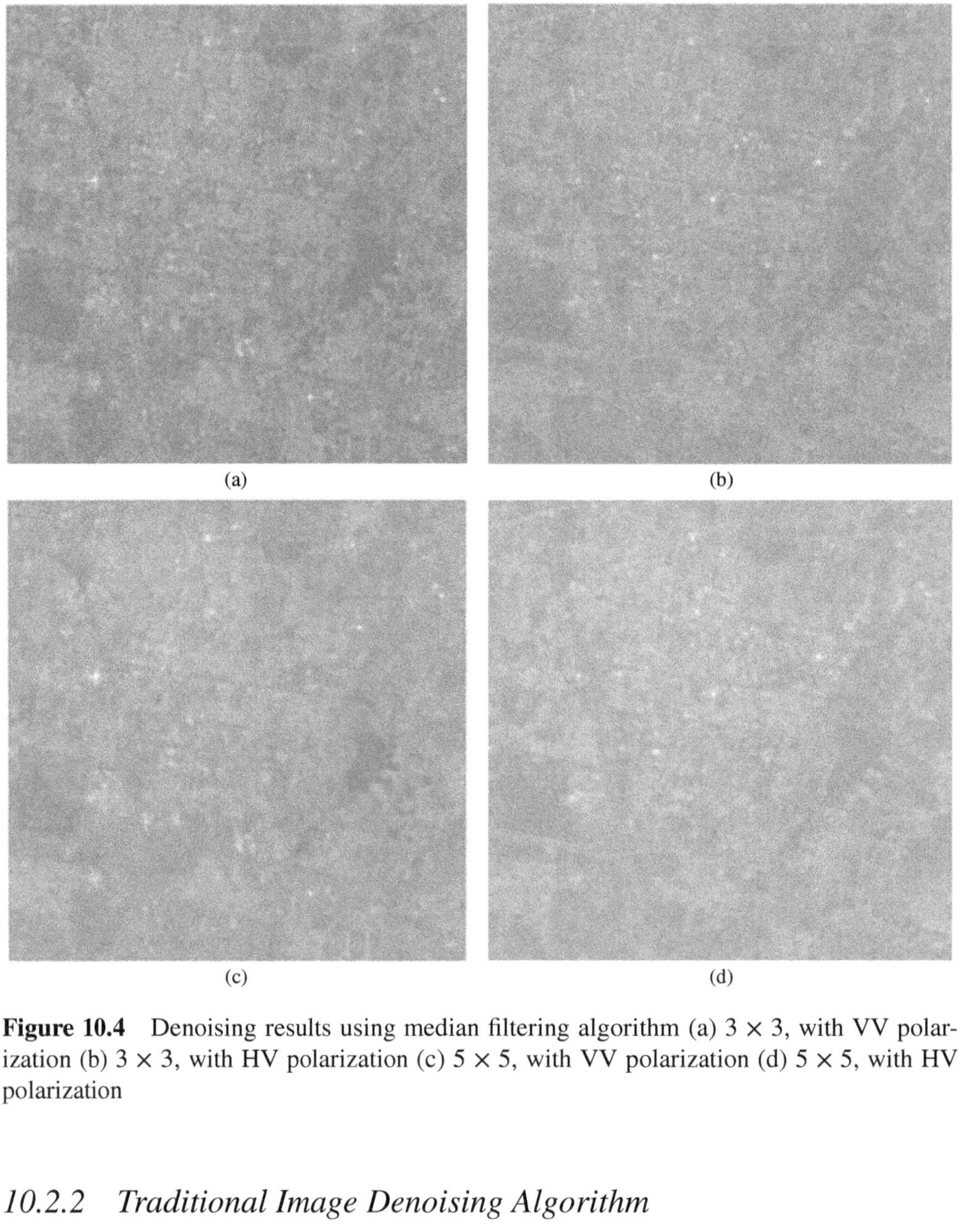

Figure 10.4 Denoising results using median filtering algorithm (a) 3 × 3, with VV polarization (b) 3 × 3, with HV polarization (c) 5 × 5, with VV polarization (d) 5 × 5, with HV polarization

10.2.2 Traditional Image Denoising Algorithm

10.2.2.1 Median Filtering

Median filtering is a nonlinear digital filter technique, often used to remove noise. It is very widely used because, under certain conditions, it preserves edges while removing noise. Denoising results by median filtering are shown in Figure 10.4.

(a) (b)

Figure 10.5 Denoising results using mean filtering algorithm (a) With VV polarization (b) With HV polarization

10.2.2.2 Mean Filtering

Mean filtering is a simple, intuitive, and easy to implement method of smoothing images, that is, reducing the amount of intensity variation between one pixel and the next. It is often used to reduce noise in images. Denoising results by mean filtering are shown in Figure 10.5.

10.2.3 ICA Denoising Algorithm

The SAR images with different polarization contain characteristics with varying levels of detail. Polarization radar applied to a scene provides a set of polarized image data, with the target characteristics dispersed in the speckle polluted images, independent of each other and the speckle noise. Suppressing speckling involves separating the desired radar image signals and noise from the observed image, which is what ICA can achieve.

A SAR image can be seen as a linear mix of the target signal and speckle noise. Because the image data generally obey a non-Gaussian distribution, such a signal can be considered to satisfy the ICA requirement of a maximum of one Gaussian source. Thus, the ICA algorithm can be used for SAR image speckle noise suppression. For this type of multiplicative noise, Arsenault and April (1976) proved that if the number of independent views is large enough (usually $N > 3$), after applying a logarithmic transformation to the Gamma distributed speckle noise, the probability density function approximates a Gaussian distribution with variance $1/N$.

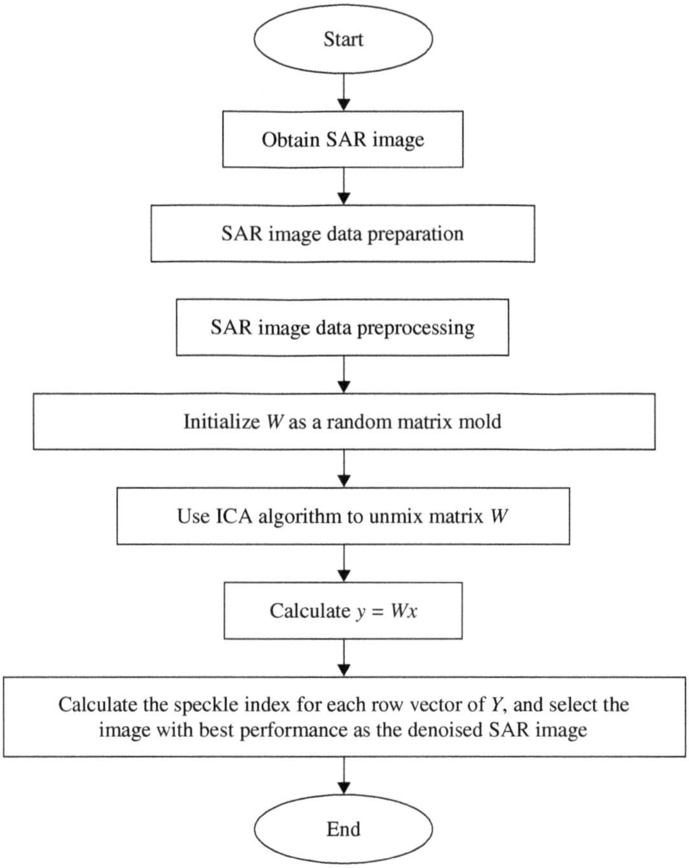

Figure 10.6 General flowchart for SAR speckle suppression based on ICA

A flowchart of SAR image speckle noise reduction using ICA is shown in Figure 10.6, while the process is described below.

Algorithm Algorithm for SAR image speckle noise reduction with ICA

Input: The original image
Output: Image after denoising
Steps:

1. **Image data preparation**: For a number of different polarized SAR images, the image data preparation process involves converting each section of the two-dimensional image data into a one-dimensional vector, so that multiple image data take the form of a matrix. In addition, a normalization process is applied to this matrix.

2. **Data preprocessing**: This involves a centering and whitening process applied to the prepared data, while at the same time reducing the dimensionality when carrying out the whitening process.
3. **ICA processing**:
   a. Initialize $\boldsymbol{W}$ as a random matrix mold.
   b. Use the ICA algorithm to unmix matrix $\boldsymbol{W}$.
   c. Calculate $\boldsymbol{y} = \boldsymbol{Wx}$, where $\boldsymbol{x}$ is a preprocessed mixed signal and $\boldsymbol{y}$ is an estimate of the unknown independent sources.
4. **Image recovery**: Transform each row vector of $\boldsymbol{y}$ into two-dimensional image data with the same size as the original observed images. Calculate the speckle index for each row vector, where the smallest speckle index is the SAR image obtained using ICA noise reduction, while the remainder is noise.

10.2.3.1 FastICA Denoising

The FastICA algorithm, also known as the fixed point algorithm, was proposed and developed by Hyvärinen and Oja (1997) from the Aalto University Institute of Technology in Finland. Denoising results by the FastICA algorithm applied to the image are shown in Figure 10.7.

10.2.3.2 ERICA Denoising

A robust ICA algorithm for a mixed additive noise signal was proposed by Cichocki *et al.* (1994, 1996, 1997, 1998). The algorithm makes use of the equivariant property,

(a) (b)

Figure 10.7 Denoised images using FastICA algorithm (a) Denoised image of IC1 (b) Denoised image of IC2

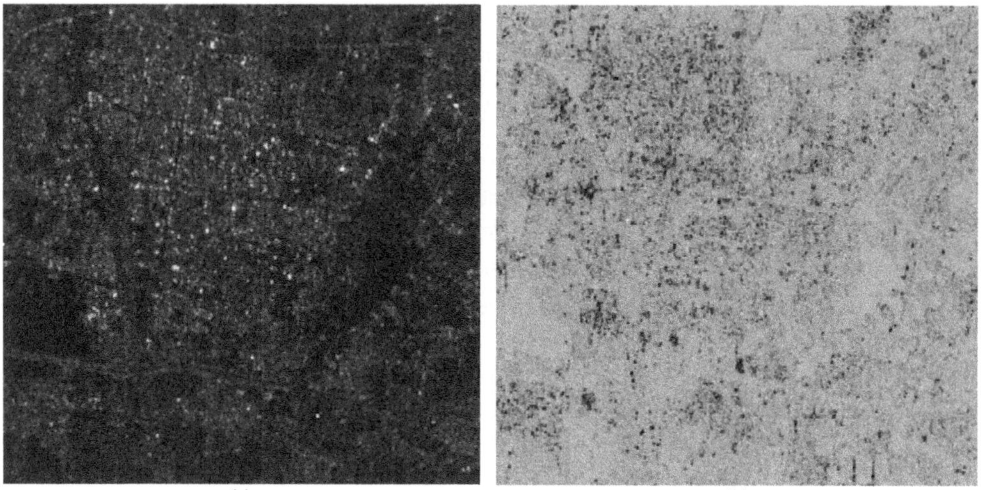

(a) (b)

Figure 10.8 Denoised images using the ERICA algorithm (a) Denoised image of IC1 (b) Denoised image of IC2

and can effectively reduce the impact of noise on the separation matrix. Image denoising results by the ERICA algorithm are shown in Figure 10.8.

10.2.3.3 Natural Gradient Flexible ICA Speckle Noise Reduction Based on the Natural Gradient

In this section, we perform natural gradient flexible ICA for the denoising of the two images. Denoising results are shown in Figure 10.9.

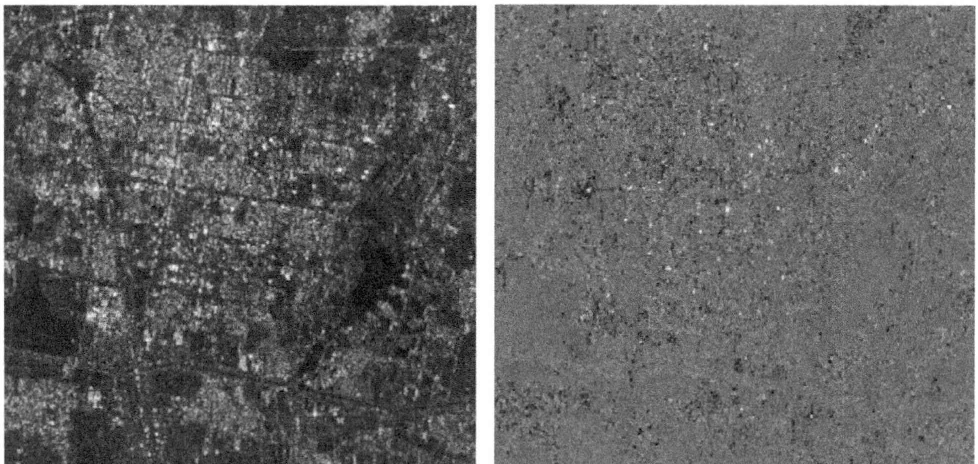

(a) (b)

Figure 10.9 Denoised images using the NG-Flexible ICA algorithm (a) Denoised image of IC1 (b) Denoised image of IC2

10.2.4 *Analysis of Results*

Typically, a good speckle filtering algorithm within a uniform region has a preferable mean holding capacity and reduced variance, while at the same time being better able to retain details on the edge. In a SAR image quality assessment system, commonly measured values, namely, the speckle index that measures the speckle intensity, and the edge retention index that measures the capacity for retaining the edge features, are used to measure the advantages and disadvantages of each algorithm. Indicators before and after filtering (namely, speckle index, equivalent number of views, edge preserving index, and mean square error) are given in Tables 10.1–10.4. A Google Earth image of the same location is shown in Figure 10.10.

It can be seen that the SAR image processed using the natural gradient flexible ICA algorithm has a smaller speckle index and higher equivalent number of views and edge retention index than the original image, which indicates that this algorithm can suppress speckle noise more effectively and also retain more texture details than the other algorithms.

Tables 10.5 and 10.6 show that the natural gradient flexible ICA algorithm has better overall performance, although there is some room for improvement with respect to its runtime.

Table 10.1 Speckle index

Algorithm	VV polarization (IC1)	HV polarization (IC2)
Original map	0.7015	0.5727
Median filtering (3 × 3)	0.5771	0.4682
Median filtering (5 × 5)	0.4466	0.3674
Mean filtering	0.5559	0.4545
FastICA	0.0250	0.0365
ERICA	0.5125	0.2089
Natural gradient flexible ICA	0.2007	0.5487

Table 10.2 Equivalent number of views

Algorithm	VV polarization (IC1)	HV polarization (IC2)
Original map	2.0321	3.0491
Median filtering (3 × 3)	3.0024	4.5624
Median filtering (5 × 5)	5.0141	7.4081
Mean filtering	3.2361	4.8405
FastICA	1599.8	751.7978
ERICA	3.8072	22.9254
Natural gradient flexible ICA	24.8230	3.3211

Table 10.3 Edge preserving index

Algorithm	VV polarization (IC1)	HV polarization (IC2)
Original map	1.0	1.0
Median filtering (3 × 3)	0.5664	0.5491
Median filtering (5 × 5)	0.3071	0.2967
Mean filtering	0.5270	0.5110
FastICA	1.1443e-016	3.9484e-016
ERICA	0.2052	0.3382
Natural gradient flexible ICA	0.7421	0.3134

Table 10.4 Mean square error

Algorithm	VV polarization (IC1)	HV polarization (IC2)
Original map	–	–
Median filtering(3 × 3)	189.5357	63.5518
Median filtering(5 × 5)	346.8465	112.7841
Mean filtering	216.4557	71.6022
FastICA	941.3433	344.6512
ERICA	879.2979	204.2355
Natural gradient flexible ICA	824.2988	240.5482

Figure 10.10 Google Earth image of the same location as that depicted in Figure 10.3

Table 10.5 Performance index of various algorithms

Algorithm	Overall performance index
FastICA	0.4926
ERICA	0.4576
Natural gradient flexible ICA	0.5927

Table 10.6 Runtime of various algorithms

Algorithm	Runtime (s)
FastICA	2.08
ERICA	6.03
Natural gradient flexible ICA	19.77

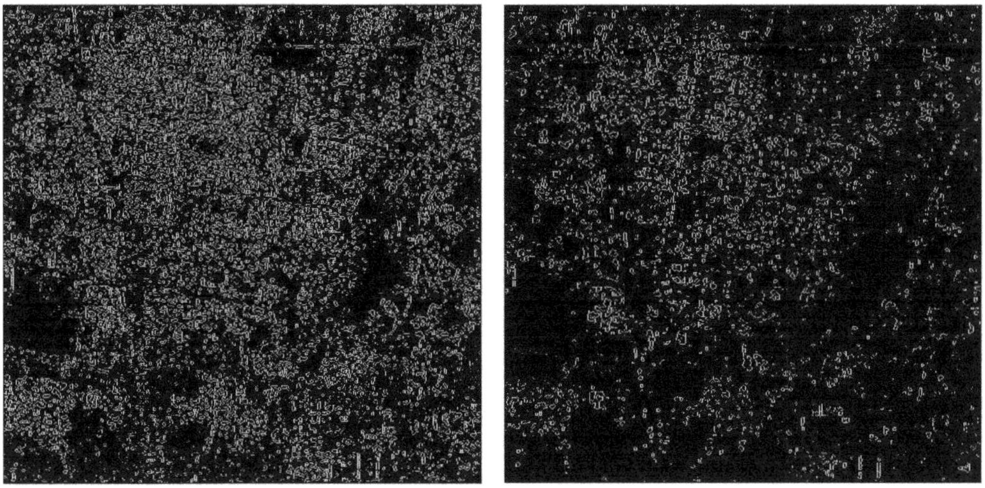

(a) (b)

Figure 10.11 Comparison of the image edge extraction results before and after denoising (a) Image edge extraction result before denoising (b) Image edge extraction result after denoising

Finally, we applied edge extraction to the images before and after natural gradient flexible ICA denoising. From Figure 10.11a,b it can be seen that the extracted edge features of the denoised image are more accurate.

In addition, we compared the image obtained as a result of natural gradient flexible ICA denoising with an image obtained from Google Earth of the same location, as shown in Figures 10.8a and 10.10, respectively. It can be seen that the processed image does inhibit the effect of noise, while the image quality has been significantly improved, compared with several other algorithms.

10.3 Chapter Summary

Natural gradient flexible ICA used for speckle suppression of SAR images with rich textured details can effectively improve the quality of the images, while retaining the polarization information. This method is simple, self-adaptable, and predictable and widely used for SAR image enhancement.

References

Amari, S.I. (1999). Natural gradient learning for over- and under-complete bases in ICA. *Neural Computation*, **11**(8), 1875–1883.

Amari, S.I.. and Cardoso, J.F. (1997). Blind source separation-semiparametric statistical approach. *IEEE Transactions on Signal Processing*, **45**(11), 2692–2700.

Amari, S.I., Cichocki, A., and Yang, H.H. (1996). A new learning algorithm for blind signal separation. Advances in Neural Information Processing Systems, Denver, CO, August 1996, pp. 757–763.

Arsenault, H.H.. and April, G. (1976). Properties of speckle integrated with a finite aperture and logarithmically transformed. *Journal of the Optical Society of America*, **66**(11), 1160–1163.

Cardoso, J.F. (1997). Infomax and maximum likelihood for blind source separation. *IEEE Signal Processing Letters*, **4**(4), 112–114.

Cardoso, J.F.. and Laheld, B.H. (1996). Equivariant adaptive source separation. *IEEE Transactions on Signal Processing*, **44**(12), 3017–3030.

Choi, S. (1999). Adaptive blind signal separation for multiuser communications: an information-theoretic approach. *Journal of Electrical Engineering and Information Science*, **4**(2), 249–256.

Choi, S., Cichocki, A.. and Amari, S.I. (2000). Flexible independent component analysis. *Journal of VLSI Signal Processing Systems for Signal Image and Video Technology*, **26**(1–2), 25–38.

Cichocki, A., Douglas, S.C.. and Amari, S. (1998). Robust techniques for independent component analysis (ICA) with noisy data. *Neurocomputing*, **22**(1), 113–129.

Cichocki, A., Thawonmas, R.. and Amari, S.I. (1997). Sequential blind signal extraction in order specified by stochastic properties. *Electronics Letters*, **33**(1), 64–65.

Cichocki, A.. and Unbehauen, R. (1996). Robust neural networks with on-line learning for blind identification and blind separation of sources. *IEEE Transactions on Circuits and Systems I-Fundamental Theory and Applications*, **43**(11), 894–906.

Cichocki, A., Unbehauen, R.. and Rummert, E. (1994). Robust learning algorithm for blind separation of signals. *Electronics Letters*, **30**(17), 1386–1387.

Hyvärinen, A.. and Oja, E. (1997). A fast fixed-point algorithm for independent component analysis. *Neural Computation*, **9**(7), 1483–1492.

11

Non-negative Independent Component Analysis and Its Application

The elements of the independent components must be either positive or negative, as determined by separation using the traditional independent component analysis (ICA) algorithm. However, in reality, many real-world signals are non-negative (non-negativity of a random signal $\boldsymbol{S}$ means that $\boldsymbol{S} \geq 0$). For example, both the gray values of image pixels in remote sensing data fusion and the chemical element content of mineral data are non-negative. If the same pre-whitening is used when processing these non-negative signals, the non-negativity information is lost.

To solve this problem, Paatero introduced the concept of positive matrix factorization (PMF), which constrains $\boldsymbol{S}$ and $\boldsymbol{A}$ to positive matrices (Paatero and Tapper, 1994). Lee and Seung (1997, 1999, 2001) proposed an effective non-negative matrix factorization (NMF) algorithm based on PMF. However, this does not guarantee that the equation has only one solution merely by using non-negativity as the constraint. Plumlbey suggested using both independence and non-negativity as constraint conditions, since this would ensure the preservation of the non-negativity information of the source signal. However, in addition to independence and non-negativity, these algorithms require that the source signal has the property of tending to zero. (The property of a random signal tending to zero refers to an arbitrary $\delta > 0$, $\boldsymbol{P}(\boldsymbol{s} < \delta) > 0$).

In fact, some signals, such as image signals, do not satisfy the property of tending to zero. To solve this problem, Wang, Yu, and Zhang (2008a) proposed a new non-negative ICA algorithm , which unlike other non-negative ICA algorithms, does not require the property of the source signals tending to zero.

Blind Source Separation: Theory and Applications, First Edition. Xianchuan Yu, Dan Hu and Jindong Xu.

11.1 Non-negative Independent Component Analysis

We propose a new method, called non-negative ICA, which is based on Lagrange constraints, and which uses high-order accumulation leptokurtosis, kurt(y), as the objective function:

$$\mathrm{kurt}(y) = \mathrm{kurt}(\boldsymbol{w}^{\mathrm{T}}\boldsymbol{x}) = \mathrm{E}\{(\boldsymbol{w}^{\mathrm{T}}\boldsymbol{x})^4\} - 3(\mathrm{E}\{(\boldsymbol{w}^{\mathrm{T}}\boldsymbol{x})^2\})^2 \quad (11.1)$$

By selecting $\mathbf{y}$ as the non-negative constraint, non-negative ICA can be transformed into a nonlinear optimization problem with constraints.

$$\begin{cases} \max(\mathrm{kurt}\,(y)) = \max(\mathrm{kurt}(\boldsymbol{w}^{\mathrm{T}}\boldsymbol{x})) \\ s.t.y = \boldsymbol{w}^{\mathrm{T}}\boldsymbol{x} \geq 0 \\ \boldsymbol{w}^{\mathrm{T}}\boldsymbol{w} = 1 \end{cases} \quad (11.2)$$

The norm of $\boldsymbol{w}$ is 1 and it is ruled by $\boldsymbol{w}^{\mathrm{T}}\boldsymbol{w} = 1$.

We can use the Lagrange multiplier method to solve these nonlinear programming problems with constraints by setting λ_{ij} as the Lagrange multiplier to restrict $\mathbf{y} = \mathbf{w}^{\mathrm{T}}\mathbf{x} \geq 0$, so that the augmented Lagrangian function of the problem is defined as

$$\mathrm{L} = \mathrm{kurt}(u) + \mathrm{tr}(\lambda u) = \mathrm{kurt}(\boldsymbol{w}^{\mathrm{T}}\boldsymbol{x}) + \mathrm{tr}(\lambda \boldsymbol{w}^{\mathrm{T}}\boldsymbol{x}) \quad (11.3)$$

By calculating $\mathrm{d}y/\mathrm{d}x$ while using the Kuhn–Tucker condition, we can transform the original planning problem into one that requires solving the following equations:

$$\begin{cases} \dfrac{\partial \mathrm{L}}{\partial \boldsymbol{w}} = 4\mathrm{E}\left[\boldsymbol{x}\left(\boldsymbol{\omega}^{\mathrm{T}}\boldsymbol{x}\right)^3\right] \text{-}3\boldsymbol{w} + \lambda\boldsymbol{x} = 0 \\ \lambda\boldsymbol{w}^{\mathrm{T}}\boldsymbol{x} = 0 \\ \boldsymbol{w} \leftarrow \dfrac{\boldsymbol{w}}{\boldsymbol{w}} \end{cases} . \quad (11.4)$$

Because the equations are non-homogeneous and there is no quick general solution for them, by using an iterative method we can only obtain an approximate solution, with the final iteration formula as

$$\begin{cases} \boldsymbol{w} \leftarrow \dfrac{4\mathrm{E}\left[\boldsymbol{x}\left(\boldsymbol{w}^{\mathrm{T}}\boldsymbol{x}\right)^3\right]}{3} \\ \boldsymbol{w} \leftarrow \dfrac{\boldsymbol{w}}{\|\boldsymbol{w}\|} \end{cases} \quad (11.5)$$

We do not need to centralize the data because a negative source does not have the characteristic of a zero mean, which is the main difference between this algorithm and classical ICA algorithms. Therefore, the mean value must be included in the entire operation. Based on the above, the iterative process for this learning algorithm can be summarized as follows:

Algorithm Non-negative ICA learning algorithm

Input: Sample set $\boldsymbol{X} = (\boldsymbol{x}_1, \boldsymbol{x}_2, \dots, \boldsymbol{x}_N)$.
Output: Separated matrices $\boldsymbol{W}_{m\times n} = (\boldsymbol{w}_1^{\mathrm{T}}, \dots, \boldsymbol{w}_n^{\mathrm{T}})^{\mathrm{T}}$.
Steps:

1. Obtain **Z** by whitening the data.
2. Choose an initial estimate m of the independent components; let $i \leftarrow 1$.
3. Choose an initialization vector $\boldsymbol{w}_i$ with unit norm; this can be randomly selected.
4. Update $\boldsymbol{w}_i$: $\boldsymbol{w}_i \leftarrow \frac{4\mathrm{E}[\mathbf{Z}(\boldsymbol{w}_i^{\mathrm{T}}\mathbf{Z})^3]}{3}$.
5. Orthogonalize $\boldsymbol{w}_i$: $\boldsymbol{w}_i \leftarrow \boldsymbol{w}_i - \sum_{j=1}^{i-1} \boldsymbol{w}_i^{\mathrm{T}}\boldsymbol{w}_j)\boldsymbol{w}_j$.
6. Standardize $\boldsymbol{w}_i$: $\boldsymbol{w}_i \leftarrow \boldsymbol{w}_i/||\boldsymbol{w}_i||$.
7. If $\boldsymbol{w}_i$ has not converged absolutely, return to step (4).
8. Let $i \leftarrow i + 1$; if $i \leq m$, return to step (3).

11.2 Application and Analysis

Since non-negative ICA retains the non-negativity information during the separation of a source signal, it can be used in a wide range of practical applications. The first experiment in this section verifies the effectiveness of the non-negative ICA algorithm, while the second presents an application thereof in image fusion.

11.2.1 Mixed-Signal Separation Experiment

To verify the effectiveness of the algorithm without requiring the property that the data tends to zero, we use a simulation signal, which is a mixture of sub-Gaussian and super-Gaussian signals as the experimental data, while adding various parameters to control the degree to which the signal tends to zero. The selected signal functions are given as

$$\boldsymbol{s} = \begin{bmatrix} s_1 \\ s_2 \\ s_3 \end{bmatrix} = \begin{bmatrix} ((\mathrm{rem}\,(x, 23) - 11)\,/9)^5 + u_1 \\ \sin(600\pi x/10000 + 6\cos(120\pi x/10000)) + u_2 \\ (2(\mathrm{rand}(500) < 0.5) - 1) \times (\log(\mathrm{rand}(500))) + u_3 \end{bmatrix} \tag{11.6}$$

In Equation 11.6, $\boldsymbol{u} = [\boldsymbol{u}_1, \boldsymbol{u}_2, \boldsymbol{u}_3]^{\mathrm{T}}$ is used to control the degree to which the non-negative signal tends to zero, function $rem(m, n)$ calculates the remainder of m/n, and function rand(m) is used to produce a random signal. The three signals tend to zero when $\boldsymbol{u} = [\boldsymbol{u}_1, \boldsymbol{u}_2, \boldsymbol{u}_3]^{\mathrm{T}} = [0, 0, 0]^{\mathrm{T}}$. To verify the effectiveness of the algorithm, we set $\boldsymbol{u}_1$, $\boldsymbol{u}_2$, $\boldsymbol{u}_3$ equal to 2.8, 1, and 20, respectively, which means that neither of these signals has the property of tending to zero. The normalized kurtosis values corresponding to these three source signals are 2.5307, −1.0453, and 3.5022, respectively, which means the first and third signals are super-Gaussian signals, while the second is a sub-Gaussian signal. The source signals are shown in Figure 11.1.

Using mixing matrix $\boldsymbol{A} = \begin{bmatrix} 1 & 53 & 47 \\ 5 & 7 & 45 \\ 3 & 45 & 67 \end{bmatrix}$ in the experiment, we obtain signal $\boldsymbol{X} = \boldsymbol{AS}$ after mixing, as shown in Figure 11.2.

Figure 11.3 illustrates the unmixed signal obtained using the non-negative ICA algorithm. As can be seen in these figures, the unmixed signal retains the non-negativity characteristics of the source signal.

To calculate the degree of separation of the signal, we denote the source signal by $(\boldsymbol{s}_1, \boldsymbol{s}_2, \boldsymbol{s}_3)^{\mathrm{T}}$ and the unmixed signal by $(\boldsymbol{x}_1, \boldsymbol{x}_2, \boldsymbol{x}_3)^{\mathrm{T}}$. Through factor analysis, the unmixed signals $\boldsymbol{x}_2, \boldsymbol{x}_1$, and $\boldsymbol{x}_3$ correspond to source signals $\boldsymbol{s}_1, \boldsymbol{s}_2$, and $\boldsymbol{s}_3$, respectively.

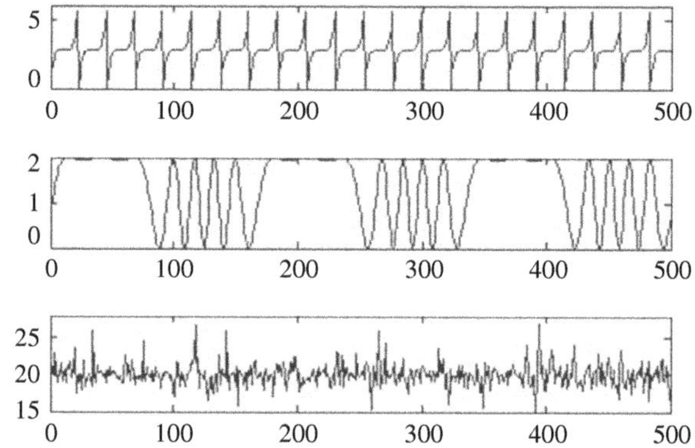

Figure 11.1 Source signals

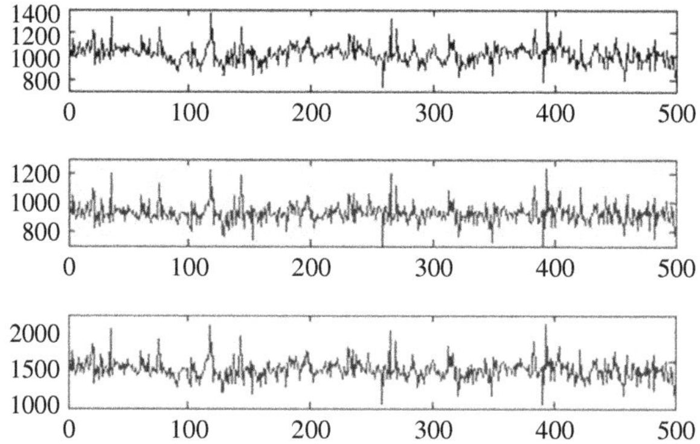

Figure 11.2 Mixed signals

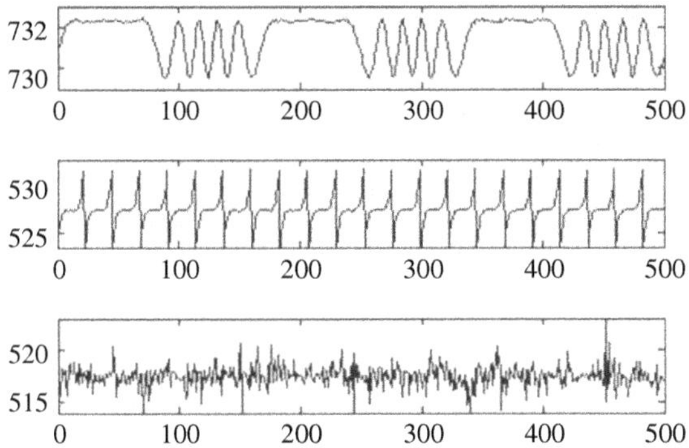

Figure 11.3 Unmixed signals

Table 11.1 Correlation coefficients of source and unmixed signals

Source signal\ unmixed signal	Unmixed signal 1	Unmixed signal 2	Unmixed signal 3
Source signal 1	0.0370	0.9999	−0.0003
Source signal 2	−0.9993	0.0108	0.0146
Source signal 3	0.0028	−0.0002	−0.9999

The correlation coefficients of the source signals and the unmixed signals are given in Table 11.1.

From Table 11.1, we see that the absolute value of the correlation coefficient that corresponds to signals $\boldsymbol{s}_1$ and $\boldsymbol{x}_2$, signals $\boldsymbol{s}_2$ and $\boldsymbol{x}_1$, as well as signals $\boldsymbol{s}_3$ and $\boldsymbol{x}_3$, is very close to 1, which indicates that the algorithm achieves good blind source separation performance without the condition of tending to zero. Moreover, it also retains the non-negativity characteristic of the signal.

11.2.2 Remote Sensing Image Fusion Experiments

11.2.2.1 Experimental Data

The enhanced thematic mapper (ETM+), manufactured by Raytheon, has a higher sensitivity than the thematic mapper (TM). The ETM+ is an eight-spectral multi-spectral scanning radiometer that works in the visible, near-infrared, shortwave, and thermal infrared spectral segments. The sensor technology indexes of the ETM+ are given in Table 11.2.

From Table 11.2, we see that Landsat-7 data can be divided into three band groups, namely, the visible/shortwave infrared (SWIR) (bands 1, 2, 3, 4, 5, and 7), thermal infrared (band 6) and panchromatic band groups, respectively.

Table 11.2 Main parameters of the eight ETM+ channels

Band no.	Band	Wavelength range (μm)	Ground resolution (m)
1	Blue	0.450–0.515	30
2	Green	0.525–0.605	30
3	Red	0.630–0.690	30
4	Near-infrared	0.775–0.900	30
5	Mid-infrared	1.550–1.750	30
6	Thermal	10.40–12.50	60
7	Mid-infrared	2.090–2.350	30
PAN	Panchromatic	0.520–0.900	15

Compared with the TM, the ETM+ has been improved in the following three areas:

1. It has an increased PAN (panchromatic) band with 15 m resolution.
2. It uses double gain technology, which improves the resolution of the far-infrared band from 6 to 60 m.
3. With the improvement to the sun scaler, the radiation calibration error of satellites is less than 5% while the accuracy is almost doubled compared with Landsat 5. In addition, the radiation calibration is also greatly improved.

SPOT4 is a resource satellite launched by France, the images from which are currently some of the most widely used in satellite remote sensing. Images from SPOT4 are grouped into panchromatic and multispectral images, with the ground resolution of these images being 10 and 20 m, respectively. The parameters for SPOT4 are given in Table 11.3.

The first Chinese resource satellite, called the "China-Brazil Earth Resource Satellite 1" (CBERS) was launched jointly by China and Brazil. The satellite is running in a nearly polar solar synchronous orbit 778 km above the ground, and circles the earth 14.3 times every day so that it can observe the Earth completely every 26 days.

Table 11.3 SPOT satellite parameter specifications

Remote sensor	HRVIR	VI
Band (μm)	0.61–0.68 (PAN) 0.50–0.59 0.61–0.68 0.79–0.89 1.58–1.75	0.43–0.47 0.61–0.68 0.79–0.89 1.58–1.75
Ground resolution (m)	10 (PAN) 20 (2–5)	1000

Table 11.4 CBERS parameter specifications

Remote sensor	CCD camera	Infrared multispectral scanner	Wide angle imager
Band (μm)	B1: 0.45–0.52 B2: 0.52–0.59 B3: 0.63–0.69 B4: 0.77–0.89 B5: 0.51–0.73	B6: 0.50–1.10 B7: 1.55–1.75 B8: 2.08–2.35 B9: 10.4–12.5	B10: 0.63–0.69 B11: 0.77–0.89
Coverage area (km)	113	119.5	890
Ground resolution (m)	19.5	77.8 (6–8), 156 (9)	256
Side-view ability	±32°	–	–

The satellite contains a five-band CCD camera with a resolution of 20 m, a four-band infrared multispectral scanner with a resolution of 78 m, a two-band wide angle imaging instrument with a resolution of 256 m, a space environmental monitor, and a data acquisition system. The satellite's high density tape recorder can be used to record the data of a region of interest that is outside the coverage of the receiving station. CBERS has attracted great interest in China's remote sensing community because of its advantages of a high viewpoint, wide angle views, fast data acquisition, repeat coverage, and continuous observation. The sensor technology indicators are given in Table 11.4.

We adopted a CBERS satellite image and a Landsat ETM+ image as the experimental data. The experimental area was located in the Doumen district of Zhuhai city in Guangdong province. The ground area consists of five types of covering, namely, urban land, water, forest, agricultural land, and unused land. The size of all images used was 256 × 256, with the multispectral image displayed as a false color image synthesized from bands 4, 3, and 2. In the experiment, we used the CBERS-1 satellite multispectral image and the Landsat ETM+ panchromatic image, registered as the reference image for the result, see Figure 11.4.

11.2.2.2 Remote Sensing Image Fusion Based on Ordered Non-negative ICA

We carried out the fusion experiment on the multispectral remote sensing image because of its abundant spectral information, while the panchromatic remote sensing image has high resolution. Traditional algorithms could not effectively eliminate the high-order redundancy among the fusion images and, therefore, for the characteristics of orderliness and non-negativity, we proposed an ordered non-negative independent component analysis (ONICA) algorithm (Wang, and Yu, 2008b) to solve the problems that the traditional algorithms failed to solve. We also improved

(a) (b)

Figure 11.4 Multispectral and panchromatic images (a) CBERS multispectral image (b) Landsat ETM+ panchromatic image

the fusion rules and obtained a good result. Next, to eliminate high-order redundancy effectively, we proposed an integer wavelet transform ordered non-negative independent component analysis (IWT-ONICA) algorithm (Wang, Yu, and Zhang, 2008c) based on the fact that the distribution of a high-frequency sub-band is similar to the Laplace distribution of a super-Gaussian. The experimental results indicate that the fusion algorithms, ONICA and IWT-ONICA, can eliminate high-order redundant information and effectively improve the speed of fusion while better retaining both the spectral and spatial detail, unlike other fusion methods, such as the discrete wavelet transform (DWT).

Idea behind the ONICA Algorithm

At present, image fusion algorithms based on ICA are rare (Bian and Zhou, 2005; Chen *et al.*, 2007; Shi *et al.*, 2008; Wang, Yu, and Zhang, 2008a). ONICA achieves the expected fusion aim of eliminating high-order redundancy. However, the algorithm does not take full account of the choice of fusion rules, and merely applies some simple superposition on the independent components after separation. This can easily lead to conflicts among the different independent components, and may also result in poor contrast of the image after fusion. In 2007, Mitianoudis *et al.* proposed a new fusion method based on ICA. In this algorithm, images are randomly divided into blocks, and then fused using the ICA algorithm. However, the approach is likely to cause excessive input signals, decrease processing efficiency, and increase operation time. Moreover, these algorithms do not consider the non-negativity property or the orderly property (Mitianoudis and Stathaki, 2007) between each band of the remote sensing images. Therefore, we introduced the ONICA algorithm for remote sensing image fusion and, at the same time, proposed new fusion rules to improve the fusion results.

For the fusion of remote sensing images, assume that there are a number of independent signal sources in a scene, which are separable in the spectrum, and thus we can use the ICA algorithms to separate the independent sources. Assume that the fusion resolution of the panchromatic and multispectral images is $m \times n$, synthesized by bands 4, 3, and 2, and labeled as $\boldsymbol{X} = [\boldsymbol{R}, \boldsymbol{G}, \boldsymbol{B}, \boldsymbol{P}]^{\mathrm{T}}$, where $\boldsymbol{R}$, $\boldsymbol{G}$, and $\boldsymbol{B}$ correspond to the three bands of the multispectral image and P represents the panchromatic image. Each independent component can be obtained by the ICA model $\boldsymbol{X} = \boldsymbol{AS}$ and the ONICA algorithm, denoted by $\boldsymbol{S} = [\boldsymbol{\mu}_1, \boldsymbol{\mu}_2, \boldsymbol{\mu}_3, \boldsymbol{\mu}_4]^{\mathrm{T}}$.

The basic idea of ONICA image fusion is as follows. First, record the image after registration as four base vectors, that is, the observation signal $\boldsymbol{X} = [\boldsymbol{R}, \boldsymbol{G}, \boldsymbol{B}, \boldsymbol{P}]^{\mathrm{T}}$. Then transform the base vectors into a non-negative independent feature subspace using the ONICA algorithm. Thereafter, apply factor analysis to the new matrix, consisting of the observation signal $\mathbf{X} = [\boldsymbol{R}, \boldsymbol{G}, \boldsymbol{B}, \boldsymbol{P}]^{\mathrm{T}}$ and the independent components $\boldsymbol{S} = [\boldsymbol{\mu}_1, \boldsymbol{\mu}_2, \boldsymbol{\mu}_3, \boldsymbol{\mu}_4]^{\mathrm{T}}$, to determine the correspondence relation between them, and execute fusion of the coefficients of the image feature in the transform domain according to certain fusion rules. The research of Olshausen and Field (1996) showed that for any observation scene, its non-deformation presentation in the ICA domain is sparser than its presentation in a single source image. In other words, larger coefficients should be enhanced by coefficient fusion, while on the other hand, the smaller ones should be constrained. Supposing that, after factor analysis, the independent components corresponding to $\boldsymbol{R}$, $\boldsymbol{G}$, $\boldsymbol{B}$, and $\boldsymbol{P}$ are, respectively, λ_1, λ_2, λ_3, λ_4, we adopt the following fusion rules:

$$\begin{cases} \lambda_1 = \max(\lambda_1, \ \lambda_4) \\ \lambda_2 = \max(\lambda_2, \ \lambda_4) \\ \lambda_3 = \max(\lambda_3, \ \lambda_4) \end{cases} \tag{11.7}$$

Finally, we obtain the fusion image by carrying out the ONICA inverse transform on the fused coefficients. The basic flow diagram is illustrated in Figure 11.5.

Further Improvement of the Algorithm (IWT-ONICA)

We know that the key to the success of the ICA algorithm is to make the correct assumptions about the probability density of the source signal through ICA. The disadvantage of ICA is that if we cannot estimate the probability density of the source signal accurately, the performance thereof is seriously affected. In fact, a remote sensing image is not entirely consistent with a non-Gaussian distribution; studies have shown that the distribution of the high-frequency sub-band of a natural image, which is almost a Gaussian distribution in the wavelet transform domain, is approximately a super-Gaussian Laplace distribution.

Remote sensing images can produce high-frequency sub-bands by DWT, which is also a kind of fusion algorithm with many excellent features. It has the ability not only to enhance the performance of the details of the space of the multispectral image effectively, but also to maintain the spectral characteristics of the image before and

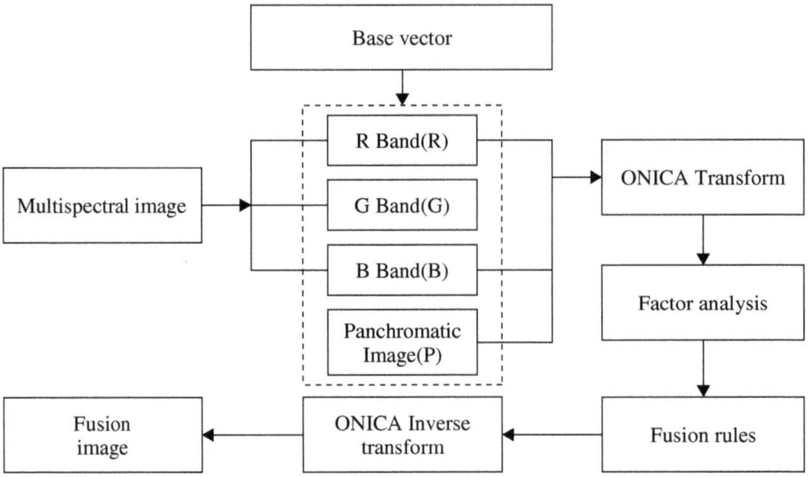

Figure 11.5 Process of orderly non-negative independent component analysis image fusion

after fusion. Because the output of the DWT filter is a floating point number, there are many problems, including slow operational speed, and massive memory and computation requirements. In the late 1990s, Daubechies and Swelden (1998) proposed constructing an integer wavelet transform (IWT) through the process of ascension. Also known as a second generation wavelet, the algorithm, which can be realized by mapping from integer to integer with a real sense of reversibility, both maintains the characteristics of the wavelet and overcomes the limitations of computational speed. IWT has the advantages of fast operation, reduced computer memory requirements, and accurate reconstruction of the image.

We can improve the ICA fusion algorithm by using probability density. Since ICA has the ability to remove high-order redundant information, while IWT has the characteristics of multi-resolution and high speed, we propose non-negative ordered ICA (IWT-ONICA) by combining ICA and IWT. The new algorithm is very fast because of the fast speed of IWT and the reduced computation of ICA.

Idea behind the IWT-ONICA Algorithm

The steps using IWT-ONICA to carry out image fusion are given below.

1. Obtain baseband data and sub-band data by applying an IWT to the three bands of the multispectral image to be registered and the panchromatic image with a decomposition level of 3. We denote the three bands of the multispectral image (bands 4, 3, and 2) and the panchromatic image by $[\boldsymbol{R}, \boldsymbol{G}, \boldsymbol{B}, \boldsymbol{P}]^{\mathrm{T}}$, where $\boldsymbol{R}$, $\boldsymbol{G}$, and $\boldsymbol{B}$ correspond to the three bands, respectively, and $\boldsymbol{P}$ represents the panchromatic

image. Figure 11.6 shows the image with the three bands corresponding to that in Figure 11.4a.

2. Carry out fusion on the high-frequency sub-band data obtained by the IWT according to the fusion rules. In the DWT domain, the high-frequency sub-band is generally decomposed into three sub-band images, namely, the sub-band image in the horizontal direction LH, the sub-band image in the vertical direction HL, and the sub-band image in the diagonal direction HH. Since the high-frequency sub-band adheres to a super-Gaussian distribution to a greater extent, we can translate the high-frequency sub-band images using ONICA, and then apply factor analysis to determine the order of the independent components. Assume that after factor analysis, the corresponding independent components of the high-frequency sub-band of ***R***, ***G***, ***B***, ***P*** are, respectively, λ_1, λ_2, λ_3, λ_4, we use the same rules as given in Equation 11.7 to fuse the obtained independent high-frequency sub-band image, namely,

$$\begin{cases} \lambda_1 = \max(\lambda_1, \lambda_4) \\ \lambda_2 = \max(\lambda_2, \lambda_4) \\ \lambda_3 = \max(\lambda_3, \lambda_4) \end{cases} \tag{11.8}$$

   Thereafter, we apply the ONICA reverse transformation to the fused non-negative independent components.

3. Apply IWT to the combination of the four fused images (***R***, ***G***, ***B***, ***P***) in the high-frequency sub-band and the corresponding low-frequency sub-band, and finally obtain the fused multispectral image.

Time Complexity Analysis of IWT-ONICA

To illustrate the impact of DWT and IWT on the fusion speed, we apply DWT and IWT to carry out decomposition/reconstruction of the panchromatic and multispectral images, respectively (Wang, Yu, and Zhang, 2008a). The experimental hardware environment consisted of a Pentium 3.00 GHz CPU with 1.00 GB memory, with VC++ 6.0 as the programming environment. The resolution of the multispectral and panchromatic images was 256 × 256, and the decomposition level of the wavelet transform was 3. The times for decomposition and reconstruction are shown in Figure 11.7.

According to Figure 11.7, compared with DWT, the reduced complexity of IWT as a percentage is

$$\frac{34.0833 + 34.0930 - 18.2202 - 18.8745}{34.0833 + 34.0930} = 45.59\%.$$

This is mainly because IWT can be achieved as an integer to integer mapping, which reduces the number of products and the complexity of the calculation at the same time. The experimental results indicate that IWT can be carried out with greater speed without affecting the fusion result.

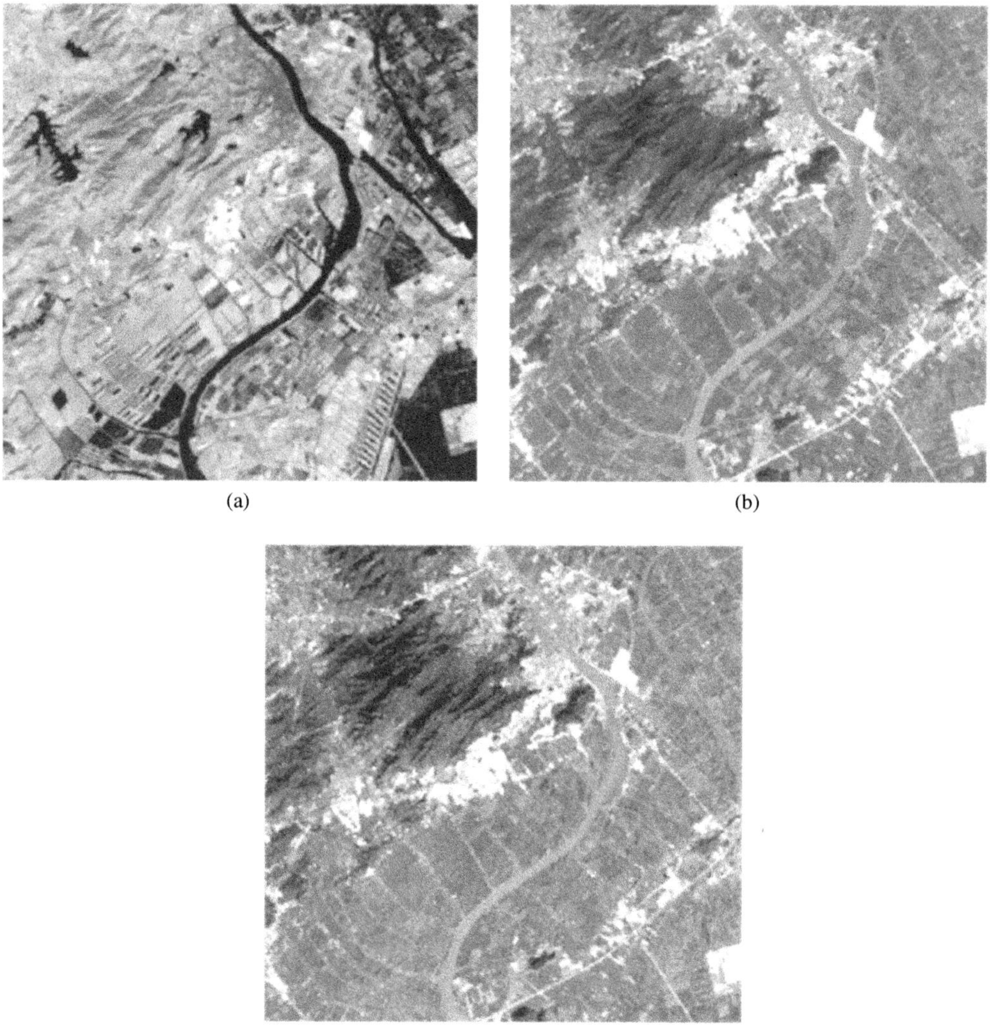

(a) (b) (c)

Figure 11.6 Three bands of a multispectral image: (a) band 4, (b) band 3, (c) band 2

11.2.2.3 Comparative Analysis of Experimental Results

To prove the validity of the fusion algorithm, we carried out the following simulation experiments. Figure 11.8a–d shows the resultant fused images using the DWT, PCA, ONICA, and IWT-ONICA algorithms, as discussed in this chapter.

To compare the fusion performance, we selected the peak signal-to-noise ratio (PNSR), which reflects the degree to which redundant information is removed; the

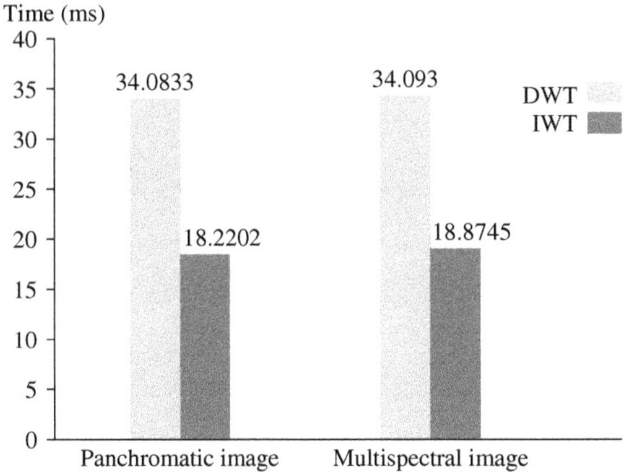

Figure 11.7 Time comparison for decomposition/reconstruction of DWT and IWT

Table 11.5 Comparison of the statistical evaluation parameters of the fusion results of different fusion methods

Band of remote Sensing image	PSNR	Correlation coefficient	Degree of distortion	Entropy
Figure 11.8a DWT 4	16.4525	0.8316	31.8932	7.6326
Figure 11.8a DWT 3	15.3591	0.7060	36.9360	7.7033
Figure 11.8a DWT 2	15.0530	0.6781	33.9408	7.6910
Figure 11.8b PCA 4	19.4903	0.8994	20.9306	7.6363
Figure 11.8b PCA 3	16.0428	0.6653	32.7310	7.4266
Figure 11.8b PCA 2	15.5584	0.6215	33.4730	7.4214
Figure 11.8c ICA 4	20.9196	0.8173	18.5833	7.7496
Figure 11.8c ICA 3	19.9235	0.8127	20.8635	7.4844
Figure 11.8c ICA 2	20.0556	0.7617	20.5433	7.5022
Figure 11.8d IWT-ONICA 4	21.7242	0.9439	15.8681	7.7408
Figure 11.8d IWT-ONICA 3	20.6423	0.8944	17.9259	7.5705
Figure 11.8d IWT-ONICA 2	20.1993	0.8812	18.9304	7.5704

correlation coefficients and distortion, which reflect the degree to which spectral information is retained; and information entropy, which can reflect the detailed spatial information as statistical parameters. These metrics give a more comprehensive evaluation of the fusion performance of ONICA and IWT-ONICA. The fusion results of a good fusion algorithm should have a larger PSNR, a larger correlation coefficient, a smaller distortion, and greater information entropy. Table 11.5 gives the statistical parameters of the fusion results using the different fusion algorithms.

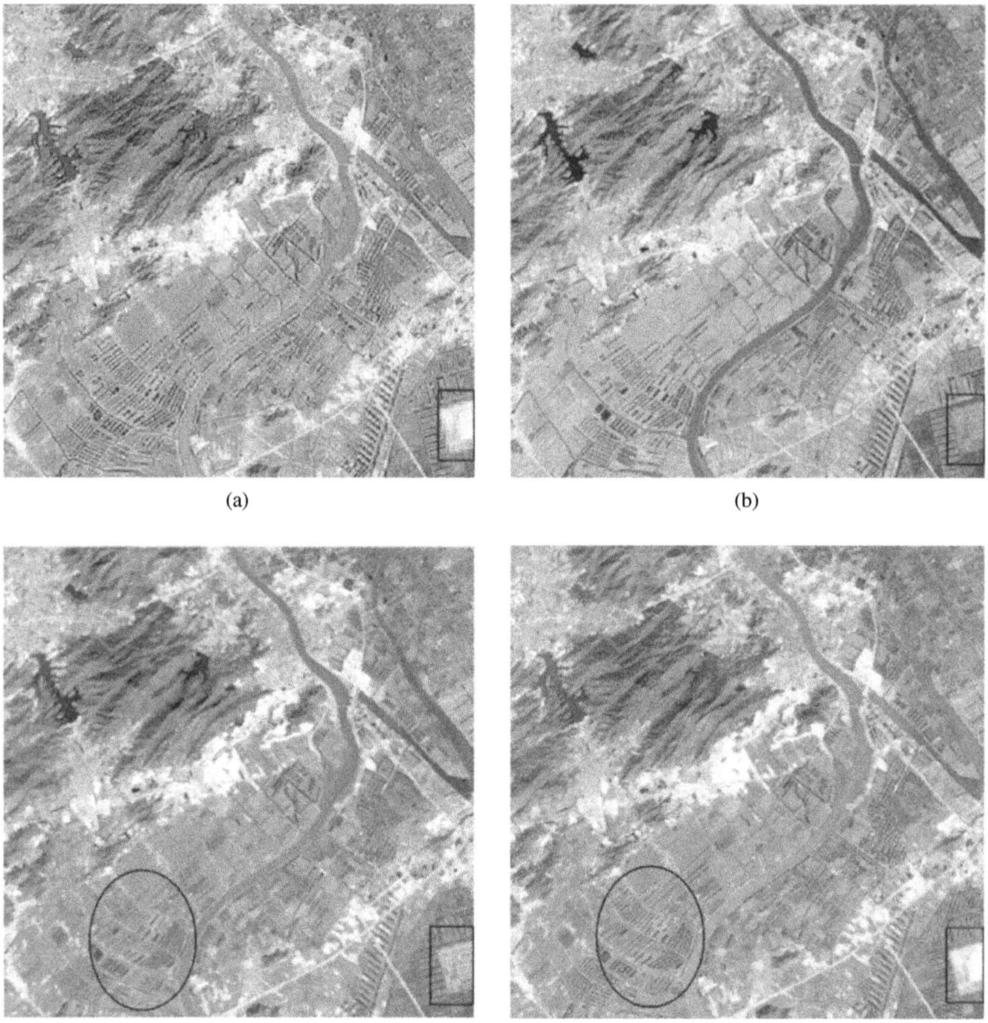
(a) (b) (c) (d)

Figure 11.8 Comparison of different fusion methods. (a) Image fusion based on DWT, (b) image fusion based on the PCA transform, (c) image fusion base on the ONICA transform, (d) image fusion based on IWT_ONICA

Comparing and analyzing the original image and the various fused images through mathematical statistics (see Table 11.5) and visual interpretation (see Figure 11.8), we reach the following conclusions:

1. From Table 11.5, we see that the PSNR of IWT-ONICA is the highest, followed by that of ONICA and PCA, with DWT having the lowest PSNR. Investigating the reasons for this phenomenon, we find that IWT-ONICA, ONICA, and PCA reduce

redundant information in the fused image, as well as the impact of noise on the fusion results, so these three methods have a larger PSNR than DWT. Moreover, since the fusion algorithm based on ICA removes the high-order information of the image data obtained from different sensors, whereas PCA only removes the second-order redundant information of the fused image data, the ICA algorithm has a larger PSNR than PCA.

2. The correlation coefficient reflects the spectral characteristics of similarity between the fusion image and the original multispectral image, which can be used to determine the spectrum retaining ability of the fusion image. From Table 11.5 we see that the correlation coefficient of IWT-ONICA is the largest, followed by that of ICA, and finally PCA. This illustrates that the fusion image adopted by IWT-ONICA retains the spectrum information of the original image commendably. From Figure 11.8, we can see that the change in the spectrum information for the PCA fusion algorithm is the largest; the visual result is consistent with the statistical results.
3. The degree of image spectral distortion reflects the degree of spectral distortion of the multispectral image: the smaller the value, the better the performance. From Table 11.5, we can see that the spectral distortion of IWT-ONICA is the lowest, followed by that of ONICA and PCA, while DWT has the highest distortion. The reason is that the fusion method based on DWT can be affected by the decomposition level, with the result that it is easy to produce a block effect and fuzzy texture (see the areas demarcated by rectangles in Figure 11.8a–c). Because it uses an IWT, the IWT-ONICA fusion algorithm also produces a block effect (see the area demarcated by a rectangle in Figure 11.8d). However, owing to the combination of the advantages of IWT and ICA, and the fact that the algorithm simultaneously adheres to the non-negativity characteristic of the data, the degree of spectral distortion is reduced.
4. The information entropy of an image is an important indicator of the amount of information. From Table 11.5, we can see that the entropy of IWT-ONICA is larger than that of PCA and ONICA. However, it is smaller than that of DWT, which means that the traditional method performs very well in terms of spectral information retention. From Figure 11.8, we can see that the fusion result of DWT is clearer than that of PCA, in terms of the texture, retention of the degree of edge profiles, and so on. Meanwhile, the fusion result of IWT-ONICA is also clearer than that of ONICA (see the circular areas in Figure 11.8c,d). This is mainly because the IWT-ONICA fusion algorithm adheres more to the assumptions of the ICA algorithm.

In summary, the ONICA and IWT-ONICA fusion algorithms presented in this chapter are superior to PCA and DWT, because both ONICA and IWT-ONICA not only take into account the non-negativity characteristic of the data, but also remove the high-order relativity of the data. As a result, the fusion results of these algorithms are much clearer with better visibility.

11.3 Chapter Summary

Removing redundant information from different sensors remains a problem in the study of image fusion, and has yet to be solved properly. In this chapter, we proposed a non-negative ICA algorithm based on the analysis of ICA theory, and the non-negativity order characteristic of remote sensing images. We proposed an image fusion solution based on the ONICA algorithm that applies the non-negative ICA algorithm to image fusion, where the source signal adheres to the non-Gaussian assumption of the ICA algorithm. By combining the advantages of IWT and ICA, we proposed the IWT-ONICA algorithm from the perspective of probability distributions. This algorithm achieved better image fusion results in our experiments. Further attempts on ICA based fusion of remote sensing images were introduced and we proposed a feasible solution for effectively resolving the conflicting (redundant) information or reducing error information among the different sensors.

References

Bian, J., and Zhou, Q.L. (2005) A fusion algorithm of multi-spectral images and panchromatic. *Microelectronics and Computer*, **22**(6), 5–11, (in Chinese).

Chen, M., Xuan, J.H., Li, D.R. *et al.* (2007) Image fusion algorithm based on independent component analysis. *Optoelectronic Engineering*, **34**, 82–87.

Daubechies, I., and Swelden, W. (1998) Factoring wavelet transform into lifting steps. *Fourier Analysis and Applications*, **4**(3), 247–269.

Lee, D.D., and Seung, H.S. (1997) Unsupervised learning by convex and conic coding. *Advances in Neural Information Processing Systems*, **9**, 515–521.

Lee, D.D., and Seung, H.S. (1999) Learning the parts of objects by non-negative matrix factorization. *Nature*, **401**, 788–791.

Lee, D.D., and Seung, H.S. (2001) Algorithms for non-negative matrix factorization. *Advances in Neural Information Processing Systems*, **13**, 556–562.

Mitianoudis, N., and Stathaki, T. (2007) Pixel-based and region-based image fusion schemes using ICA bases. *Information Fusion*, **8**(2), 131–142.

Olshausen, B.A., and Field, D.J. (1996) Emergence of simple-cell receptive field properties by learning a sparse code for natural images. *Nature*, **381**, 607–609.

Paatero, P., and Tapper, U. (1994) Positive matrix factorization: a non-negative factor model with optimal utilization of error estimates of data values. *Environmetrics*, **5**(2), 111–126.

Shi, Y.Q., Yu, X.C., Cheng, X.C. *et al.* (2008) A new BSS algorithm based on the data fusion and ICA. 7th IEEE International Conference on Cybernetic Intelligent Systems, London, UK, pp. 1–6.

Wang, Z.N., Yu, X.C., and Zhang, L.B. (2008a) A remote sensing image fusion algorithm based on integer wavelet transform. 7th World Congress on Intelligent Control and Automation, Chongqing, People's Republic of China, pp. 5950–5954.

Wang, Z.N., and Yu, X.C. (2008) A remote sensing image fusion algorithm based on non-negative ordinal independent component analysis by using lagrange algorithm. International Conference on Computer Science and Software Engineering, Wuhan, People's Republic of China, pp. 610–613.

Wang, Z.N., Yu, X.C., and Zhang, L.B. (2008b) A novel remote sensing image fusion algorithm based on IWT-ICA. 7th International Conference on Advanced Language Processing and Web Information Technology, Dalian, People's Republic of China, pp. 187–192.

12

Constraint Independent Component Analysis Algorithms and Applications

12.1 Overview

Currently, most independent component analysis (ICA) algorithms attempt to estimate all source signals, although in many practical applications, the number of components needed is often less than the number of source signals. For example, in biomedical signal processing, there are a large number of sensor signals, but we are typically only interested in a small number of these (Wu *et al.*, 2006, whereas in geochemical data, separation of mixed-signal data is important to predict mineral resources, and thus we observe the mixed-signal data, which is the superimposed function result of a past historical event (Yu *et al.*, 2007; Liu *et al.*, 2007). Thus, it is necessary to reduce the dimensions in the standard ICAs input and output processes. Although many one-unit ICA algorithms based on compression are able to extract a small number of signals, these algorithms have various shortcomings, including inefficiency and the fact that the order in which components are extracted is random. In addition, although the nonlinear principal component analysis (PCA) (Oja, 1997; Woo and Sali, 2002; Shi, 2008) and fast independent component analysis (FastICA) (Hyvärinen and Oja, 1997, 2000; Hyvärinen, 1998, 1999; Hyvärinen, Karhunen, and Oja, 2001) algorithms provide parallel extraction methods, they too have issues. Their main problems are that in many algorithms and applications of global expansion ICA, the extraction of a component mainly relies on the selection of the target function, and pre-whitening or related processes used in the process interfere with the original data, thus causing some of the useful information in the data to be lost.

In fact, we often have some prior information about the source signals, and taking full advantage of this information can help us to improve the algorithm's processing capacity. CICA (constrained independent component analysis) takes this into consideration; by combining traditional ICA with prior information, it reduces

Blind Source Separation: Theory and Applications, First Edition. Xianchuan Yu, Dan Hu and Jindong Xu.

the dimensions of the output signals to obtain the signals of interest. The difference between CICA and traditional ICA is that the latter algorithm has substantive characteristics, that is, uncertainties in the order of separating signals, symbols, and ranges. Without additional assumptions and constraints, these uncertainties cannot be reduced. For example, the FastICA algorithm only defines the direction of the independent components, while the independent components' separation range and matrix norms are still variable. On the other hand, the CICA algorithm can reduce the uncertainty of ICA. Certain restrictions of the CICA algorithm are given below:

1. To avoid order arbitrariness of the output composition, we can use statistics to provide order; commonly used statistics are variance and standard kurtosis, among others.
2. Since a standardized separation matrix is used to mitigate the effects of the output components' range, we obtain a learning algorithm with a stable numerical value (Choi, 2006).

Application of these conditions transforms the ICA issue into a constrained optimization problem, with the inequality and equality constraints used as prior information and additional conditions.

The success of CICA relies on providing a systematic and flexible method that embeds prior information or certain assumptions in the objective function, so that it can change ill-posed ICA issues into well-posed ones. This makes it more conducive to practical application. By adding certain constraints, the dimensions of the ICA output can be reduced. In addition, by embedding prior information in the source signals (such as statistical properties or reference source signals), the local minimum problem can be avoided, thus improving the properties of the algorithms. Lu and Rajapakse (2001, 2005) first presented an overall framework for CICA models, and pointed out how constraints can be imported into traditional ICA models. They solved the uncertainty problems of traditional ICA, and also proposed two other applications of CICA – less complete ICA and ICA with reference (ICA-R). For these less complete ICA algorithms, they used the uncorrelation of the output components as a constraint, which is compatible with the independence criterion and also solves the problem of the possibility of repeated composition. For the ICA-R algorithm, as long as the choice of threshold is appropriate, and the degree of closeness between the extracted signal and the reference signal is measured correctly, we are able to extract any period of the signal distribution or non-periodic signals, including Gaussian signals. In the ICA-R algorithm, negative entropy is used as the target function. Compared with the previous method, which is based on a second-order statistical property, the source signal extracted is independent. Both of these applications show that, if we add certain appropriate constraints, CICA not only enhances the processing power of the ICA algorithm, reduces the dimensions of the problem, and increases stability of convergence, but also reduces the process after detecting the source signals and noise features (Hyvärinen, Hoyer, and Oja, 1998). Compared with the traditional

ICA algorithm, which separates and outputs all source signals, CICA's computation is reduced (Rasheed *et al.*, 2006). Therefore, how to combine prior information or certain assumptions with the traditional ICA method and turn "blind" issues into semi-blind issues so as to solve the blind source separation (BSS) problem, has important practical significance.

12.2 CICA Algorithm

12.2.1 CICA Algorithm Based on Negative Entropy

The basics of the CICA and FastICA algorithms are the same, that is, to extract an independent composition at each iteration. FastICA uses negative entropy as the criterion, where the independent components that are extracted first have the largest negative entropy of all the components we need to extract. Thus, the order of extracting signals and observation signals differs, and it is difficult to extract certain specific components first.

The CICA algorithm solves the above problems by introducing a one-dimensional constraint reference vector $\boldsymbol{r} = (r_1, r_2, \ldots, r_n)^{\mathrm{T}}$ (n is the sample length), which contains feature step information of the target signal source to be extracted. This information changes according to the actual application, but is not exactly the same as the source signal. During the computation, the constraint vectors can cause the target signal to have non-Gaussian maximum utility, which makes the source signal reach its maximum negative entropy under the negative entropy criterion. This allows the CICA algorithm to extract the target source first, and sort the independent components by the size of the statistical measure, which solves the disorder problem of the FastICA algorithm when it separates signals. An overview of the algorithm is illustrated in Figure 12.1.

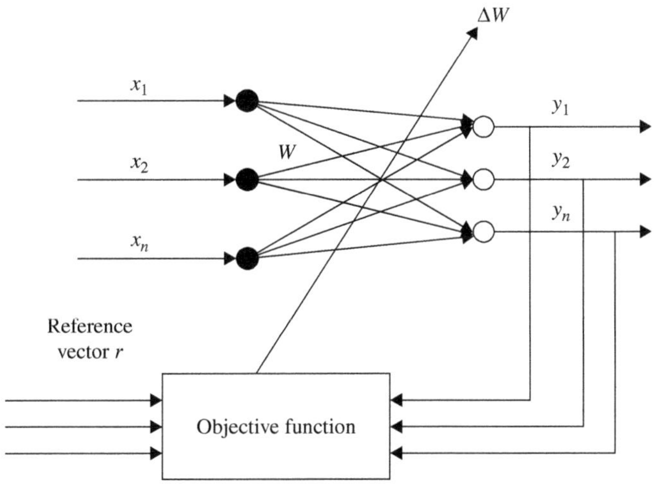

Figure 12.1 CICA algorithm schematic

12.2.1.1 Target Functions

The aim of the CICA algorithm is given as follows: After a series of iterative training, we obtain a one-dimensional weight vector $\boldsymbol{w}$, which is a row of the unmixing matrix $\boldsymbol{W}$, so the one-dimensional output component $\boldsymbol{y} = \boldsymbol{w}^{\mathrm{T}}\boldsymbol{X}$ approximates $\boldsymbol{s}^*$, where $\boldsymbol{s}^*$ is a one-dimensional target signal obtained by using reference vector $\boldsymbol{r}$. Similar to the FastICA algorithm, the CICA algorithm is described below.

First, we introduce the negative entropy's approximation comparison function $J(\boldsymbol{y})$:

$$J(\boldsymbol{y}) \approx \rho[E\{G(\boldsymbol{y})\} - E\{G(\boldsymbol{v})\}]^2 \tag{12.1}$$

where ρ is a positive constant, $\boldsymbol{v}$ is a Gaussian variable with mean 0 and variance 1, and $G(\cdot)$ is a non-quadratic function. The above principle is basically the same as that for FastICA.

In the general literature, the mean square error norm is given by

$$\varepsilon(\boldsymbol{y},\boldsymbol{r}) = E\{(\boldsymbol{y}-\boldsymbol{r})^2\} \tag{12.2}$$

In calculating the error norm, when $\boldsymbol{y}$ approaches the target source signal $\boldsymbol{s}^*$ completely, $\varepsilon(\boldsymbol{y},\boldsymbol{r})$ has its minimum value. If $\varepsilon(\boldsymbol{y},\boldsymbol{r})$ satisfies Equation 12.3, we get output $\boldsymbol{y}$, which is the source signal.

$$g(\boldsymbol{w}) = \varepsilon(\boldsymbol{y},\boldsymbol{r}) - \xi \leq 0 \tag{12.3}$$

where ξ is the threshold. Let $g(\boldsymbol{w})$ be the feasibility constraint of Equation 12.1; then it can project the algorithm's solutions onto feasibility constraint functions, thereby further limiting the extracted components. Thus, the CICA algorithm can be described by the following objective function:

$$\max\ J(y) \approx \rho[E\{G(y)\} - E\{G(\boldsymbol{v})\}]^2 \tag{12.4}$$

$$\text{s.t. } h(\boldsymbol{w}) = E\{y^2\} - 1 = 0 \tag{12.5}$$

where $h(\boldsymbol{w})$ is the limit function of contrast function $J(\boldsymbol{y})$ and weight vector $\boldsymbol{w}$. With Equations 12.4 and 12.5 established, the CICA algorithm terminates and $\boldsymbol{y}$ is the solution vector.

12.2.1.2 Optimization Algorithm

From the previous section, we know that Equations 12.4 and 12.5 define the target function of the CICA algorithm: solving the maximum of $J(\boldsymbol{y})$ can be transformed into solving the extremum problem of Lagrangian function L_w, the definition of which is given as

$$L(\boldsymbol{w},\mu,\lambda) = J(y) - \frac{1}{2\gamma}[\max{}^2\{\mu + \gamma g(\boldsymbol{w}), 0\} - \mu^2] - \lambda h(\boldsymbol{w}) - \frac{1}{2}\gamma\|h(\boldsymbol{w})\|^2 \tag{12.6}$$

Using the quasi-Newton method, we get $\boldsymbol{w}$'s iterative formula

$$\boldsymbol{w}_{k+1} = \boldsymbol{w}_k - \eta \boldsymbol{R}_{xx}{}^{-1} L'_{w_k} / \delta(\boldsymbol{w}_k) \tag{12.7}$$

where k is the number of iterations, η is the learning rate, $\boldsymbol{R}_{xx}$ is the covariance matrix of the observational data matrix $\boldsymbol{X}$, and L'_w is the Lagrangian function L_w's first derivative of $\boldsymbol{w}$.

$$L'_w = \rho E\{\boldsymbol{X}G'_y(\boldsymbol{y})\} - 0.5\mu E\{\boldsymbol{X}g'_y(\boldsymbol{w})\} - \lambda E\{\boldsymbol{X}\boldsymbol{y}\} \tag{12.8}$$

$$\delta(\boldsymbol{w}) = \rho E\{G'_{yy}(\boldsymbol{y})\} - 0.5\mu E\{g'_{yy}(\boldsymbol{w})\} - \lambda \tag{12.9}$$

where $G'_y(\boldsymbol{y})$ and $G''_{yy}(\boldsymbol{y})$ are non-quadratic function $G(\boldsymbol{y})$'s first and second derivatives of $\boldsymbol{y}$, respectively, and $g'_y(\boldsymbol{w})$ and $g'_{yy}(\boldsymbol{w})$ are function $g(\boldsymbol{w})$'s first and second derivatives of $\boldsymbol{y}$, respectively. μ and λ are Lagrange multiplication factors, with their negative entropy learning formula given as

$$\mu_{k+1} = \max\{0, \mu_k + \gamma g(\boldsymbol{w}_k)\} \tag{12.10}$$

$$\lambda_{k+1} = \lambda_k + \gamma h(\boldsymbol{w}_k) \tag{12.11}$$

where γ is a scalar processing parameter.

The CICA algorithm, which is based on negative entropy, is defined below.

Algorithm CICA algorithm based on negative entropy

Input: Whitening matrix $\boldsymbol{X} = (\boldsymbol{x}_1, \boldsymbol{x}_2, \ldots \boldsymbol{x}_n)$
Output: Independent component $\boldsymbol{Y} = (\boldsymbol{y}_1, \boldsymbol{y}_2, \ldots, \boldsymbol{y}_n)$
Steps:

1. Extract constraint reference vector $\boldsymbol{r}$, which has source signal characteristics
2. Initialize weight vector $\boldsymbol{w}$ (using a unit random matrix)
3. Calculate L'_w and $\delta(w)$
4. Adjust $\boldsymbol{w}$, repeat steps (3) to (4) until $\boldsymbol{w}$ converges
5. Obtain independent component $\boldsymbol{y}_i = \boldsymbol{w}^{\mathrm{T}}\boldsymbol{X}$.

The greatest advantage of the CICA algorithm is adding a reference vector as a constraint, which restricts the target function.

12.2.2 CICA Algorithm Based on Kurtosis

The independent components can be sorted by the determined statistical significance. Assuming that the statistical measure of the signal is $J(\boldsymbol{u})$, we can use the size of the statistical measure to sort the independent components. Statistical measures that are often used include variance and kurtosis. Under the constraints of sorting independent components and separating the standardized matrix, the likelihood function is optimized, and then we obtain the following CICA problem:

$$\begin{aligned} \max \quad & L(\boldsymbol{u}, \boldsymbol{W}) = E\{\sum_{i=1}^{N} \log p_i(\boldsymbol{w}_i^{\mathrm{T}}\boldsymbol{x}(t))\} + \log|\det(\boldsymbol{W})| \\ \text{s.t.} \quad & G(\boldsymbol{W}) = (G_1(\boldsymbol{W}), \ldots, G_{N-1}(\boldsymbol{W}))^{\mathrm{T}} \geq 0, \\ & H(\boldsymbol{W}) = (H_1(\boldsymbol{W}), \ldots, H_N(\boldsymbol{W}))^{\mathrm{T}} = 0 \end{aligned} \tag{12.12}$$

where $G_i(\boldsymbol{W}) = J(\boldsymbol{u}_{i+1}) - J(\boldsymbol{u}_i), i = 1, 2, \ldots, N$.

$H_j(\boldsymbol{W}) = \boldsymbol{w}_j^{\mathrm{T}}\boldsymbol{w}_j - 1, j = 1, 2, \ldots, N$ specifies that $\boldsymbol{W}$'s row norm is 1, while $\boldsymbol{w}_j^{\mathrm{T}}$ is $\boldsymbol{W}$'s j-th row vector.

Combining the projection method with the Lagrange multiplier method, we present a solution process for the CICA problem. Because the last constraint condition of the constraint problem, which is merely the standardization of the row norm, is relatively simple, we can use the projection strategy on the constraint set. This means we can first use the Lagrange multiplier method to solve the following inequality constrained optimization problems:

$$\begin{aligned} \max \quad & L(\boldsymbol{u}, \boldsymbol{W}) = E\{\sum_{i=1}^{N} \log p_i(\boldsymbol{w}_i^{\mathrm{T}}\boldsymbol{x}(t))\} + \log|\det(\boldsymbol{W})| \\ \text{s.t.} \quad & G(\boldsymbol{W}) = (G_1(\boldsymbol{W}), \ldots, G_{N-1}(\boldsymbol{W}))^{\mathrm{T}} \geq 0 \end{aligned} \tag{12.13}$$

After each iteration, we project the solution onto the constraint set to satisfy the equation constraints in Equation 12.13. The augmented Lagrangian function is defined as

$$L(\boldsymbol{W}, \lambda) = -L(\boldsymbol{u}, \boldsymbol{W}) + \frac{1}{2\gamma}\sum_{i=1}^{N-1}\{[\max\{0, \tilde{G}_i(\boldsymbol{W})\}]^2 - \lambda_i^2\} \tag{12.14}$$

where $\boldsymbol{\lambda} = (\lambda_1, \ldots, \lambda_{N-1})^{\mathrm{T}}$ is the Lagrange multiplier, γ is a penalty parameter, and $\tilde{G}_i(\boldsymbol{W}) = \lambda_i + \gamma G_i(\boldsymbol{W})$. Using gradient descent methods we get the following iteration rule.

$$\Delta \boldsymbol{W} \propto \boldsymbol{W}^{-\mathrm{T}} - \Psi(\boldsymbol{u})\boldsymbol{x}^{\mathrm{T}} \tag{12.15}$$

where

$$\Psi(\boldsymbol{u}) = \begin{pmatrix} -g_1(u_1) - \lambda_1 J'(u_1) \\ -g_2(u_2) + (\lambda_1 - \lambda_2)J'(u_2) \\ \vdots \\ -g_{N-1}(u_{N-1}) + (\lambda_{N-2} - \lambda_{N-1})J'(u_{N-1}) \\ -g_N(u_N) + \lambda_{N-1}J'(u_N) \end{pmatrix} \tag{12.16}$$

$J'(.)$ is the first derivative of $J(.)$ while the selection of $g_i(u_i)$ is the same as in the Infomax algorithm. The iterative formula for the Lagrange multiplier λ_i is given as

$$\lambda_i(k+1) = \max\{0, \lambda_i(k) + \gamma[J(u_{i+1}(k)) - J(u_i(k))]\} \tag{12.17}$$

The algorithm contains a matrix inversion process, which causes the algorithm to have numerical problems. We can incorporate the natural gradient method to improve the algorithm:

$$\Delta \boldsymbol{W} \propto (\boldsymbol{W}^{-\mathrm{T}} - \Psi(\boldsymbol{u})\boldsymbol{x}^{\mathrm{T}})\boldsymbol{W}^{\mathrm{T}}\boldsymbol{W} = (\boldsymbol{J} - \Psi(\boldsymbol{u})\boldsymbol{u}^{\mathrm{T}})\boldsymbol{W} \tag{12.18}$$

Including the projection algorithm, we finally get the gradient algorithm for the optimization problem:

$$\begin{aligned} &\boldsymbol{W} \leftarrow \boldsymbol{W} + \alpha(\boldsymbol{J} - \Psi(\boldsymbol{u})\boldsymbol{u}^{\mathrm{T}})\boldsymbol{W} \\ &\boldsymbol{w}_i \leftarrow \frac{\boldsymbol{w}_i}{\|\boldsymbol{w}_i\|}, \quad i = 1, 2, \dots, N \end{aligned} \tag{12.19}$$

where α is the learning rate, for which we generally use kurtosis as the statistical measure.

$$J_{\mathrm{kur}}(u_i) = \frac{E\{u_i^4\}}{E\{u_i^2\}^2} - 3 \qquad J'_{\mathrm{kur}}(u_i) = \frac{4u_i^3}{E\{u_i^2\}^2} - \frac{4E\{u_i^4\}u_i}{E\{u_i^2\}^3} \tag{12.20}$$

The CICA algorithm based on kurtosis is given below.

Algorithm CICA algorithm based on kurtosis

Input: Albino matrix $\boldsymbol{X} = (\boldsymbol{x}_1, \boldsymbol{x}_2, \dots \boldsymbol{x}_n)$
Output: Independent component $\boldsymbol{Y} = (\boldsymbol{y}_1, \boldsymbol{y}_2, \dots, \boldsymbol{y}_n)$
Steps:

1. Initialize weight vector $\boldsymbol{w}$ (using a unit random matrix)
2. Calculate $\lambda_i(k)$
3. Adjust $\boldsymbol{w}$, repeat steps (2) to (3) until $\boldsymbol{w}$ converges
4. Obtain independent component $\boldsymbol{y} = \boldsymbol{w}^{\mathrm{T}}\boldsymbol{X}$.

12.3 Application and Analysis

12.3.1 CICA of Voice Signals

We selected four speech signals with kurtosis values 1.2106, 0.6827, 2.7844, and 2.0755, respectively, and mixed them with a randomly generated hybrid 4×4 matrix, to obtain mixed signals. Then, we used the CICA algorithm to separate the mixed signals, the experimental results of which are shown in Figure 12.2.

From the experimental results, we can see that the CICA method can separate the mixed speech signals well, while the kurtosis of the separated independent components can be arranged in ascending order.

12.3.2 CICA for Prediction of Mineral Resources

Without any prior knowledge, CICA cannot really restore the corresponding arrangement of independent components; it can only guarantee the consistency of

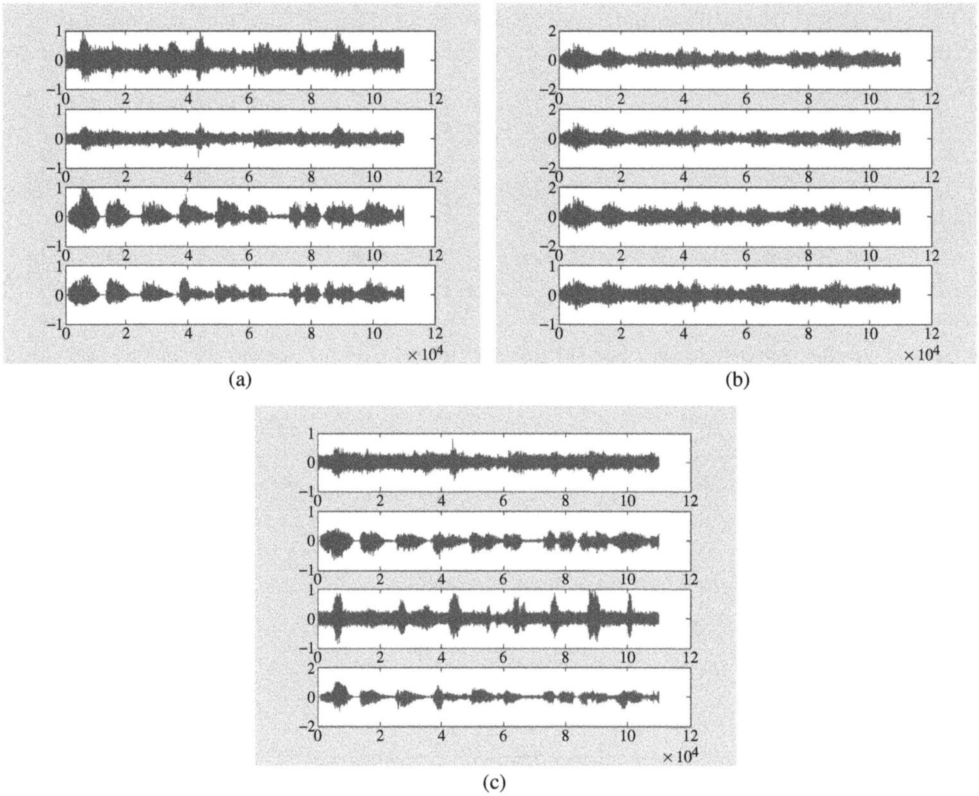

Figure 12.2 Application example: CICA used in the separation of a super-Gaussian mixture of voice signals. (a) Original super-Gaussian speech signals. (b) Linear mixed speech signals. (c) CICA separation results

each experimental result. Therefore, for mineral prediction, it does not have much practical significance (Zhang, Yu, and Lu, 2007).

As previously emphasized, the ICA algorithm has certain limitations, including the fact that the separated signal sequences do not have a one-to-one correspondence with the source signals. Based on the particularity of the ICA model, if there is no prior knowledge, it is impossible to resolve the order uncertainty problem.

This problem has not attracted the attention of many researchers. The reason for this is that although ICA has been widely used in numerous kinds of blind source analysis problems for a long time, in these types of problems the order of the estimated separated data is undefined and immaterial to the solution of the problem. For example, in an application to speech signal separation, we only need to identify the voices of the different people; we do not need to match each person to a voice (in other words, the separation signals may not have a one-to-one correspondence with the original signals).

However, in the field of geochemical data analysis, the sequence of the separated signals is important. Table 12.1 shows some of the data of the gold peripheral 1/50 000 stream sediment in Tuanjiegou in Heilongjiang Province.

Table 12.1 Some of the data for the gold peripheral 1/50 000 stream sediment in Tuanjiegou in Heilongjiang Province

Sample number	x-axis	y-axis	Cu	Pb	Zn	Hg	Bi	As	Sb	Au	Ag	W	Mo
1	97.8	69.3	34	32.5	114.8	11.71	69.4	12.55	40	207	652	1.4	1.81
2	97.9	69.8	22	24.7	89.4	16.3	15.1	5.13	190	10.3	46	1.17	1.41
3	98.4	70.7	31	44.7	85.7	9.4	11.6	5.39	490	1.8	169	0.93	1.11
4	98.7	70.9	27	38.3	72.6	11.1	61.4	8.53	950	1.6	160	1.28	1.95
5	100	70.7	17	17.7	60.8	11.5	45	4.94	180	0.1	165	0.7	1.95
6	98.3	70.1	35	14.8	78.5	9.7	56	11.17	1110	4.9	51	1.17	2.17
7	99.1	70.3	43	31.8	98.5	8.1	29.4	4.93	490	5.3	47	1.17	1.61
8	99.3	70.3	27	11.3	70.6	5.6	29.4	4.52	290	0.6	50	1.17	1.44

In this table, the first column gives the sample number while the second and third columns denote the spatial coordinates of a sample point on the earth's surface (with the x-axis as the abscissa and the y-axis as the ordinate). From the fourth column onwards, each column represents the content value of a metal element at that point. In the experiment, we first obtained the distribution of the source data, and then analyzed it using ICA. After preliminary processing of the signals, we separated the mixture, and used the signals to estimate the evolution and development of the mineral resources in the region. For instance, if we can ascertain ore-caused abnormals in the data, we can indicate the location of the mineral (Figure 12.3).

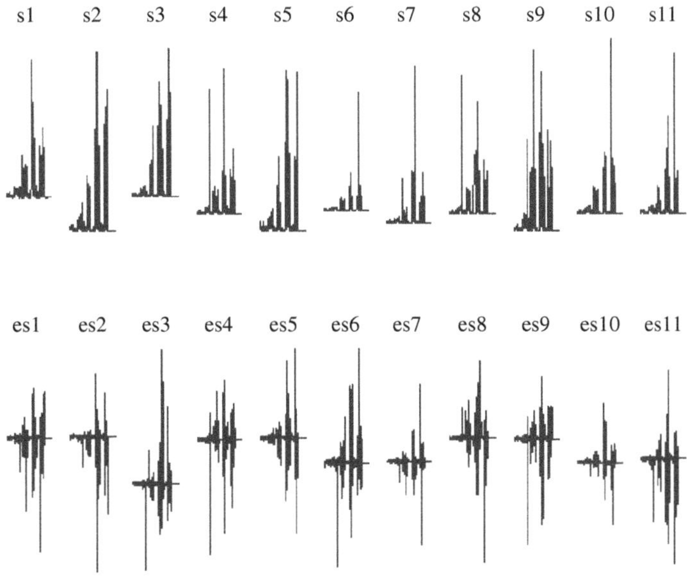

Figure 12.3 Original content and values after ICA processing of the gold peripheral stream sediments in Tuanjiegou in Heilongjiang Province

In Figure 12.3, the first row represents the original content of various elements at the sampling points, while the second row of the image represents the evolution results by ICA estimation. According to the explanation given above about the uncertainty of the order of ICA, we cannot determine whether the mineral represented by es1 corresponds to the mineral represented by s1, since it could represent any of the mineral evolution results from s1 to s11. In this case, it is particularly important to determine the corresponding relation of the 11 minerals; otherwise, evolution and development would be impossible to research.

To solve this problem, we introduce another multivariate statistical analysis method, namely, factor analysis. The basic aim of this method is to use a few random variables to describe the covariance relationships between many variables. Several of these random variables can be unobserved, and are usually referred to as factors. In essence, factor analysis is based on the following idea: divide the variables into groups according to the size of correlation, so that the same group has a high correlation between the variables but lower correlation between different groups of variables. Each group's variable represents a basic structure, or factor, which reflects the correlation that has been observed.

Suppose the mean value of the p-dimensional random vector x is μ and the covariance matrix is Σ. The factor model assumes that x depends on a few unpredictable random variables $f_1, f_2, \ldots, f_m (m < p)$ and p additional variance sources $\varepsilon_1, \varepsilon_2, \ldots, \varepsilon_p$. We usually call $f_1, f_2, \ldots, f_m$ common factors and $\varepsilon_1, \varepsilon_2, \ldots, \varepsilon_p$ special factors or errors. Then, the matrix expression of the orthogonal factor model is

$$\boldsymbol{X} = \mu + \boldsymbol{L}f + \varepsilon \tag{12.21}$$

where $\boldsymbol{L} = (l_{ij})_{p\times m}$, with l_{ij} the load of factor j for variable i.

To establish a factor model, first we need to estimate the factor load matrix and special variance, for which the main parameter estimation methods commonly used are PCA, the principal factor method, and the maximum likelihood method. After we have obtained the load matrix of the factor model, we also need to give a reasonable explanation for the public factor; this interpretation requires certain expertise and experience, as we need to give a name with a practical meaning to each factor, which can be used to reflect the importance of the common factor when each observable original variable is predicted; this is equivalent to the load of this factor. The interpretation of the factor often includes some subjectivity, which can be reduced by rotating the factor. After rotation, the absolute value of the elements in each column of the load matrix spreads the size distance as far as possible; in other words, we make some elements as close to 0 as possible, and other elements close to ± 1. In general, we use the greatest variance rotation method.

If all public factors are uncorrelated, factor load l_{ij} is the correlation coefficient between the i-th original variable and the j-th public factor, and therefore, the greater the absolute value l_{ij}, the stronger the relationship between the public factor f_j and the original variable x_i.

In practical applications, the known mixed observed signals are not a simple mix of several independent signal sources. These independent signals are influenced or disturbed by other signals, so that the observed signals are often mixed, but these signals are the main influential factors in the corresponding mixed component. For example, in geochemical data related to mineral deposits, concomitant element conditions often exist. To restore the source signals corresponding to the independent components, we regard the mixed signals and independent components as the observation data and apply factor analysis. After obtaining the public factors of the number of source signals, we analyze the factor load matrix. The two largest components of the same public factor load can be thought of as having a corresponding relationship with each other.

We used geochemical experimental data of an unspecified location in Inner Mongolia, containing 10 855 sample points and 18 different elements (Ag, As, Au, Bi, Cd, Co, Cr, Cu, Fe_2O_3, Hg, Mn, Mo, Ni, Pb, Sb, Sn, W, and Zn). The area is also rich in polymetallic minerals. In fact, we were mainly concerned with five elements, namely, Ag, Au, Cu, Pb, and Zn, as well as polymetallic ores.

Our aim was to obtain the correlation between the independent components and their linear combinations, in order to predict the areas in which minerals could possibly occur. Figure 12.4 shows an isogram of the raw data, while Figure 12.5 shows the distribution of minerals that were identified. In this experiment, we used the former diagram as the background for ease of comparison of the results.

To display the results more clearly, we added the background to Figure 12.6. The red circles denote that the raw data for the region are consistent with the actual data,

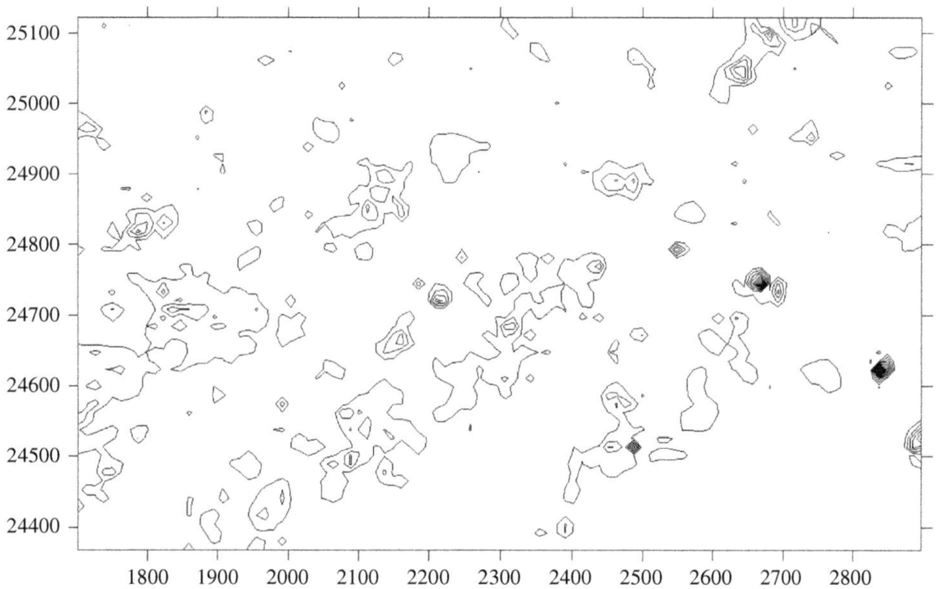

Figure 12.4 Geochemical element data of a location in Inner Mongolia

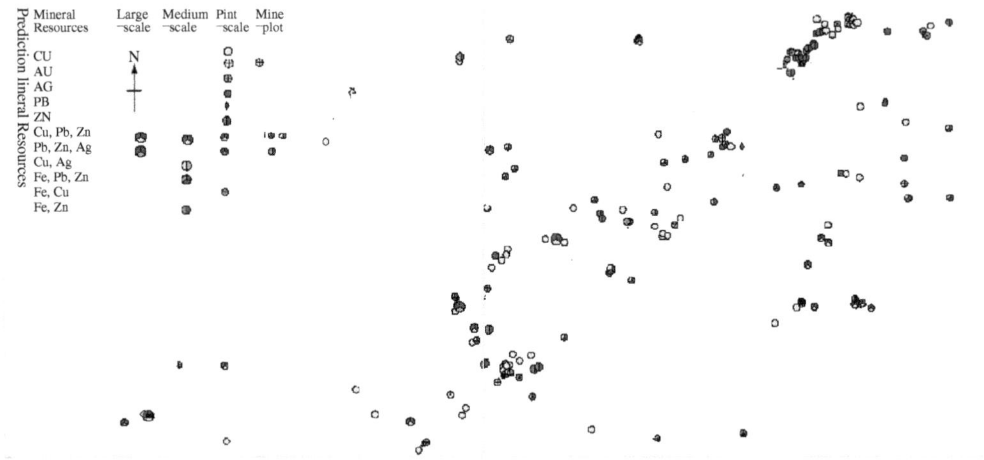

Figure 12.5 Spatial distribution of polymetallic ore at a location in Inner Mongolia

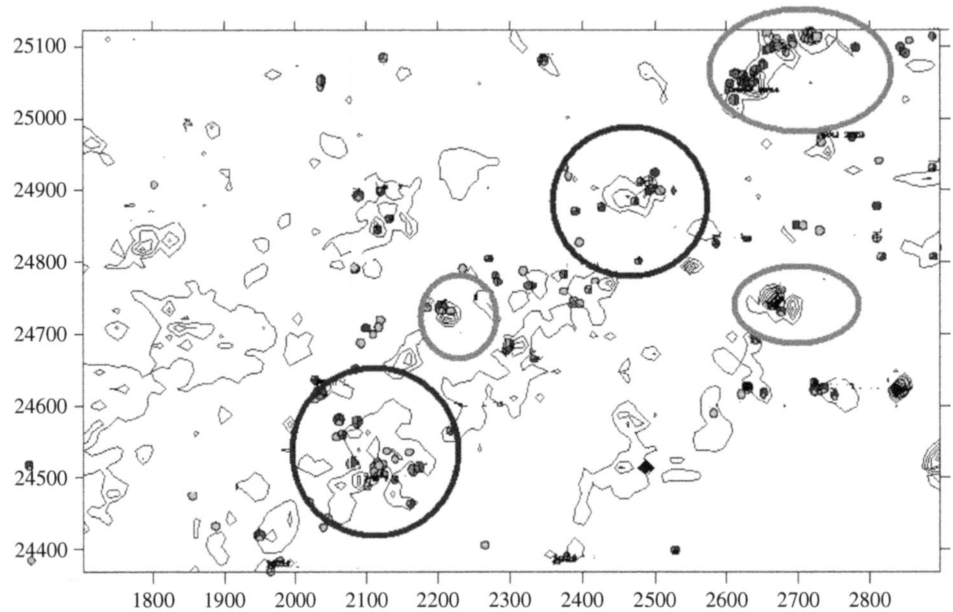

Figure 12.6 Geochemical raw data graph with known mines included

while the blue circles denote that the raw data for the region are inconsistent with the actual data.

We used the CICA algorithm to calculate the raw data, which had been preprocessed. The ICA results with the same background are shown in Figure 12.7.

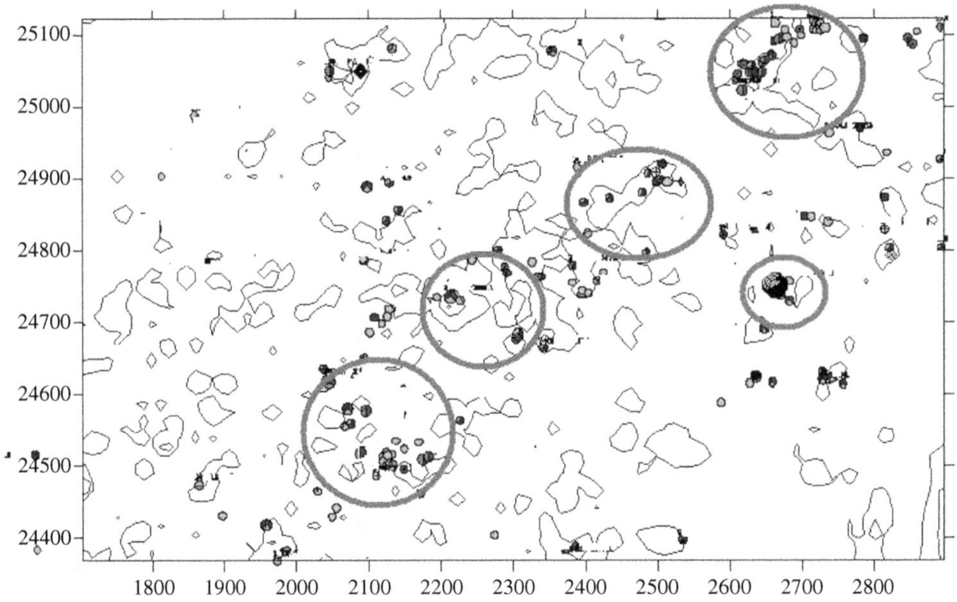

Figure 12.7 Independent components of polymetallic ore with known mines included

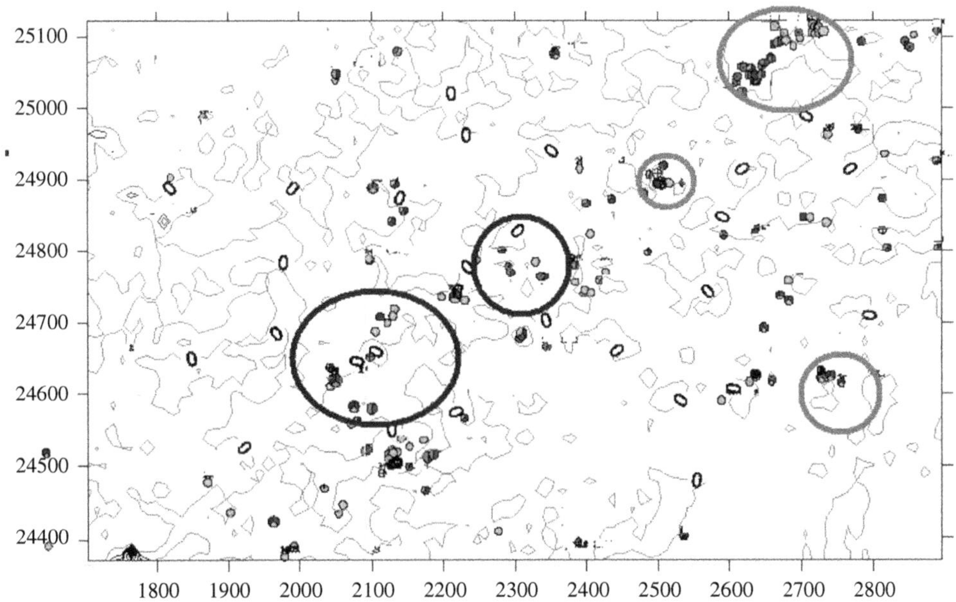

Figure 12.8 PCA processing results

A comparison of Figures 12.6 and 12.7 shows that, after removing specific values, the results are closer to the actual situation in the mining area. Compared with the results of PCA in Figure 12.8, it is obvious that the CICA results coincide better with the actual regional environment.

12.4 Chapter Summary

The CICA algorithm was proposed to solve the connatural problems of ICA, such as the instability of the amplitude and the nondeterminacy of the order. In this algorithm we use kurtosis as a statistical measure to sort the independent components. By incorporating a projection method to normalize the independent component amplitude to 1, the uncertainty of the ICA algorithm can be eliminated. When applied to mineral prediction, the results of this algorithm more closely match the actual situation. Thus, CICA has very important practical significance in many situations where the order of the variables must be considered.

References

Choi, P.S. (2006) Differential learning algorithms for decorrelation and independent component analysis. *Neural Networks*, **19**(10), 1558–1567.

Hyvärinen, A. (1998) New approximations of differential entropy for independent component analysis and projection pursuit. *Advances Neural Information Processing System*, **10**(NIPS*97), 273–279.

Hyvärinen, A. (1999) Fast and robust fixed-point algorithms for independent component analysis. *IEEE Transactions on Neural Networks*, **10**(3), 626–634.

Hyvärinen, A., Hoyer, P. and Oja, E. (1998) Sparse code shrinkage for image denoising. *Neural Networks Proceedings on IEEE World Congress on Computational Intelligence*, Vol. **2**, IEEE Press, pp. 859–864.

Hyvärinen, A., Karhunen, J. Oja, E. (2001) *Independent Component Analysis*, John Wiley & Sons, Inc.

Hyvärinen, A., and Oja, E. (1997) A fast fixed-point algorithm for independent component analysis. *Neural Computation*, **9**(7), 1483–1492.

Hyvärinen, A., Oja, E. (2000) Independent component analysis: algorithms and applications. *Neural Networks*, **13**(4–5), 411–430.

Liu, S., Huang, Z., Yu, X., H., Leng, (2007) An improved mineral prediction method based on independent component analysis. *Journal of Mineral*, **27**(z1), 450–451, (in Chinese).

Lu, W., Rajapakse, J.C. (2001) Constrained independent component analysis. *Advances in Neural Information Processing Systems*, **13**(NIPS2000), 570–576.

Lu, W., Rajapakse, J.C. (2005) Approach and application of constrained ICA. *IEEE Transactions on Neural Networks*, **16**(1), 203–212.

Oja, E. (1997) The nonlinear PCA learning rule in independent component analysis. *Neurocomputing*, **17**(1), 25–45.

Rasheed, T., Kim, T.S., and Lee, Y.K. *et al.* (2006) Constrained ICA–based ballistocardiogram and electro-oculogram artifacts removal from EEG signals acquired inside MRI. International Society for Magnetic Resonance in Medicine, Seattle, WA, pp. 255–263.

Shi, X. (2008) *Blind Signal Processing-Theory and Practice*. Shanghai Jiaotong University Press, ShangHai (in Chinese).

Woo, W.L., Sali, S. (2002) General multilayer perceptron demixer scheme for nonlinear blind signal separation. *IEEE Proceedings-Vision Image and Signal Processing*, **149**(5), 253–262.

Wu, S., Qi, H., Liu, G., Yang, X. (2006) Online palmprint recognition base. *Computer Engineering and Design*, **27**(18), 3433–3435, (in Chinese).

Yu, X., Liu, S., Ren, J., and Zhang, T. (2007) Robust fast independent component analysis applied to mineral resources prediction. Proceedings of the IAMG'07: Geomathematics and GIS Analysis of Resources, Environment and Hazards, Beijing, China, pp. 94–97.

Zhang, T., Yu, X.C., and Liu, L.W. (2007) Constrained fast independent component analysis applied to mineral resources prediction. Proceedings of the IAMG'07: Geomathematics and GIS Analysis of Resources, Environment and Hazards, Beijing, China, pp. 535–540.

13

Optimized Independent Component Analysis Algorithms and Applications

13.1 Overview

A typical ICA model usually assumes the same or a similar distribution of the independent signal sources. For example, in the separation process of the traditional ICA algorithm we refer to second- or fourth-order, which is an implicit assumption that the data source satisfies a highly symmetric distribution. However, this assumption is not usually satisfied in actual data analysis. If the original signal does not satisfy the implicit assumption of a symmetric distribution, an ICA algorithm based on fourth-order statistics can realize an ideal separation result. Since no traditional ICA method exists that can be applied to source signals with an asymmetric distribution (Stone *et al.*, 2002), we propose a method based on density estimation of the source signals to optimize the separation results of such signal sources using traditional ICA.

The ICA approach is divided into two stages, namely, criterion optimization and algorithm optimization. The basic independence criterion should be introduced in the probability density function. However, since the probability density function is often unknown, it is difficult to estimate this. Thus, the target function is taken as the optimized criterion function. Selection criteria for choosing a suitable target function are given below:

1. It must be easy for the algorithm to deal with any kind of question.
2. In practical applications, (such as application to functional magnetic resonance imaging (fMRI) data), the selected target function should be a best fit to the data.
3. The differences that exist between various functions (or families of functions) in practical applications should be considered.

Blind Source Separation: Theory and Applications, First Edition. Xianchuan Yu, Dan Hu and Jindong Xu.

Since the traditional ICA algorithm is not able to solve problems involving asymmetric source signals very well, Stone *et al.* (2002) introduced the following function

$$f'(x; a, b) \propto \exp\left(\frac{a-b}{2}x - \frac{a+b}{2}\sqrt{x^2+1}\right) \tag{13.1}$$

As an optimal nonlinear function, it successfully solves the class specifically targeted at separating an asymmetric source. It can be seen that the asymptote of the function is

$$\lim_{x \to -\infty} f' = \exp(ax) \text{ and } \lim_{x \to +\infty} f' = \exp(-bx) \tag{13.2}$$

In addition, the exponential decay rate ensures that its robustness is the same as that for the simpler hyperbolic tangent probability density function

$$\text{sech}^2 x = \frac{\mathrm{d} \tanh}{\mathrm{d}x} \tag{13.3}$$

This method has been proven to improve the accuracy of the separation results in some specific cases. However, by using only a fixed non-linear function, it still cannot achieve optimal separation of a variety of signal source distributions.

Taking into account different regions of interest in a brain diagram associated with different local properties, even for a single test, various composite signal sources (such as a signal source related to a task, a physiological signal source, and so on) are likely to produce different distributions. So having only a fixed nonlinear function is usually not enough.

Stone *et al.* presented a new way of thinking. Based on their initial study, models with an asymmetric distribution must satisfy the following conditions:

1. The distribution should be non-symmetrical (skew).
2. By changing the relevant parameters, a large number of different shapes can be determined for the model.
3. The weighted iterative and deviation modulation equations should have a relatively good shape.

13.2 Optimized ICA Algorithm

Taking the output of the conventional ICA algorithm as the new source signal, we can estimate the parameters of the probability density function of the signal source based on their statistical characteristics. Furthermore, we can find the density function with the best fit for each channel of the signal source. Thereafter, the next task is to determine the weighted iterative and deviation modulation equations of the density function in the ICA algorithm. The second ICA estimation is then conducted, which results in a more accurate approximation of the signal source than the first ICA separation. This realizes an optimized ICA algorithm.

13.2.1 *The Weighted Iterative Algorithm for Optimized ICA*

To solve the target function, the ICA weighted iterative algorithm based on extreme value distributions is obtained as follows:

$$\boldsymbol{y} = D(\boldsymbol{u}) = \exp\left(-\exp\left(\frac{a-\boldsymbol{u}}{b}\right)\right) \tag{13.4}$$

$$\frac{\partial \boldsymbol{y}}{\partial \boldsymbol{x}} = \frac{\partial}{\partial \mathbf{x}}\left(\exp\left(-\exp\left(\frac{a-\boldsymbol{u}}{b}\right)\right)\right)\frac{\partial \boldsymbol{u}}{\partial \boldsymbol{x}} = \frac{\boldsymbol{W}}{b}\exp\left(\frac{a-\boldsymbol{u}}{b} - \exp\left(\frac{a-\boldsymbol{u}}{b}\right)\right) \tag{13.5}$$

$$\frac{\partial}{\partial \boldsymbol{W}}\left(\frac{\partial \boldsymbol{y}}{\partial \boldsymbol{x}}\right) = \frac{1}{b}\exp\left(\frac{a-\boldsymbol{u}}{b} - \exp\left(\frac{a-\boldsymbol{u}}{b}\right)\right) \times \left(\boldsymbol{W}\frac{\boldsymbol{x}}{b}\exp\left(\frac{a-\boldsymbol{u}}{b} - 1\right)\right) \tag{13.6}$$

By substituting the weighted iterative equation of the ICA algorithm in Equation 13.7, we get Equation 13.8.

$$\Delta \boldsymbol{W} \propto \frac{\partial H(\boldsymbol{y})}{\partial \boldsymbol{W}} = \left(\frac{\partial \boldsymbol{y}}{\partial \boldsymbol{x}}\right)^{-1}\frac{\partial}{\partial \boldsymbol{W}}\left(\frac{\partial \boldsymbol{y}}{\partial \boldsymbol{x}}\right) \tag{13.7}$$

$$\frac{\partial H(\boldsymbol{y})}{\partial \boldsymbol{W}} = \left(\frac{\partial \boldsymbol{y}}{\partial \boldsymbol{x}}\right)^{-1}\frac{\partial}{\partial \boldsymbol{W}}\left(\frac{\partial \boldsymbol{y}}{\partial \boldsymbol{x}}\right) = \boldsymbol{W}^{-1} + \frac{\boldsymbol{x}}{b}\left[\exp\left(\frac{a-\boldsymbol{u}}{b}\right) - 1\right] \tag{13.8}$$

So, the weighted iterative value is

$$\Delta \boldsymbol{W} \propto [\boldsymbol{W}^{\mathrm{T}}]^{-1} + \frac{1}{b}\left[\exp\left(\frac{a-\boldsymbol{u}}{b}\right) - 1\right]\boldsymbol{x}^{\mathrm{T}} \tag{13.9}$$

Accordingly, we can obtain the deviation modulated value as

$$\Delta \boldsymbol{w}_0 \propto \frac{\mathrm{e}^{(a-u)/b} - 1}{b} \tag{13.10}$$

Similarly, corresponding ICA weighted iterative algorithms for three of the most commonly used symmetric distribution models, namely the Gaussian, Laplace, and logistic functions, are given. For example, in a Laplace distribution, without loss of generality, we suppose that $\mu = 0, \sigma = 1$ to obtain the nonlinear function

$$\boldsymbol{y} = F(\boldsymbol{u}) = \begin{cases} 1 - \frac{1}{2}\exp(-\boldsymbol{u}), \mathbf{u} \geq 0 \\ \frac{1}{2}\exp(\boldsymbol{u}), \ \boldsymbol{u} < 0 \end{cases} \tag{13.11}$$

For $\boldsymbol{u} \geq 0$, we have

$$\frac{\partial \boldsymbol{y}}{\partial \boldsymbol{x}} = \frac{\partial}{\partial \boldsymbol{x}}\left(1 - \frac{1}{2}\exp(-\boldsymbol{u})\right)\frac{\partial \boldsymbol{u}}{\partial \boldsymbol{x}} = \frac{\boldsymbol{W}}{2}\exp(-\boldsymbol{u}) \tag{13.12}$$

$$\frac{\partial}{\partial \boldsymbol{W}}\left(\frac{\partial \boldsymbol{y}}{\partial \boldsymbol{x}}\right) = \frac{\exp(-\boldsymbol{u})}{2}(1 - \boldsymbol{W}\boldsymbol{x}) \tag{13.13}$$

$$\frac{\partial H(\boldsymbol{y})}{\partial \boldsymbol{W}} = \boldsymbol{W}^{-1} - \boldsymbol{x} \tag{13.14}$$

For $\boldsymbol{u} < 0$, we have

$$\frac{\partial y}{\partial \boldsymbol{x}} = \frac{\boldsymbol{W}}{2}\exp(\boldsymbol{u}) \tag{13.15}$$

$$\frac{\partial H(\boldsymbol{y})}{\partial \boldsymbol{W}} = \boldsymbol{W}^{-1} + \boldsymbol{x} \tag{13.16}$$

We obtain the ICA iterative algorithm for estimating weights as

$$\Delta \boldsymbol{W} \propto [\boldsymbol{W}^{\mathrm{T}}]^{-1} - (\mathrm{sgn}(\boldsymbol{u}))\boldsymbol{x}^{\mathrm{T}} \tag{13.17}$$

$$\Delta \boldsymbol{w}_0 \propto -\mathrm{sgn}(\boldsymbol{u}) \tag{13.18}$$

Similarly, the cumulative distribution function (cdf) of the Gaussian distribution and the iterative equation for estimating the weights are given as

$$\Delta \boldsymbol{W} \propto [\boldsymbol{W}^{T}]^{-1} - \boldsymbol{u}\boldsymbol{x}^{\mathrm{T}} \tag{13.19}$$

$$\Delta \boldsymbol{w}_0 \propto -\boldsymbol{u} \tag{13.20}$$

In addition, the weighted iterative equation for the logistic function is

$$\Delta \boldsymbol{W} \propto [\boldsymbol{W}^{\mathrm{T}}]^{-1} + (1 - 2\boldsymbol{y})\boldsymbol{x}^{\mathrm{T}} \tag{13.21}$$

$$\Delta \boldsymbol{w}_0 \propto 1 - 2\boldsymbol{y} \tag{13.22}$$

where $\boldsymbol{y} = (1 + \mathrm{e}^{-u})^{-1}$.

To avoid solving the inversion of a matrix, reduce the computational cost, and speed up convergence, the natural gradient method is applied to simplify the iterative equation for the model weights given above, and calculate the deviation modulation equation. The resulting equations are given in Table 13.1.

In the optimized ICA algorithm, the standard Infomax algorithm is first applied to all information channels. After convergence, we take the estimate $\boldsymbol{u}$ of the ICA signal as the new source. Finally, based on different statistical characteristics of the source signals, we adopt the density evaluation skill to estimate the various information channel probability densities. We adaptively choose the best fitting nonlinear function, and calculate parameters a and b corresponding to random variables that satisfy the skew distributions.

Unlike the traditional Infomax algorithm, optimized ICA utilizes a non-deterministic nonlinear function to update weights. So it is possible to adaptively

Table 13.1 Weighted iterative equation and deviation modulation equation for different models

Model	Weighted iterative equation	Deviation modulation equation
Extremum	$\boldsymbol{W}(n+1) = \boldsymbol{W}(n) + \mu\left(I + \frac{\mathrm{e}^{(a-u)/b}-1}{b}\boldsymbol{u}^{\mathrm{T}}\right)\boldsymbol{W}(n)$	$\boldsymbol{w}_0(n+1) = \boldsymbol{w}_0(n) + \mu\sum\left(\frac{\mathrm{e}^{(a-u)/b}-1}{b}\right)^{\mathrm{T}}$
Gaussian	$\boldsymbol{W}(n+1) = \boldsymbol{W}(n) + \mu(I - \boldsymbol{u}\boldsymbol{u}^{\mathrm{T}})\boldsymbol{W}(n)$	$\boldsymbol{w}_0(n+1) = \boldsymbol{w}_0(n) + \mu(-\boldsymbol{u})^{\mathrm{T}}$
Laplace	$\boldsymbol{W}(n+1) = \boldsymbol{W}(n) + \mu(I - \mathrm{sgn}(\boldsymbol{u})\boldsymbol{u}^{\mathrm{T}})\boldsymbol{W}(n)$	$\boldsymbol{w}_0(n+1) = \boldsymbol{w}_0(n) + \mu[-\mathrm{sgn}(\boldsymbol{u})]^{\mathrm{T}}$
Logistic	$\boldsymbol{W}(n+1) = \boldsymbol{W}(n) + \mu[\boldsymbol{I} + (1 - 2\boldsymbol{y})\boldsymbol{u}^{\mathrm{T}}]\boldsymbol{W}(n)$	$\boldsymbol{w}_0(n+1) = \boldsymbol{w}_0(n) + \mu(1 - 2\boldsymbol{y})^{\mathrm{T}}$

select an optimum nonlinear function based on the different channel conditions, and ultimately find an optimal weight matrix $\boldsymbol{W}$.

13.2.2 Optimized ICA Weighted Iterative Algorithm

The central idea of the optimized ICA algorithm is given below as separation steps in the following three stages:

1. Apply the standard Infomax ICA to evaluate the initial statistically independent source signals.
2. Adopt the nuclear estimates to determine the probability density of these source signals.
3. Finally, obtain the "best" nonlinear function based on the source density in the separation stage.

The basic framework for the optimized ICA algorithm is given below.

Algorithm Optimized ICA weighted iterative algorithm

Input: Mixed signal source $\boldsymbol{X}$
Output: Independent component $\boldsymbol{Y}$
Steps:

1. Preprocess data (e.g., whitening).
2. Apply standard ICA algorithm (e.g., Infomax algorithm) to update $\Delta \boldsymbol{W}$. If the weight is less than a threshold, stop the training.
3. Take the separation result from the second step as the new signal source, the density function of which must be estimated:
   a. Perform kernel density estimation on the new signal source $\boldsymbol{u} = \boldsymbol{W}\boldsymbol{x}$.
   b. Determine the nonlinear function that achieves optimal matching according to the minimized Kullback–Leibler divergence between the probability density function of the raw data and the estimated density function.
4. According to the selected matching nonlinear function, update the corresponding channel again until convergence is achieved.
5. If convergence is not reached, go back to step (3).

An overview of the optimized ICA model framework is illustrated in Figure 13.1.

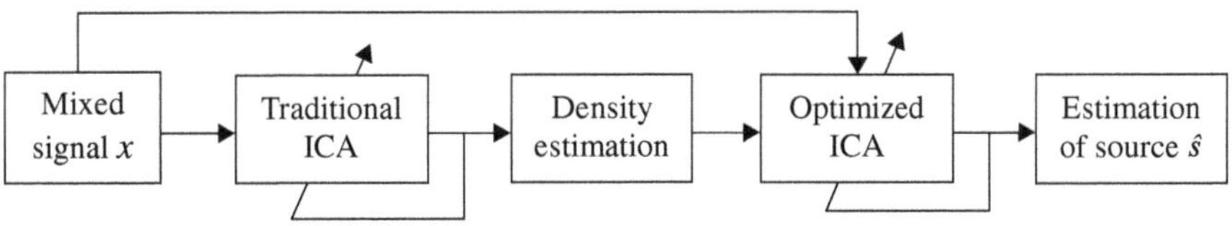

Figure 13.1 Optimized ICA algorithm model framework

13.2.3 Choice of Number of Independent Components

The minimum description length (MDL) (Fishler and Messer, 1999, 2000; Li, Cichocki, and Amari, 2003) criterion is based on the source estimation criterion for eigenvalues. The MDL criterion from the given model library finds the shortest encoding length and the best model to describe the data, and works as follows:

Assuming that P geometrically structured arbitrary arrays are obtained from $Q(\leq P-1)$ narrow-band signals from different directions $\theta_1, \theta_2, \cdots \theta_Q$. Then the complex envelope of the received signal of the array is given as

$$x(t_i) = \sum_{k=1}^{Q} \boldsymbol{\alpha}(\boldsymbol{\theta}_k) s_k(t_i) + n(t_i) = A(\boldsymbol{\theta}) s(t_i) + n(t_i) \; i = 1, 2, \cdots, M \tag{13.23}$$

where the $x(t_i)$ are the received observation signals at moment t_i; $\boldsymbol{\alpha}(\boldsymbol{\theta}_k)$ is the response vector (derivative vector) in direction $\boldsymbol{\theta}$ for the array; $\boldsymbol{A}(\boldsymbol{\theta}) = [\boldsymbol{\alpha}(\boldsymbol{\theta}_1), \boldsymbol{\alpha}(\boldsymbol{\theta}_2), \cdots \boldsymbol{\alpha}(\boldsymbol{\theta}_Q)]$ is a Vandermonde column full rank matrix; and $s_k(t_i)$ is the complex envelope of the kth signal, $\boldsymbol{S}(t_i) = [s_1(t_i), s_2(t_i), \cdots s_3(t_i)]^{\mathrm{T}}$. Here, $n(t_i)$ is an independent and identically distributed Gaussian noise vector with zero mean and $\sigma^2 \boldsymbol{I}$ is the covariance matrix. M is the sampling length. The covariance matrix of $\boldsymbol{x}(\boldsymbol{t})$ is given as

$$\boldsymbol{R} = \boldsymbol{E}[\boldsymbol{x}(\boldsymbol{t})\boldsymbol{x}^{\mathrm{H}}(\boldsymbol{t})] = \boldsymbol{A}(\boldsymbol{\theta})\boldsymbol{R}_s\boldsymbol{A}^{H}(\boldsymbol{\theta}) + \boldsymbol{\sigma}^2\boldsymbol{I} \tag{13.24}$$

where $\boldsymbol{H}$ is the conjugate transpose matrix and $\boldsymbol{I}$ is a $P \times P$ unit matrix.

The eigenvalues are sorted in descending order from large to small:

$$\lambda_1 \geq \lambda_2 \geq \cdots \lambda_Q \geq \lambda_{Q+1} = \lambda_{Q+2} = \cdots = \lambda_P = \sigma^2 \tag{13.25}$$

If the length of the observation data is infinite, the number of repetitions of the smallest eigenvalue of $\boldsymbol{R}$ is directly adopted to determine the number of sources. In practical applications, it is possible to obtain the maximum likelihood estimation from $\boldsymbol{R}$:

$$\widehat{\boldsymbol{R}} = \frac{1}{M} \sum_{i=1}^{M} \boldsymbol{x}(t_i)\boldsymbol{x}^{\mathrm{H}}(t_i) \tag{13.26}$$

Owing to the existence of noise, Equation 13.25 can be replaced by the following equation:

$$\lambda_1 \geq \lambda_2 \geq \cdots \lambda_Q \geq \lambda_{Q+1} \geq \lambda_{Q+2} \geq \cdots \geq \lambda_P \tag{13.27}$$

The P eigenvalue probabilities are different. The question then arises how we can utilize the P eigenvalues of the "sampling covariance matrix" from the output of the array to correctly estimate the number of sources Q in the array based on limited sampling points. Assuming that there are k signal sources reaching the array, a source number estimation method based on the MDL criterion is needed to minimize the following log-likelihood function:

$$LLF(k) = M(P-k)\log\left[\frac{\frac{1}{(P-k)}\sum_{i=k+1}^{P}\lambda_i}{\left(\prod_{i=k+1}^{P}\lambda_i\right)^{\frac{1}{P-k}}}\right] \tag{13.28}$$

Recorded as

$$P_{\mathrm{MDL}}(M,P,k) = \frac{1}{2}k(2P-k)\log M \tag{13.29}$$

the MDL criterion used in the determination is:

$$k_{\mathrm{MDL}} = \arg\min\{LLF(k) + P_{\mathrm{MDL}}(M,P,k)\} \tag{13.30}$$

When the MDL measure is at its minimum, $k = K$ is the estimate of the number of sources. The MDL measure function is monotonically decreasing before the arrival of K signals, and is monotonically increasing after K. If $M \to \infty$, $k_{\mathrm{MDL}} = Q$.

13.3 Application and Analysis

13.3.1 Separation Simulation for an Artificial Mixed Signal

Through simulation experiments, we investigated the algorithm performance in separating artificially mixed signal sources (including symmetric and asymmetric/skew signal sources, as well as sub-Gaussian and super-Gaussian signal sources). Six signal sources (Table 13.2) were used in the experiments, with 2000 sample points per signal source. Together these signal sources include Gaussian, Laplace, hyperbolic secant, Rayleigh, Gamma, and exponential distributions. A full rank matrix is randomly generated to mix the signal sources. To ensure the reliability of the experiments, we performed a Monte Carlo experiment 100 times. The experimental runtime environment was set up with the Microsoft Windows XP Professional SP2 operating system running on an 1.8 GHz AMD Athlon 64 Processor 3000+ CPU with 1 GB memory, while programming was done in MATLAB® 7.0.2.

Table 13.2 High-order accumulated values of the simulated data

High-order accumulated value	Rectangular wave	Attenuation wave	Pulse wave	Exponential wave	Gamma wave	Rayleigh wave
Kurtosis	−1.9994	−0.41523	0.40289	1.782	2.4775	6.8687
Gaussianity	Sub-Gaussian	Gaussian	Gaussian	Super-Gaussian	Super-Gaussian	Super-Gaussian
Skewness	−0.024	0.048515	1.2727	1.1992	−0.01389	2.2749
Symmetry	Symmetric	Symmetric	Asymmetric	Asymmetric	Symmetric	Asymmetric

For comparison with the experimental results of the optimized ICA algorithm, traditional ICA algorithms were applied to the simulation data, including the Infomax algorithm (Bell and Sejnowski, 1995), the extended Infomax algorithm (Lee, Girolami, and Sejnowski, 1999), the FastICA (fixed-point) algorithm (Hyvärinen and Oja, 1997; Hyvärinen, 1999a, 1999b), and the JADE(joint approximate diagonalization of eigen-matrices) algorithm. The source signal is shown in Figure 13.2, the mixed signal in Figure 13.3, and the separation results of the optimized ICA and various traditional ICA algorithms in Figures 13.4 and 13.5, respectively, with 500 points in each example. In Figures 13.2 to 13.5, the horizontal and vertical axes of the left figure denote the sampling moment and the signal amplitude, while the corresponding axes of the right figure denote the signal amplitude and the sample number corresponding to the amplitude, respectively.

As can be seen in Figure 13.5, optimized ICA has a significant advantage in separating data with a skew distribution, while the results of traditional ICA are not very satisfactory. These figures provide rough intuitive displays for an initial evaluation of the results of optimized ICA; below we provide more accurate indicators describing the performance of the algorithms.

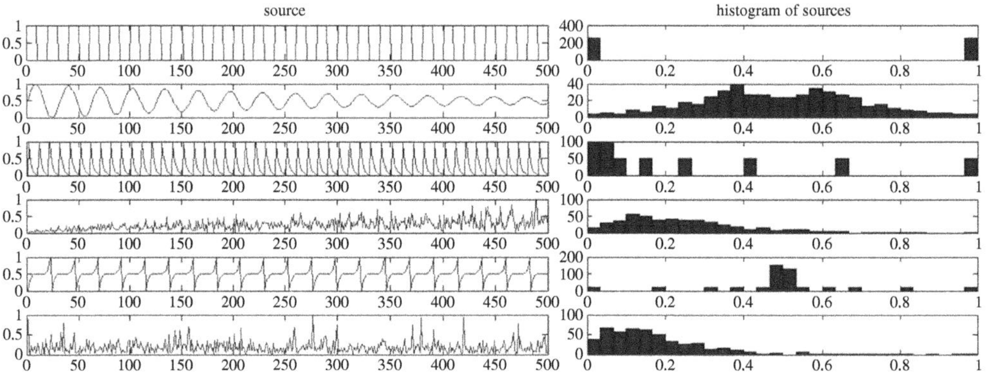

Figure 13.2 Signal source and its corresponding histogram

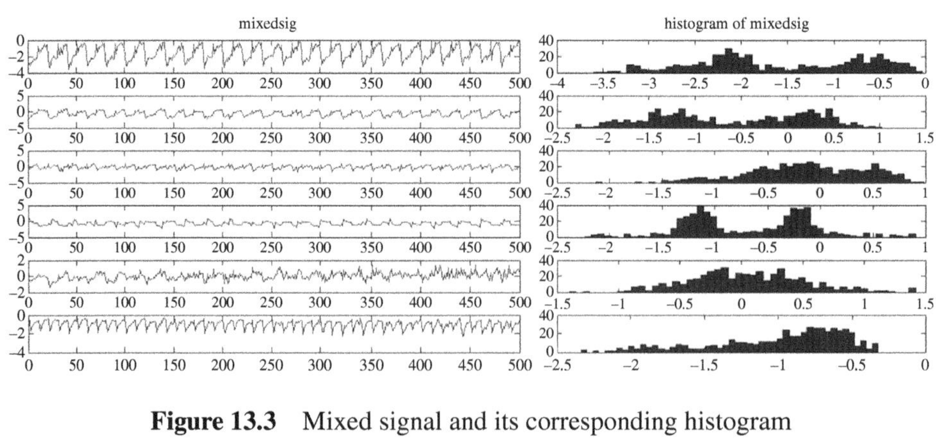

Figure 13.3 Mixed signal and its corresponding histogram

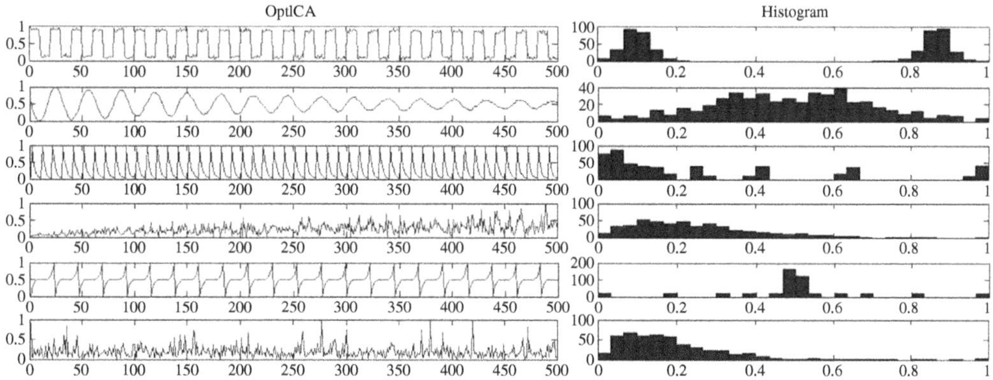

Figure 13.4 Optimized ICA and its corresponding histogram

$\boldsymbol{G} = \boldsymbol{WA}$ is called the global system matrix. If $\boldsymbol{G} = \boldsymbol{I}$ ($\boldsymbol{I}$ is an $n \times n$ unit matrix) after learning, the goal of separating (or restoring) the source signal has been achieved. In fact, as long as only one element in each row and column of $\boldsymbol{G}$ is close to 1 and the other elements are close to zero, we can consider that the signal has been separated successfully.

$$\boldsymbol{G}_{\text{optica}} = \begin{bmatrix} \mathbf{1.04400} & 0.00087 & -0.03135 & 0.00426 & 0.00171 & -0.0095 \\ 0.00164 & \mathbf{1.03100} & -0.00513 & 0.0007 & 0.00028 & -0.00156 \\ -0.00157 & -0.00014 & \mathbf{1.03300} & -0.00067 & -0.00027 & 0.0015 \\ 0.0278 & 0.00242 & -0.0869 & \mathbf{1.06300} & 0.00474 & -0.02637 \\ 0.00239 & 0.00021 & -0.00746 & 0.001 & \mathbf{1.08400} & -0.00226 \\ 0.00434 & 0.00038 & -0.0135 & 0.00184 & 0.00074 & \mathbf{1.12200} \end{bmatrix} \tag{13.31}$$

After performing normalization, according to the signal-to-noise ratio (SNR) equation

$$SNR = 10\lg \frac{E(\mathbf{s}(t)^2)}{E\{[\mathbf{s}(t) - \widehat{\mathbf{s}}(t)^2]^2\}} \tag{13.32}$$

we calculate the mixed-signal observations and SNR between the results of the ICA algorithm and the original signal source. A comparison of the separation performance of the optimized ICA and traditional ICA algorithms using an artificial mixed-signal source is given in Table 13.3.

From the analysis results shown in Figure 13.6 (based on performing 500 Monte Carlo simulations), it can be concluded that for separating an asymmetric source signal, optimized ICA has a significantly better SNR than the traditional ICA algorithms. The SNR of a separated pulse signal source is improved by 5 dB, that of a separated exponential signal source by between 5 and 30 dB, and that of a separated Rayleigh signal source by between 10 and 30 dB. In particular, as the number of sampling points increases, the advantage becomes more obvious.

For a symmetrical signal source, the optimized ICA algorithm has only a slight advantage, with an increased SNR of between 1 and 2 dB; this is notable for the

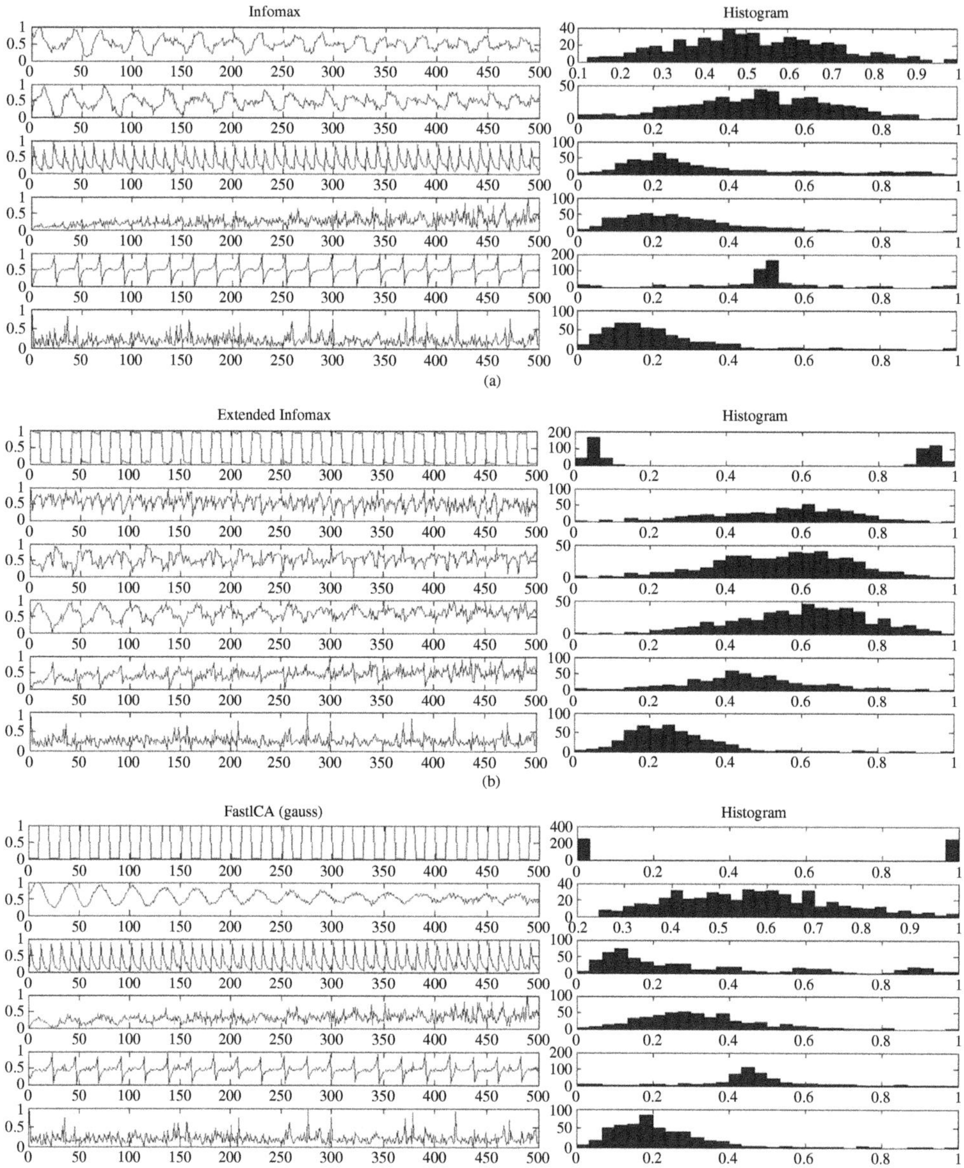

Figure 13.5 Separation results of the traditional ICA algorithms and their corresponding histograms (a) Separation results of Infomax ICA and its corresponding histogram (b) Separation results of extended Infomax ICA and its corresponding histogram (c) Separation results of FastICA (Gauss) and its corresponding histogram (d) Separation results of FastICA (tanh) and its corresponding histogram (e) Separate results of JADE and its corresponding histogram

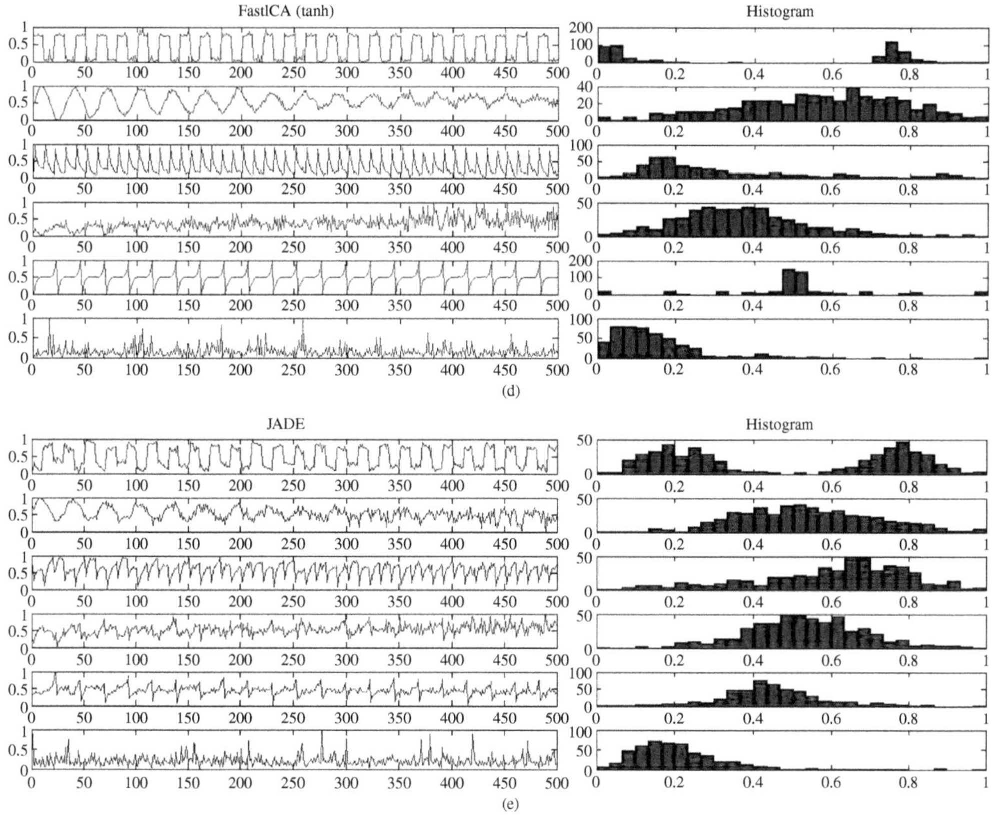

Figure 13.5 *Continued*

Gamma wave signal source, where the separation performance is almost the same as that of the traditional methods. Both the optimized ICA algorithm and traditional ICA algorithms for the Gamma wave signal source use a logistic function as the nonlinear fitting function, which results in similar performance.

The experimental results of the simulation data show that our proposed optimized ICA algorithm based on density estimation can be applied adaptively to any source signal, including sub- and super-Gaussian, and symmetric and asymmetric source signals. In some sense, this satisfies the actual full "blind" requirement of the probability density of the source signal.

13.3.2 Optimized ICA for fMRI Data Analysis

In this section, we discuss application of the optimized ICA algorithm to fMRI data analysis (Zhang, Yu, and Ding, 2006; Yu *et al.*, 2007). Developed recently, fMRI is a new imaging technology that allows non-invasive studies of the brain function (Friston, Jezzard, and Turner, 1994; Clare *et al.*, 1999; Shing *et al.*, 2001; Lysetskiy,

Table 13.3 Performance comparison of traditional ICA and traditional optimized ICA algorithms

Method	Original signal source	Optimized ICA	Infomax	Extended infomax	FastICA (Gauss)	FastICA (tanh)	JADE
Kurtosis							
Rectangular	−1.8008	−1.7944	−0.48572	−1.7921	−1.7599	−1.9994	−1.5861
Exponential	−0.19782	−0.23118	−0.03114	−0.65621	−0.66112	−0.41523	−0.30716
Attenuation	−0.14924	−0.15105	0.019441	−0.48968	−0.45711	−0.40289	−0.12935
Gamma	1.0558	1.0833	0.93709	−0.1036	1.0311	0.76369	0.67469
Rayleigh	3.0668	2.9819	3.4077	1.1557	3.5865	2.4775	3.4707
Impulse	3.7257	3.7258	3.6932	3.6892	3.6283	3.0589	3.5882
Skewness							
Rectangular	−0.01183	−0.012678	0.2501	0.001486	−0.009	−0.024	−0.04707
Exponential	0.78297	0.74035	0.46527	0.21136	0.26131	0.048515	0.007267
Attenuation	0.028857	0.018695	0.065559	0.090582	0.16625	−0.0139	−0.35883
Gamma	1.4887	1.4974	1.2588	0.64211	−1.4402	0.97976	1.0364
Rayleigh	1.5183	1.5019	1.447	0.28049	−0.08649	2.4732	0.071688
Impulse	−0.03176	−0.03162	0.042483	−0.03892	1.393	1.2727	−1.422
SNR (db)							
Rectangular	0.19278	16.0535	0.7594	18.6825	12.1565	15.878	1.417375
Exponential	−0.85313	21.226	3.474175	6.96625	22.556	20.405	2.7138738
Attenuation	−2.3542	19.7515	17.5	17.7411	17.037575	18.9875	3.77075
Gamma	−0.35254	40.805375	13.8365	10.3	22.0285	36.02875	3.7918
Rayleigh	−3.7354	23.073175	13.523	10.780375	11.4395	8.8695	10.141
Impulse	−0.61429	21.877	16.3285	12.741525	10.7575	18.242375	10.423263
CPU-time (s)		10.438	0.96875	5.5	0.23438	0.33977	0.07815

Note: The first column of SNR indicates the SNR between the observation signal value and the signal source.

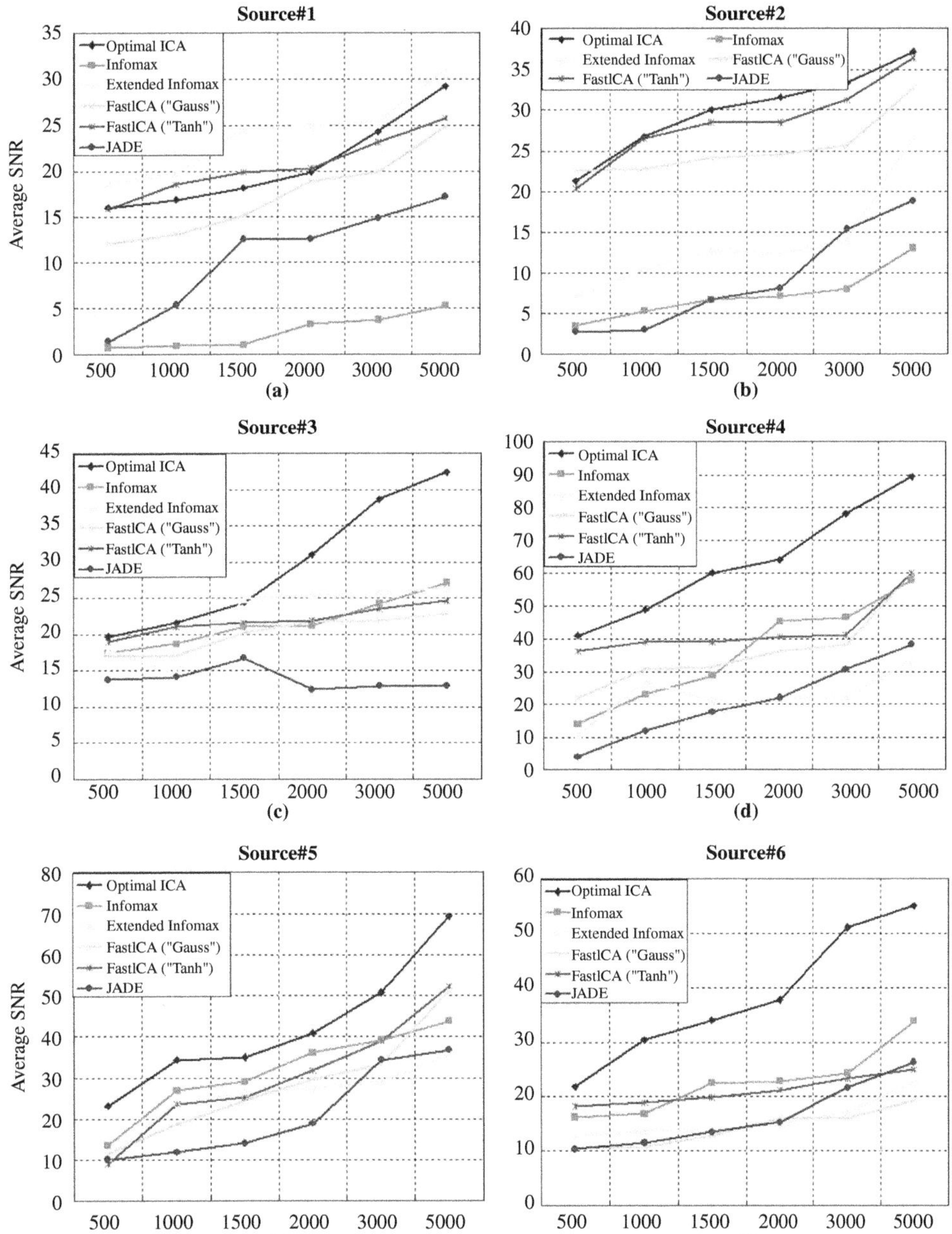

Figure 13.6 Comparison of average SNR of decomposed source signals using ICA algorithms listed in Table 13.3

Lozowski, and Zurada, 2002). When a specific region of the cerebral cortex receives sensory stimulation, carries out some intellectual activity, or is affected by drugs, fMRI measures the magnetic susceptibility change signal of the local organization in the brain caused by a change in the ratio of oxyhemoglobin and deoxyhemoglobin (i.e., the oxygen content), thus reflecting the related activity change in brain function. This technique is mainly applied to the study of the location of brain functions. The ICA algorithm is an important means of studying brain signals (McKeown *et al.*, 1998; Arfanakis *et al.*, 2000; Calhoun *et al.*, 2001; Suzuki, Kiryu, and Nakada, 2001; Chen *et al.*, 2002, 2003; Svensen, Krnggel, and Benali, 2002; Fan, Tang, and Tang, 2003; James and Gibson, 2003; Zhong, Tang, and Tang, 2003; Beckmann and Smith, 2004; Wu, Tang, and Tang, 2004).

An experiment was set up as a single block design task that explored brain plasticity induced by the learning of vocabulary. The subjects in the experiment, namely, 12 healthy, right-handed students with normal vision, were asked to focus on unlearned vocabulary, thereby being exposed to language stimulation. In the model, "a" represents the stimulated state as a result of the vocabulary, while "*" represents the baseline state, which is the state without any associated activities. Volume = 145, TR = 2 s, continue to scan for 4 mi and 50 s, this can be visually represented as follows:

14	12	11	12	13	12	12	12	11	12	12	12
*	a	*	a	*	a	*	a	*	a	*	a

In accordance with the experimental design, we utilized SPM5b (http://www.fil.ion.ucl.ac.uk/spm/) to correct head movement data, carry out standardization, and spatial smoothing preprocessing, and construct a design matrix model, which is shown in Figure 13.7.

Scan parameters: scanning instrument 3 T Siemens Magnetom Trio;

T2*(EPI): TR/TE = 2 s/35; FOV = 220×220 mm; matrix $64 \times 64 \times 25$; layer thickness 4.5 mm;

T1 (SPGR): FOV = 256×256 mm; matrix $512 \times 512 \times 128$; layer thickness 1 mm.

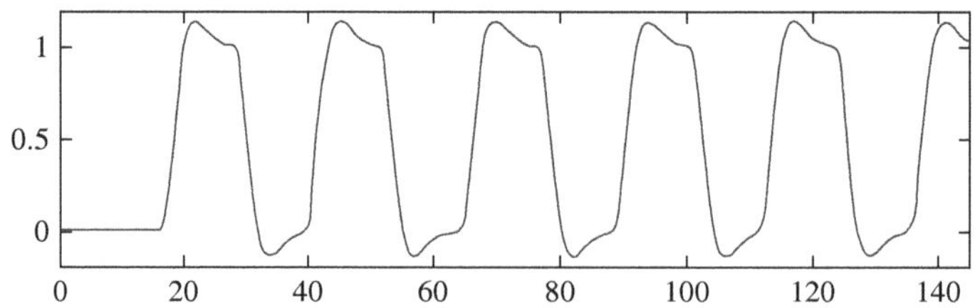

Figure 13.7 fMRI cognitive experimental paradigm

In the experiment we collected 12 sets of fMRI signals, with each set consisting of a total of 145 sequences (i.e., $M = 145$), and each sequence containing 102 400 individuals (i.e., $T = 102\,400$). We applied four ICA decomposition operations (namely, the optimized ICA algorithm, extended Infomax algorithm, FastICA (tanh) algorithm, and JADE algorithm) to the fMRI signals, from which we extracted 145 spatially independent components and obtained the separation matrix $\boldsymbol{W}_{145\times 145}$. The results show space independent component figures and their corresponding time course curves (i.e., each column of $\boldsymbol{w}^{-1}$ represents a time course curve). Different space independent components can be obtained, including an independent component related to instantaneous tasks, a head movement independent component, an independent component of the noise signal, and one of the class periodic signals. We compared the results of the various ICA algorithms applied to the fMRI data, by demonstrating the different space independent components and their corresponding time sequence curves. Red in each independent component spectrum indicates the activated state, with the z-value showing the significance of each element ($z \geq 1.0$) in the independent component diagram; in other words, it indicates the activation threshold. The typical independent components are analyzed here.

Taking the fMRI signal of the region of interest into account, the signal component with physiological significance in the region of interest is controlled by only a small part of the whole brain. That is, an interesting image source usually comprises a small proportion of all voxels, which includes local characteristics, such as the activation related to the task. The fMRI data contains a lot of background information of low importance, which makes the interesting local features darker in the activation images. These characteristics reflect a right-skewed distribution in the histogram, with the background information obscuring the importance of the voxels in the distribution tail. Traditional ICA is able to locate the local features (or signals) of interest in the first step. Since this method is not sensitive enough to detect small local characteristics, optimized ICA was adopted to restore the mixed signal, which contains both a super- or sub-Gaussian distribution as well as a symmetrical or non-symmetrical one. This completes the estimation of the optimized signal source. Owing to the use of a priori information, especially the voxel importance information represented by the distribution tail of the signal source, the optimized ICA algorithm is able to separate the fMRI signal sources more accurately.

13.3.2.1 Independent Component Related to the Task

The success of the experiment is measured by the fact that each group of fMRI signals contains an independent component related to the task. An important task of fMRI signal analysis is to find the brain activation area corresponding to the task stimulus. A brain atlas of task stimulated activation areas and task-related curves can be obtained directly from ICA. We found that the task activated areas were mainly concentrated in parts of the lingual gyrus of the occipital lobe and cerebellar regions. The lingual gyrus is mainly involved in stimulated visual information processing functions with lingual gyrus activity often seen in tasks of looking at and reading words. From the comparison of the four ICA algorithms given in Table 13.4, we can see that although similar independent components related to tasks can be obtained from the ICA, the corresponding time sequence of the composition obtained from the optimized ICA algorithm separation has the largest correlation coefficient of the design paradigm, while the amount of activation is also the most significant.

13.3.2.2 Instantaneous Task-Related Independent Component

During the course of the experiment, some brain areas are involved in the task over a period of time, while in the remaining time they do not participate in the task, which is only partially consistent with the active region associated with the task-related independent component. Such a component is called an instantaneous task-related component. We can see from the time series curves that the independent component is involved in the task for the period represented by the block, while at other times it does not participate in the task. From the independent components depicted in the figures, the instantaneous task-related component is the same for all four ICA methods used in the experiment to find the Brodmann6 zone activation (Table 13.5).

13.3.2.3 Signal Independent Component of Head Movements

Head movements in the experiments can be divided into two categories: instantaneous head movements and slow head movements. Because the head movement signal reflected in the brain atlas is an offset of the activated region, it is clear that if the head movement signal is too strong, it will affect the correct analysis of the fMRI signal. Through application of ICA we can easily find such a signal, which determines the reliability of the experimental data.

1. Independent component of instantaneous head movements (Table 13.6): During the experiment, if the subjects made head movements with large amplitude in a very short time, the time sequence curve would reflect an abrupt dip or peak over the period.
   By observing the time sequence curve, we can ascertain when the subjects made large head movements in any one direction from the time period in the diagram block. This is reflected in the independent component diagram.
2. Independent component of slow head movement (Table 13.7): If the experiment continued for a long time, the subjects might make only slight head movements owing to fatigue. It can be seen from the general trend of the time series curve whether the subjects make slow head movements. A component with the spatial structure of a "ring" in the activation figure is generally considered to be a slow head movement component, and this slow change is a non-task-related component.

Table 13.4 Brain spectra and task-related curves for four ICA algorithms

ICA	Activation composition diagram	Corresponding time sequence curves
Optimized ICA	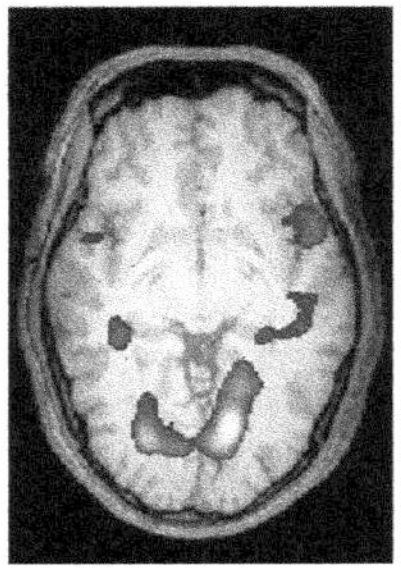	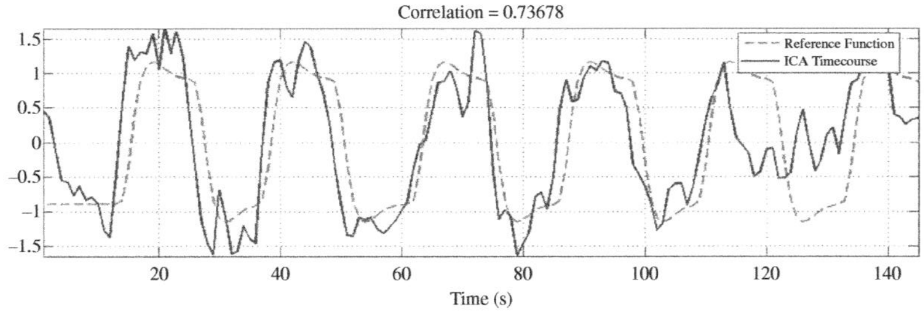 Correlation coefficient = 0.73678; maximum activated voxel value = 8.9561
Extended Infomax	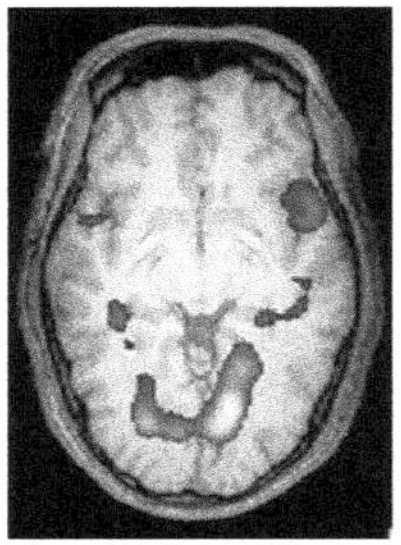	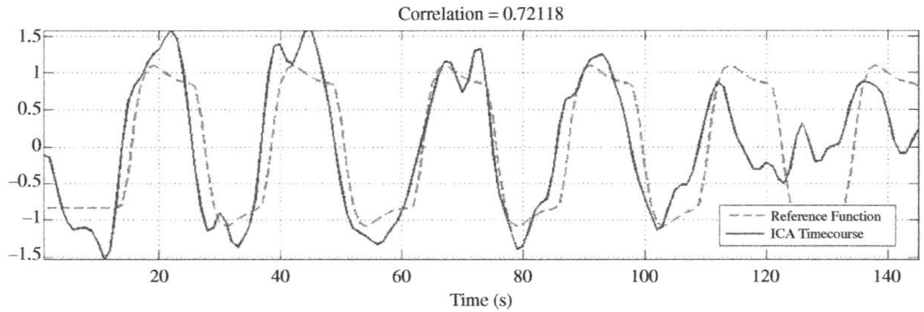 Correlation coefficient = 0.72110; maximum activated voxel value = 8.7565

(*continued overleaf*)

Table 13.4 (*continued*)

ICA	Activation composition diagram	Corresponding time sequence curves
FastICA-tanh	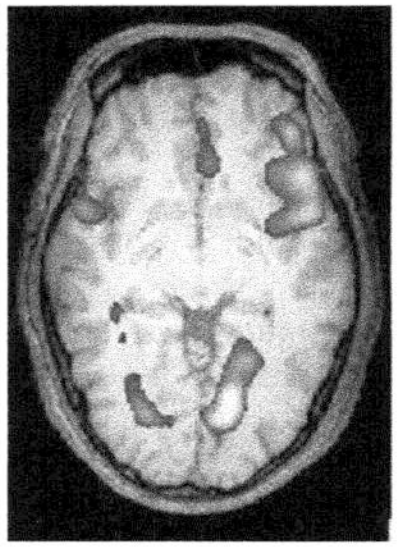	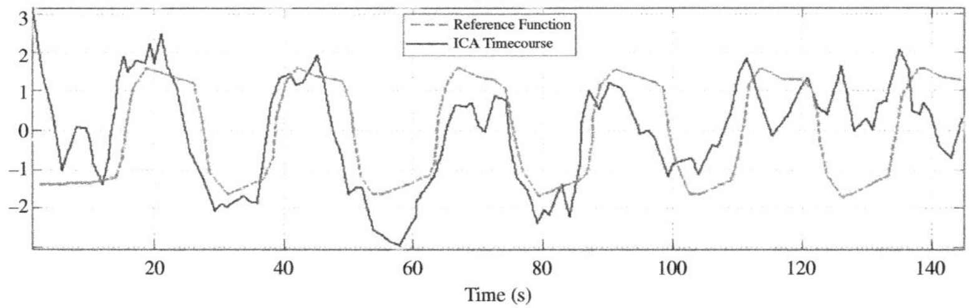 Correlation coefficient = 0.6362; maximum activated voxel value = 7.2711
JADE	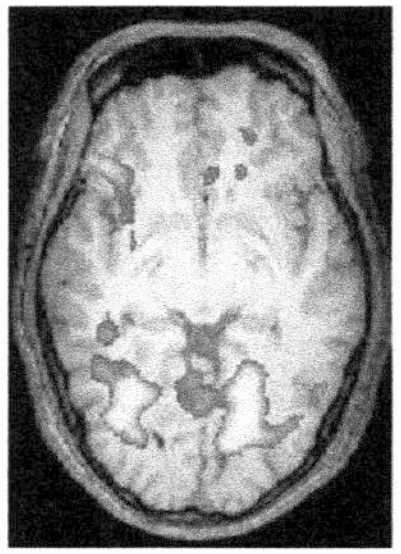	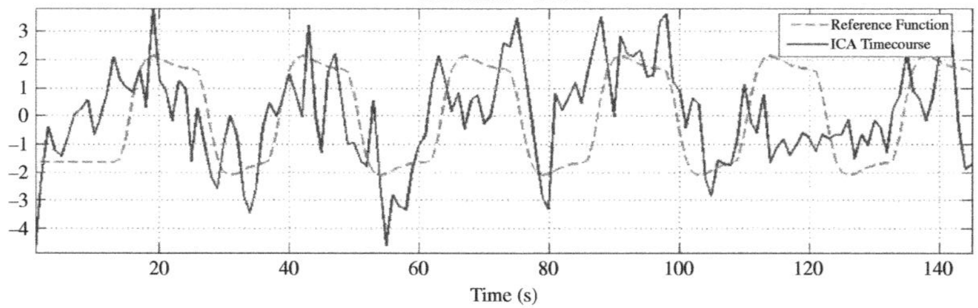 Correlation coefficient = 0.4566; maximum activated voxel value = 8.6262

Table 13.5 Brodmann6 zone activation

Independent component activation figures	Corresponding time sequence curves
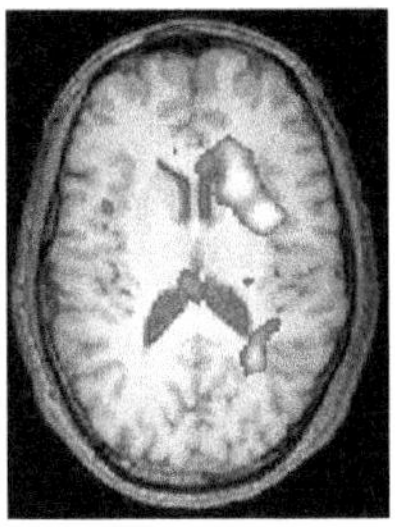	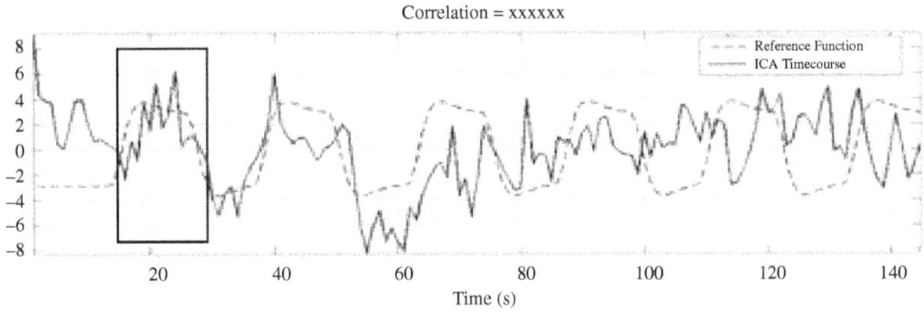
The main activation referred to as the Brodmann6 district, spans the supplementary motor and premotor areas. This location is more likely to be the premotor area, which usually has activities in tasks produced by speech. However, when subjects complete a speech recognition task, we can also see activity in this part. Therefore, some people believe that there is a link between language understanding and speech production	

Table 13.6 Independent component for substantial head movements

Independent component activation figures	Corresponding time sequence curves
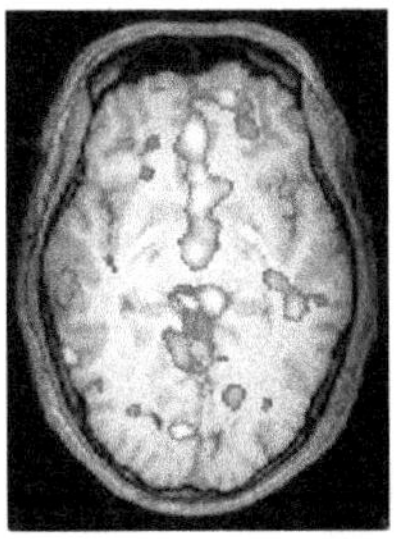	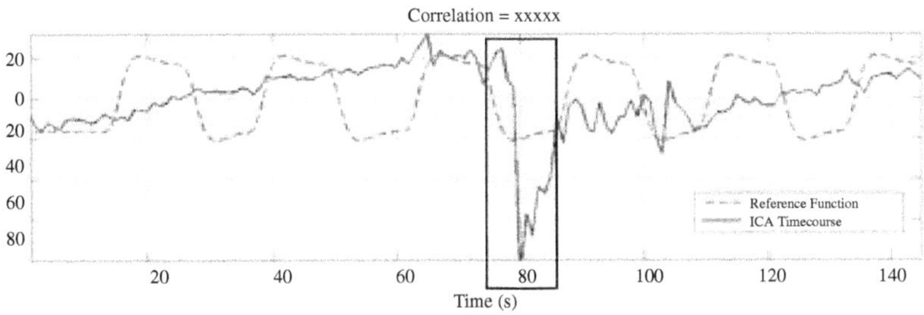
Instant head movement component. The subjects made sudden head movements approximately between 78 and 82 s	

13.3.2.4 Independent Component of Class Periodic Signals

Part of the time series process curves of the independent components approximately represents a cyclic state. The spatial distribution of such components has a high recurrence rate in various experiments. A few examples of independent components of a class periodic signal, whose time series curve reaches a peak approximately every

Table 13.7 Independent components of slow head movement

ICA	Activation composition diagram	Corresponding time sequence curves
Optimized ICA	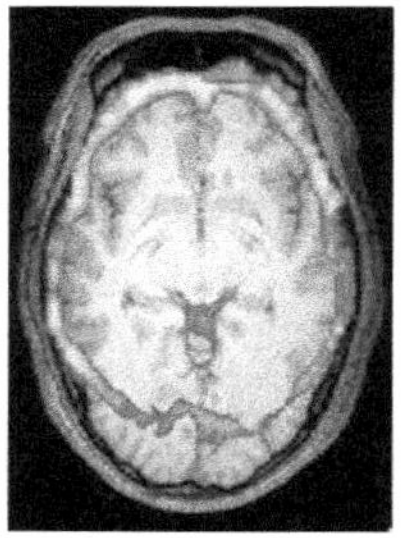	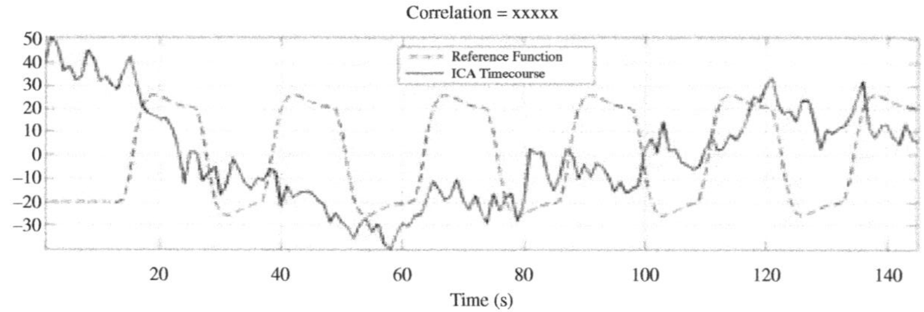 The optimized ICA algorithm clearly and completely separates the spatial structure of the “ring” line with a slow head movement component, which is not connected with other brain regions. Thus, this determines the existence of a slow head movement component.
Extended infomax	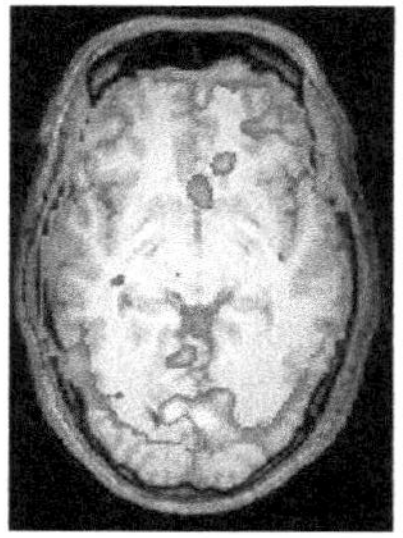	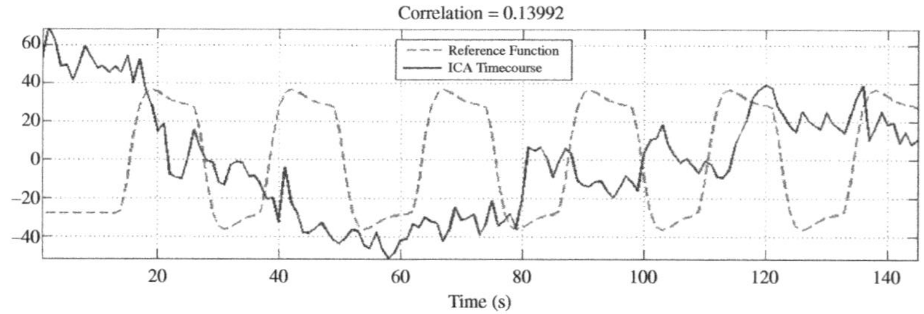 The extended Infomax algorithm is able to find the spatial structure of the ring, but the clarity is inferior to the space component separated by optimized ICA. Meanwhile, there are connections with other brain regions, which indicates poor separation performance.

FastICA-tanh

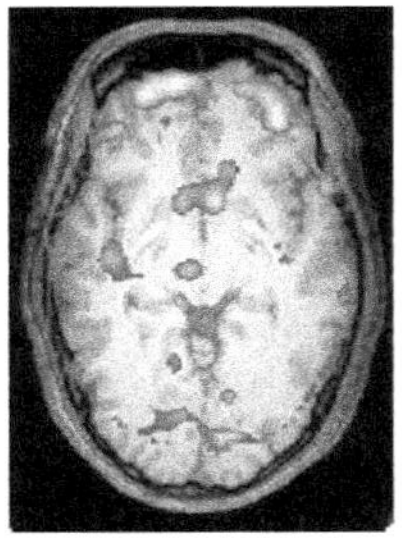

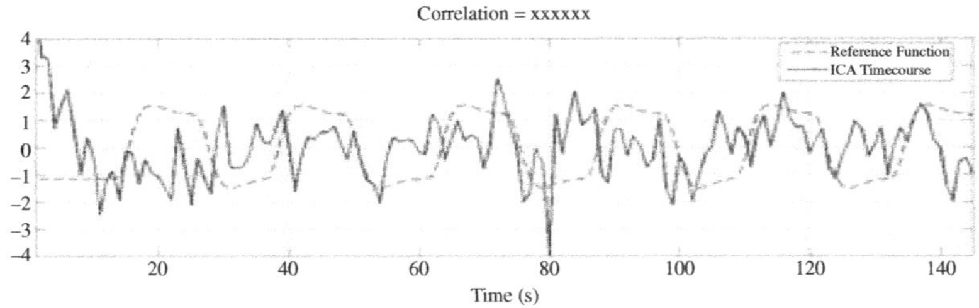

The spatial structure of the “ring” from FastICA is very clear, and is almost depicted in the form of scattered points. In addition, activated points of other brain regions also appear in the component activation figures, which indicates poor separation performance.

JADE

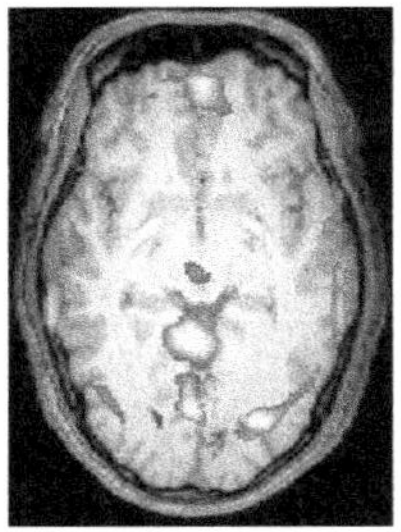

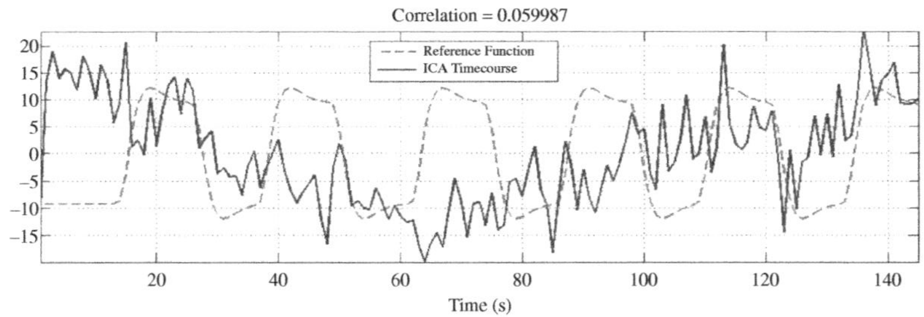

From the results of the JADE algorithm, it is almost impossible to find any activation component corresponding to the spatial structure of a “ring”.

Table 13.8 Independent components of class periodic signals

<table>
<tr><th>Activation composition diagram</th><th>Corresponding time sequence curves</th></tr>
<tr><td>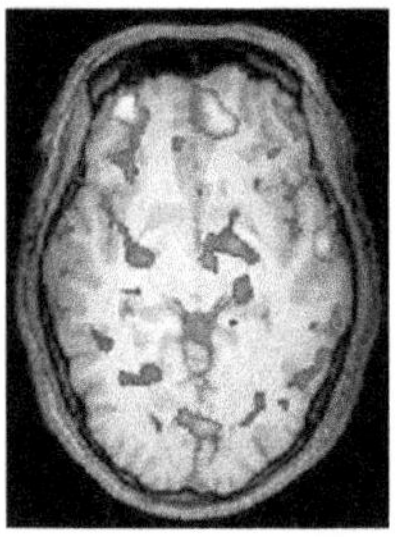</td><td>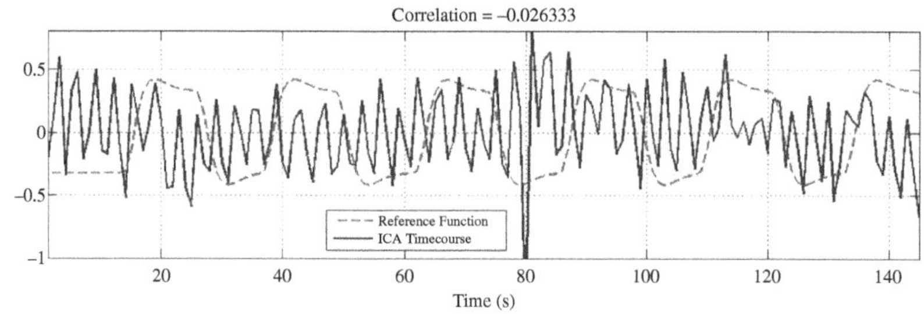</td></tr>
<tr><td>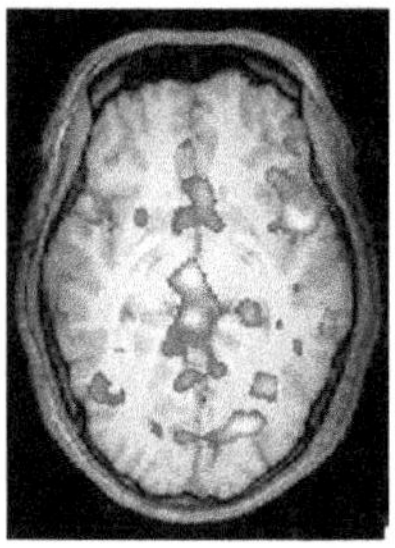</td><td>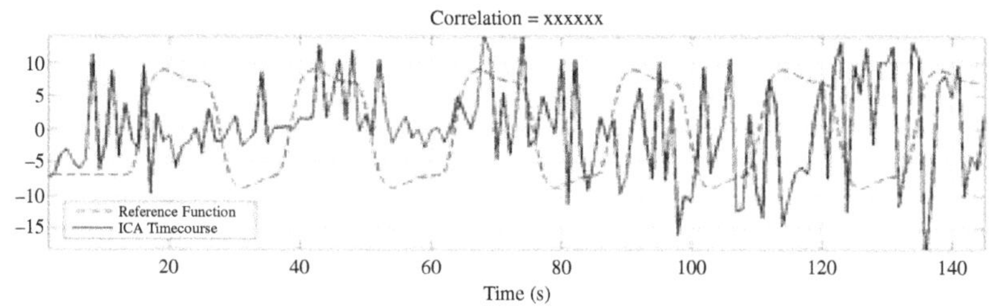</td></tr>
<tr><td>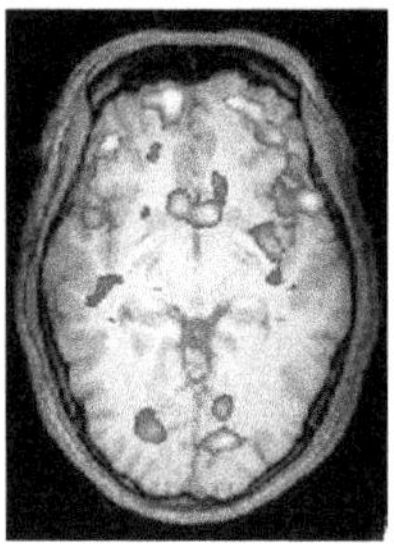</td><td>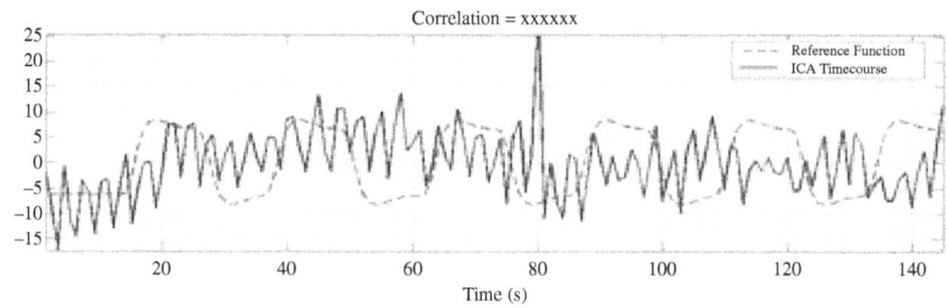</td></tr>
</table>

3–4 s, are shown in Table 13.8. The independent components' contribution to the fMRI signal shows a swing state, with such signals possibly caused by heartbeats, breathing, and blood flow.

13.3.2.5 Independent Component of the Noise Signal

A large proportion of all the independent components are noise signals, two examples of which are shown in Table 13.9, and their active regions are randomly distributed over the brain regions. In addition, their time sequence curves are also disorganized.

Table 13.9 Independent components of noise signals

Activation composition diagram	Corresponding time sequence curves
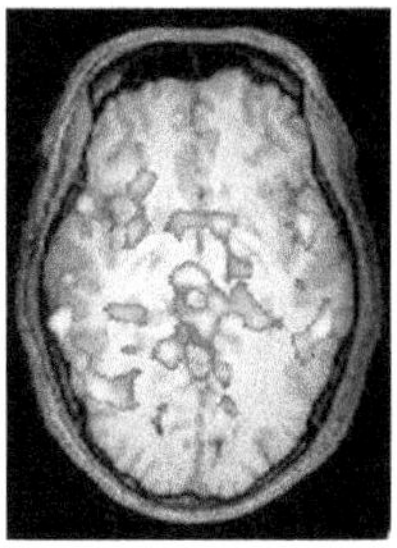	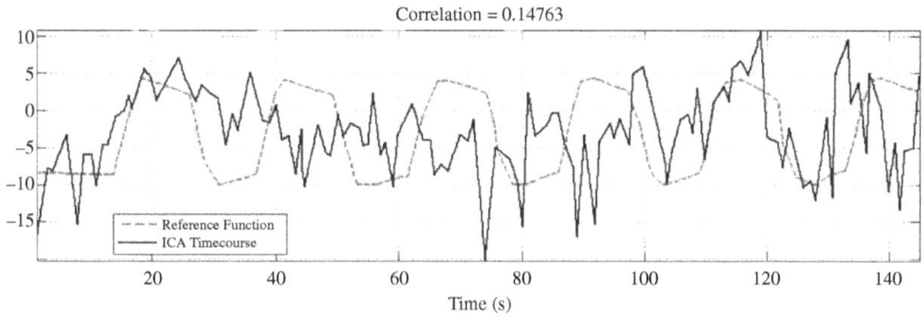
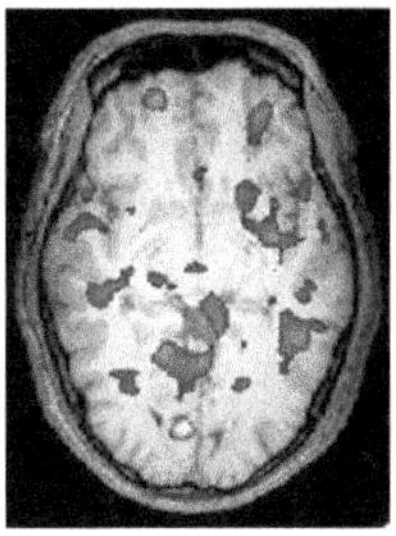	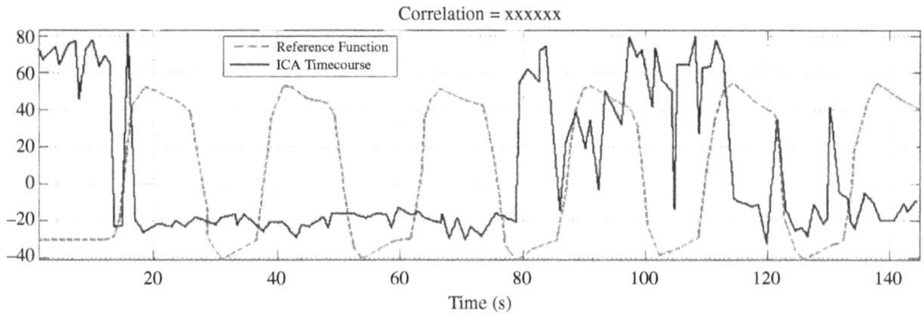

13.3.2.6 Component with a Special Meaning

During the process of extracting components using the optimized ICA algorithm, a special kind of component is obtained, which is different from the results obtained from the application of other methods, as shown in Table 13.10.

13.4 Chapter Summary

The decomposition of fMRI data by ICA produced task-related independent components, instantaneous task-related independent components, head movement independent components, and independent components of class periodic signals and noise signals. The corresponding time series curves of each were also obtained. These indicate that ICA is a good tool for analyzing fMRI data.

Since there are asymmetric distribution signals in fMRI data, we proposed the application of the optimized ICA algorithm to the fMRI data. Compared with the traditional ICA algorithms for fMRI data analysis, the optimized ICA method has greater sensitivity and accuracy in detecting activation areas. More relevant components can be obtained by optimizing the ICA algorithm for the detection of task-related components; simultaneously, the detected activation areas became more significant. To detect

Table 13.10 Component with a special meaning

Activation composition diagram	Corresponding time sequence curves
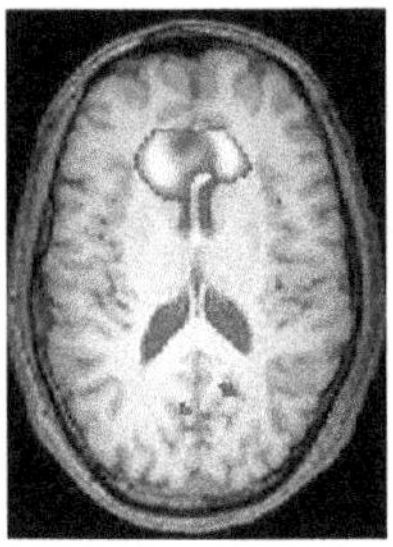	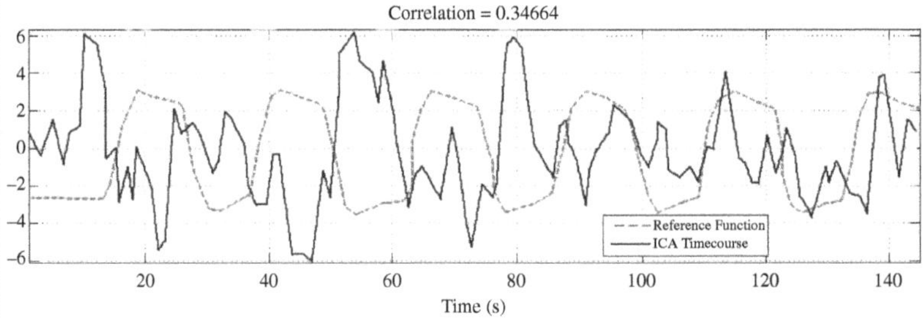
The main activation region is the Brodmann32 district, whose function is thinking and planning related to the learning task	
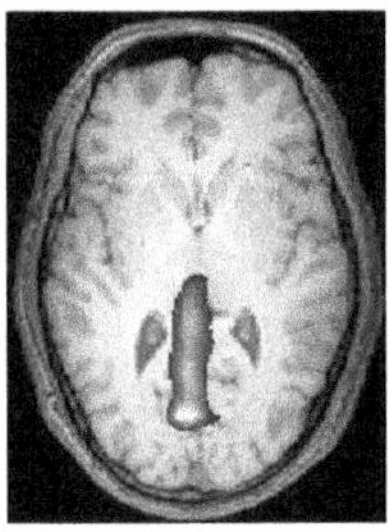	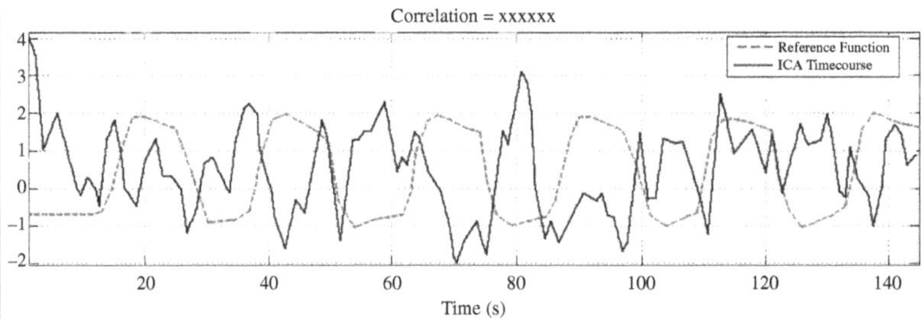
The major activation area is the precuneus, whose main function is processing visuo-spatial tasks. It may be related to the processing of glyphs and configuration in the case of characters because the inferior parietal lobule and the nearby region are more involved in the analysis of attention and visuo-spatial tasks. In some Hanzi learning experiments, a relationship between this area and the effect of word rules was found.	

independent components of slow head movements, the optimized ICA algorithm was able to separate a very clear and complete component of a "ring" spatial structure, which is not connected to other functional areas in the brain. In addition, we also found that the precuneus activation region was related to the processing of glyphs and configurations, while activation of the Brodmann32 district was associated with learning tasks and planning. Thus, richer brain activation information and more accurate results of blind separation were obtained using the optimized ICA algorithm.

References

Arfanakis, K., Cordes, D., Haughton, V.M. *et al.* (2000) Combining independent component analysis and correlation analysis to probe interregional connectivity in fMRI task activation datasets. *Magnetic Resonance in Medicine Magnetic Resonance Imaging*, **18**(8), 921–930.

Beckmann, C.F., and Smith, S.M. (2004) Probabilistic independent component analysis for functional magnetic resonance imaging. *IEEE Transaction on Medicine Imaging*, **23**(2), 137–152.

Bell, A.J., and Sejnowski, T.J. (1995) An information maximization approach to blind separation and blind deconvolution. *Neural Computation*, **7**(6), 1129–1159.

Calhoun, V.D., Adali, T., Pearlson, G.D. *et al.* (2001) A method for making group inferences from functional MRI data using independent component analysis. *Human Brain Mapping*, **14**(3), 140–151.

Chen, H.F., Rao, D.Z., Zhuo, Y. *et al.* (2002) A new fMRI data processing method: neighborhood independent component method and its preliminary application. *Science in China Series E-Technological Sciences*, **32**(5), 686–692.

Chen, H.F., Yao, D.Z., Chen, L. *et al.* (2003) Analysis of fMRI data by blind separation of data in a time spatial domain into independent temporal component. *Brain Topography*, **15**(4), 223–232.

Clare, S., Humberstone, M., Hykin, J. *et al.* (1999) Detecting activations in event-related fMRI using analysis of variance. *Magnetic Resonance in Medicine*, **42**(6), 1117–1122.

Fan, L.W., Tang, H.W., Tang, Y.Y. (2003) The application of independent component analysis to fMRI data analysis. *Journal of Dalian University of Technology*, **43**(4), 399–402.

Fishler, E., Messer, H. (1999) Order statistics approach for determining the number of sources using an array of sensors. *IEEE Signal Processing Letters*, **6**(7), 179–182.

Fishler, E., Messer, H. (2000) On the use of order statistics for improved detection of signals by the MDL criterion. *IEEE Transactions on Signal Processing*, **48**(8), 2242–2247.

Friston, K.J., Jezzard, P., and Turner, R. (1994) The analysis of functional MRI time-series. *Human Brain Mapping*, **1**(2), 153–171.

Hyvärinen, A. (1999a) Fast and robust fixed-point algorithm for independent component analysis. *IEEE Transactions on Neural Networks*, **10**(3), 626–634.

Hyvärinen, A. (1999b) The fixed-point algorithm and maximum likelihood estimation for independent component analysis. *Neural Processing Letters*, **10**(1), 1–5.

Hyvärinen, A., Oja, E. (1997) A fast fixed-point algorithm for independent component analysis. *Neural Computation*, **9**(7), 1483–1492.

James, C.J., Gibson, O.J. (2003) Temporally constrained ICA: an application to artifact rejection in electromagnetic brain signal analysis. *IEEE Transactions on Biomedical Engineering*, **50**(9), 1108–1116.

Lee, T.W., Girolami, M., Sejnowski, T.J. (1999) Independent component analysis using an extended Infomax algorithm for mixed sub-Gaussian and super-Gaussian sources. *Neural Computation*, **11**(2), 417–441.

Li, Y., Cichocki, A., and Amari, A. (2003) Sparse component analysis for blind source separation with less sensors than sources. Proceedings of the 4th International Symposium on Independent Component Analysis and Blind Signal Separation, pp. 89–94.

Lysetskiy, M., Lozowski, A., Zurada, J.M. (2002) Temporal-to-spatial dynamic mapping, flexible recognition, and temporal correlations in an olfactory cortex model. *Biological Cybernetics*, **87**(1), 58–67.

McKeown, M.J., Makeig, S., Brown, G.G. *et al.* (1998) Analysis of fMRI data by blind separation into spatial independent component analysis. *Human Brain Mapping*, **6**(3), 160–188.

Shing, C.N., William, F.A., Shantanu, S. *et al.* (2001) Activation detection in event-related fMRI data based on spatio-temporal properties. *Magnetic Resonance Imaging*, **19**(9), 1149–1158.

Stone, J.V., Porrill, J., Porter, N.R. *et al.* (2002) Spatiotemporal independent component analysis of event-related fMRI data using skewed probability density functions. *NeuroImage*, **15**(2), 407–421.

Suzuki, K., Kiryu, T., Nakada, T. (2001) Fast and precise independent component analysis for high field fMRI time series tailored using prior information on spatiotemporal structure. *Human Brain Mapping*, **15**(1), 54–66.

Svensen, M., Krnggel, F., Benali, H. (2002) ICA of fMRI group study data. *NeuroImage*, **16**(3), 551–563.

Wu, Z.H., Tang, H.W., Tang, Y.Y. (2004) The new independent component analysis algorithm to achieve the blind source separation of magnetic resonance imaging signals. *Acta Biophysica Sinica*, **20**(3), 188–192.

Yu, X.C., Ren, J.M., Zhang, N., *et al.* (2007) An optimal ICA algorithm applied to fMRI data. Proceedings of 2007 International Conference on Machine Learning and Cybernetics, Vol. 6, pp. 3565–3570.

Zhang, N., Yu, X.C., and Ding, G.S. (2006) An optimal independent component analysis approach for functional magnetic resonance imaging data. Proceedings 2006 International Conference on Intelligent Information Hiding and Multimedia Signal Processing, pp. 163–166.

Zhong, M.J., Tang, H.W., Tang, Y.Y. (2003) Space independent component analysis for blind source separation of fMRI signals. *Acta Biophysica Sinica*, **19**(1), 79–83.

14

Supervised Learning Independent Component Analysis Algorithms and Applications

Independent component analysis (ICA) is a common blind source separation (BSS) method. There exist in the mixed pixel decomposition problem constraint conditions that the sum of the decomposition results must equal 1 and the decomposition results for each land cover type must range between 0 and 1, hence, ICA cannot be directly applied to this problem. Considering the drawback of traditional ICA, we propose a new algorithm, supervised learning independent component analysis (SL-ICA) to solve the problem of mixed pixels in synthetic aperture radar (SAR) images. Adding supervised learning restrictive conditions to the negentropy objective function, we constrain the negentropy and restrictive conditions in a unified objective function and optimize the objective function by applying a new dual-gradient descent algorithm iteratively, which accelerates the computation speed. The results from applying SL-ICA, principal component analysis (PCA), and ICA to computer simulated SAR images and ENVISAT-ASAR images of Beijing show that SL-ICA yields more precise results than PCA and ICA.

14.1 Overview

Mixed pixels are commonly found in all types of remote sensing images, especially in regions with complex ground features (Chang *et al.*, 2004). The mixed-pixel problem has become a key factor influencing the accuracy of remote sensing image classification. Existing mixed-pixel models focus on optical remote sensing images and can be divided into two major categories: linear spectra hybrid models (Shaw and Burke, 2003) and nonlinear spectra hybrid models (Chang *et al.*, 1998). Linear

Blind Source Separation: Theory and Applications, First Edition. Xianchuan Yu, Dan Hu and Jindong Xu.

models, which are the most popular, include some classic methods, such as Fuzzy C-Means (k-Means) (Giles and Manoj, 1996), soft maximum likelihood estimation classifier (Bastin, 1997), the method based on Bayesian probability theory (Bressan, Guillamet, and Vitria, 2003), multi-layer perceptron neural network (Kanellopoulos *et al.*, 1991), and so on. Experiments have proved that due to the large differences between SAR images and optical remote sensing images (Inglada and Adragna, 2002), many of the mature theories and methods for mixed pixel decomposition of optical remote sensing images are not suitable for mixed pixel decomposition of SAR images.

ICA is a recently developed method for multi-signal processing and BSS. The method has been applied extensively to isolate independent signals from a mixed-signal source and in-depth to speech signals. In recent years, ICA has received increasingly wide use in functional magnetic resonance imaging (Loghmari, Naceur, and Boussema, 2002), remote sensing imagery classification (Hyvdrinen, Karhunen, and Oja, 2001; Shah, Varshney, and Arora, 2007), the extraction of remote sensing imagery information (Si and Zhang, 2002; Szu and Kopriva, 2001), telecommunications (Ristaniemi and Joutsensalo, 1999), and pattern recognition (Tsai *et al.*, 2006).

However, ICA has not been used in the area of mixed pixel decomposition of SAR images. The reason for this is that the objective function of the existing ICA algorithm is unable to satisfy some of the restrictive conditions required by the mixed pixel decomposition of remote sensing images. Examples of these restrictive conditions are that the sum of the decomposition results must equal 1, and the decomposition results for each land cover type must range between 0 and 1.

Here, we propose a new ICA algorithm, SL-ICA, with particular emphasis on the objective function and optimization algorithm. We apply SL-ICA, PCA, and ICA to artificially simulated SAR images and real SAR images. Experimental results show that SL-ICA is more precise than PCA and ICA.

14.2 Mathematical Model

14.2.1 Mathematical Model for Mixed Pixels in SAR Images

Assume that there are m (we adopt $m = 2$ in the second experiment with ENVISAT-ASAR images) remote sensing receivers. The receiver vector for each pixel is expressed as $\boldsymbol{X} = (x_1, x_2, \ldots, x_m)^{\mathrm{T}}$. Since the measurement model for these data is linear and instantaneous, we can assume that the data satisfy the model for linear vector multiplication (Shaw and Burke, 2003):

$$\boldsymbol{X} = \sum_{i=1}^{n} a_i \boldsymbol{s}_i, \boldsymbol{x}_j = \sum_{i=1}^{n} a_{ji} \boldsymbol{s}_i, \boldsymbol{X} = \boldsymbol{A}\boldsymbol{S} \tag{14.1}$$

where $\boldsymbol{A} = (a_1, a_2, \ldots, a_n)$ is the backscattered coefficient matrix, which is unknown, and A changes as the pixel changes. $\boldsymbol{S} = (s_1, s_2, \ldots, s_n)^{\mathrm{T}}$ denotes the various ground resources composed of different signal sources. The upper bound of the sum of all elements of $\boldsymbol{S} = (s_1, s_2, \ldots, s_n)^{\mathrm{T}}$ is 1. In other words, all elements of $\boldsymbol{S} = (s_1, s_2, \ldots, s_n)^{\mathrm{T}}$ satisfy the constraint below (Inglada and Adragna, 2002).

$$\sum_{i=1}^{n} s_i = 1 \tag{14.2}$$

The goal of mixed pixel decomposition of SAR images is to calculate the independent component vector $\boldsymbol{S}$ using $\boldsymbol{X}$ and assuming that $\boldsymbol{A}$ and $\boldsymbol{S}$ are unknown.

14.2.2 ICA Model

Assuming that each ground component is a random variable (random signal source), SAR remote sensing images can be considered as a multi-source mixed signal. In addition, because the precise composition of the ground is usually unknown, the mixed pixel decomposition of SAR remote sensing images can be seen as a BSS problem of the multi-source mixed signal. Equation 14.1 represents the standard model for ICA, with the main task being to estimate the independent sources from observed data $\boldsymbol{X}$; that is, to find a linear transformation matrix $\boldsymbol{W}$ about the correlation signal $\boldsymbol{X}$ to make the output as independent as possible:

$$U(t) = \boldsymbol{WX} = \boldsymbol{WAS} \tag{14.3}$$

where vector $\boldsymbol{U}$ is the estimated value of the source vector. In a linear mixed or nonmixed model, we assume that the number of sensors should be greater than or equal to the number of sources, $n \geq m$, the sources for each moment are independent of each other, and there is at most one source with a Gaussian distribution.

The model for mixed pixel decomposition of SAR images satisfies the general assumptions of ICA, enabling us to solve mixed pixels decomposition of SAR images based on image signal separation.

At the same time, from Section 14.2.1, the mixed-pixel model for SAR images not only satisfies Equation 14.1, but also Equation 14.2, so merely using the original traditional ICA is not sufficient to solve the decomposition problem of mixed pixels in SAR images. Sections 14.3 and 14.4 discuss the principles and implementation of SL-ICA, which is used to decompose mixed pixels in SAR images.

14.3 Principles of SL-ICA

SL-ICA is actually an optimization problem, which is implemented in three steps. First step: centering and whitening; second step: defining the objective function; third step: choosing a learning algorithm to optimize the objective function.

14.3.1 Centering

SL-ICA assumes that all the mixed vector and independent components have a zero mean. If not, we adjust the mean by centering. The specific process is as follows.

Preprocess the original vector x': $x = x' - E\{x'\}$, so that the independent component $\boldsymbol{S}$ also has a zero mean. After estimating the mixed matrix $\boldsymbol{A}$ and independent component $\boldsymbol{S}$, we add $A^{-1}E\{x'\}$ to the independent component S to add the cut mean value.

14.3.2 Whitening

Whitening means linearly transforming a mixed vector x to obtain a new vector v, where the components of v are uncorrelated and have a variance of 1. In other words, the covariance matrix of v is an identity matrix, $E\{vv^{\mathrm{T}}\} = \boldsymbol{I}$. A common method for whitening is $v = \boldsymbol{D}^{-1/2}\boldsymbol{E}^{\mathrm{T}}$, where $\boldsymbol{E}$ is an orthogonal matrix composed of the eigenvectors of $E\{vv^{\mathrm{T}}\}$ and $\boldsymbol{D}$ is a diagonal matrix composed of eigenvalues (Szu and Kopriva, 2001).

14.3.3 Objective Function of SL-ICA

Optimizing the objective function (comparison function) achieves the SL-ICA separation process. The SL-ICA design maximized objective function is based on negentropy as follows:

$$\max_{\lambda,A} L(\boldsymbol{S},\boldsymbol{A}) = J(\boldsymbol{S}) - E(\boldsymbol{S},\boldsymbol{A}) \tag{14.4}$$

The first term of Equation 14.4 is the negentropy of the independent component S. The concept of negentropy is used to describe the non-Gaussian signal's entropy, with negentropy defined as: $J(\boldsymbol{S}) = H(\boldsymbol{S}_{\mathrm{Gauss}}) - H(\boldsymbol{S})$, where $\boldsymbol{H}$ denotes the entropy and $\boldsymbol{S}_{\mathrm{Gauss}}$ is a Gaussian random vector with the same variance as $\boldsymbol{S}$. Assuming that $\boldsymbol{X}$ has zero mean and unit variance, we can adopt a method based on the principle of maximum entropy (Chang, 2002) to estimate the negentropy as

$$J(S) \propto [E\{G(S)\} - E\{G(S_{\mathrm{Gauss}})\}]^2 \tag{14.5}$$

given G: $G(u) = -\frac{1}{a_1}\exp(a_2u^2/2), (a_2 \approx 1)$.

The second term of Equation 14.4 is the error function $E(S,A)$. Since $E(S,A)$ should satisfy the restrictive conditions of mixed pixel decomposition of SAR images, we define the following error function:

$$E(S,A) = \lambda^{\mathrm{T}}(AS - X) + (\lambda_0 - 1)\left(\sum_{i=1}^{n} s_i - 1\right) + \sum_{i=1}^{m}[s_i^{2h} + (1 - s_i)^{2h}] \tag{14.6}$$

Focusing on the right-hand side of Equation 14.6, the first term $\lambda^{\mathrm{T}}(AS - X)$ is used to constrain the decomposition results so that it satisfies Equation 14.1. The second term $(\lambda_0 - 1)\left(\sum_{i=1}^{n} s_i - 1\right)$ is used to ensure that the sum of the decomposition results is 1, thereby satisfying Equation 14.2. And the third term $\sum_{i=1}^{m}[s_i^{2h} + (1 - s_i)^{2h}]$ is used to limit the scope of the decomposition results, so as to restrict them to the range 0 to 1, with h being a large number. If $0 \leq s_i \leq 1$, s_i^{2h} and $(1 - s_i)^{2h}$ are both approximately 0. If $s_i < 0$ or $s_i > 1$, either s_i^{2h} or $(1 - s_i)^{2h}$ is much greater than one, so our iterative operation can help to limit s_i to be between 0 and 1.

Combining Equations 14.5 and 14.6 into Equation 14.4, and making use of Equation 14.1 we get the objective function of each pixel. The objective function of SL-ICA unifies the negentropy and the error in an objective function framework and forms a combined objective function with constraint conditions. When the objective function reaches its highest value, the negentropy of the independent component $\boldsymbol{S}$ is at its maximum, thus causing the error of the restrictive conditions to be at its smallest.

14.3.4 Optimized Algorithm

To maximize the objective function through iterative calculation, we must select an optimization algorithm. Common optimization algorithms include the random gradient method and natural gradient method. SL-ICA uses a dual random gradient algorithm based on the random gradient method.

The basic idea of the random gradient algorithm is as follows. First, determine the initial value $\boldsymbol{A}_0$ of the separation matrix $\boldsymbol{A}$. Next, calculate the gradient of the objective function $L(S, A)$ at point $\boldsymbol{A}_0$, and then move an appropriate step in the gradient direction to obtain a new separation matrix $\boldsymbol{A}_1$. Repeating this process, we obtain the iterative formula for $\boldsymbol{A}$:

$$A(k+1) = A(k) + a(k)\frac{\partial L(S,A)}{\partial A}|_{A=A(k)} \tag{14.7}$$

where k is the number of iterations and $a(k)$ is the step length (we adopt $a(k) = 1.2$ in the experiment in Section 14.5). Defining $\Delta A = A(k+1) - A(k)$, Equation 14.7 can be rewritten as:

$$\Delta A = a\frac{\partial L(S,A)}{\partial A} \tag{14.8}$$

This is a common iterative formula for the random gradient method. We can obtain different iterative formulas from different objective functions.

In the objective function, $\lambda, \boldsymbol{A}, \boldsymbol{S}$ are unknown vectors. From Equation 14.1 we know that $\boldsymbol{S}$ can be uniquely represented by $\boldsymbol{A}$ and $\boldsymbol{X}$, and thus the objective function is equivalent to a dual objective function. By improving the random gradient method, we obtain the dual gradient descent method:

$$\Delta A = a\frac{\partial \underset{A,\lambda}{\mathrm{L}}(S,A)}{\partial A}, \Delta\lambda = a\frac{\partial \underset{A,\lambda}{\mathrm{L}}(S,A)}{\partial \lambda} \tag{14.9}$$

This iterative process is repeated until $\Delta A, \Delta\lambda$ are both less than a specified value.

14.4 SL-ICA Implementation Process

14.4.1 Iterative Process of SL-ICA

The iterative process of the algorithm is given below.

14.4.1.1 Initialization

1. Initialize Lagrange multiplier vector λ.
2. Initialize mixed matrix $\boldsymbol{A}$.
   Because of the convergence of the iterative process, λ can be selected randomly. However, to use the a priori affiliation of $\boldsymbol{A}$ and $\boldsymbol{X}$, we have Equation 14.10 (Shao and Deng, 1989):

$$a_{ji} \approx x_j \frac{1}{1 + \mathrm{e}^{-x_i}} \tag{14.10}$$

3. Centralize using the method discussed in Section 14.3.1.
4. Carry out whitening using the method given in Section 14.3.2.

14.4.1.2 Recursive Operation

1. Using Equation 14.5, calculate $\underset{A,\lambda}{\mathrm{L}}(S, A)$.
2. Calculate $\frac{\partial \underset{A,\lambda}{\mathrm{L}}(S,A)}{\partial A}$.
3. Calculate $\frac{\partial \underset{A,\lambda}{\mathrm{L}}(S,A)}{\partial \lambda}$.
4. Using Equation 14.9, compute $\Delta\lambda$.
5. $\lambda + = \Delta\lambda$.
6. Using Equation 14.9, compute ΔA.
7. $A + = \Delta A$.
8. If $\Delta\lambda$, ΔA are both less than a specified value, terminate the algorithm. Otherwise, jump to (1).

14.4.1.3 Post Processing

Based on the method given in Section 14.3.1, add $\boldsymbol{S}$'s mean to $\boldsymbol{S}$, which was subtracted during the centering.

14.4.2 Flowchart of SL-ICA Algorithm

A flowchart of the process is shown in Figure 14.1.

14.5 The Experiment

To test the effectiveness and benefits of SL-ICA, and to compare the performance of SL-ICA with that of PCA and ICA, we first use computer simulated SAR images

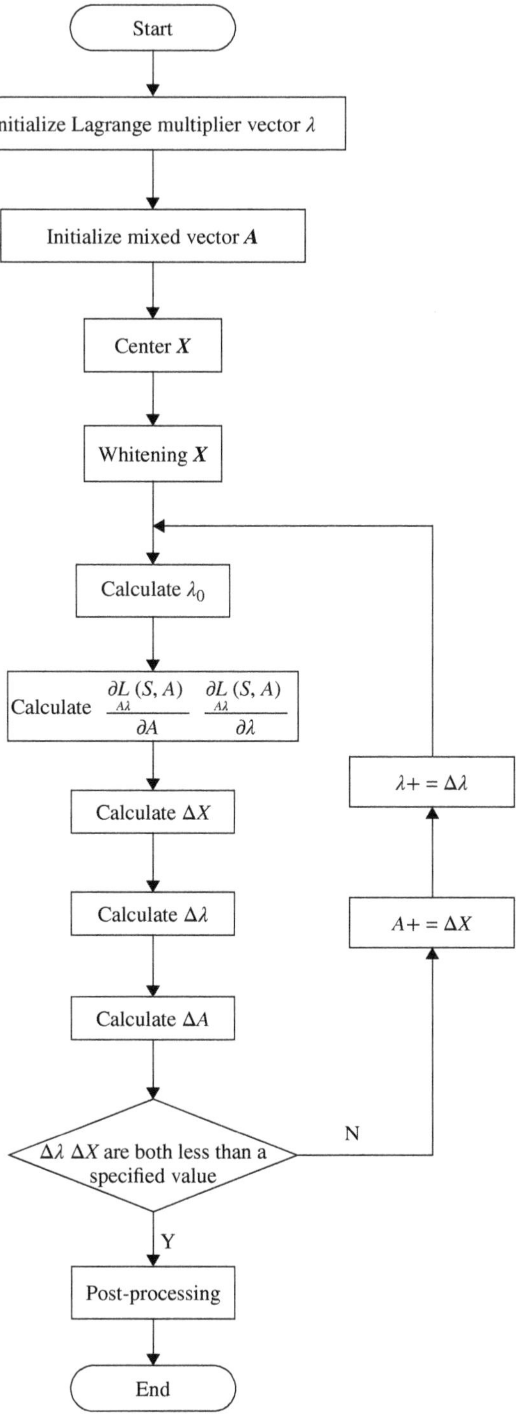

Figure 14.1 Flowchart for SL-ICA

with multi-polarization channels as experimental data. Thereafter, we use real SAR images (Figures 14A.1 and 14A.2) with two polarization (VV, HV) channels as experimental data.

14.5.1 Computer Simulated Multi-polarization SAR Images

In this section, we describe computer simulations conducted to demonstrate the performance of SL-ICA, PCA, and ICA in the mixed pixel decomposition. Computer simulated SAR images can be obtained as follows:

1. Independent component vectors $s_i, i \in (1, n)$ for 256×256 pixels are generated based on Rayleigh distribution, satisfying the condition:

$$s_i \in (0, 1), \sum_{i=1}^{n} s_i = 1, i \in (1, n) \tag{14.11}$$

   The independent component vector s_i is taken as the standard, to compare the effects of SL-ICA, PCA, and ICA.

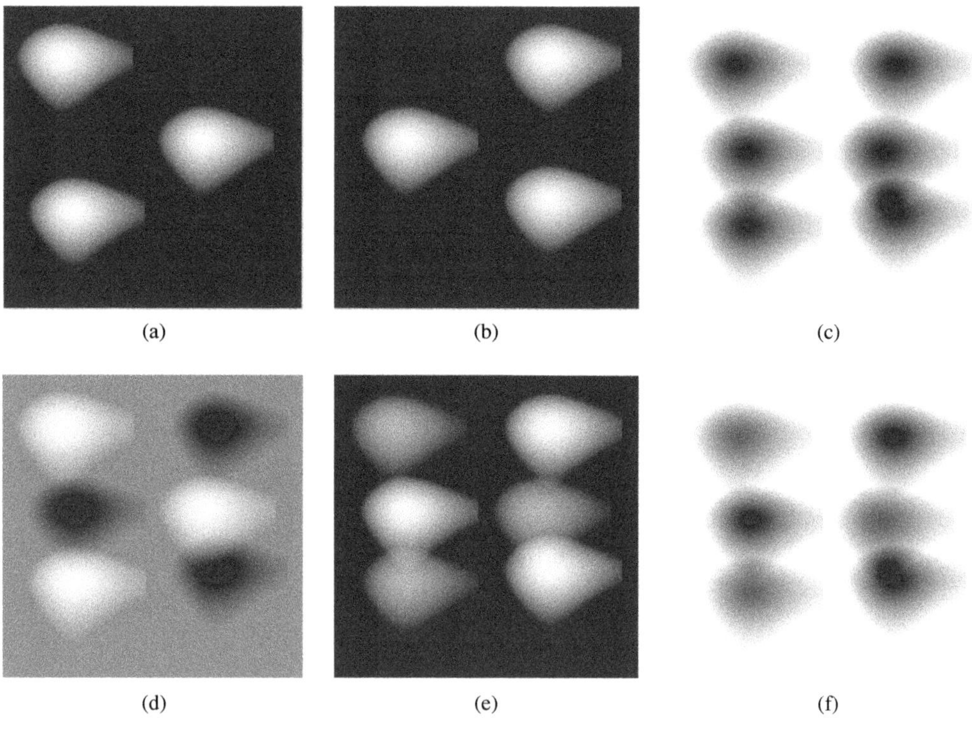

Figure 14.2 Computer simulated SAR remote sensing images: (a) pure pixel image from channel 1, (b) pure pixel image from channel 2, (c) pure pixel image from channel 3, (d) mixed pixel image of the three channels from receiver 1, (e) mixed pixel image of the three channels from receiver 2, and (f) mixed pixel image of the three channels from receiver 3

2. Mixed matrix A with 256×256 pixels is generated randomly.
3. Compute the receiver vector X for the 256×256 pixels according to Equation 14.1, and then transform this into m images containing 256×256 pixels. This yields computer simulated SAR images with multi-polarization channels.

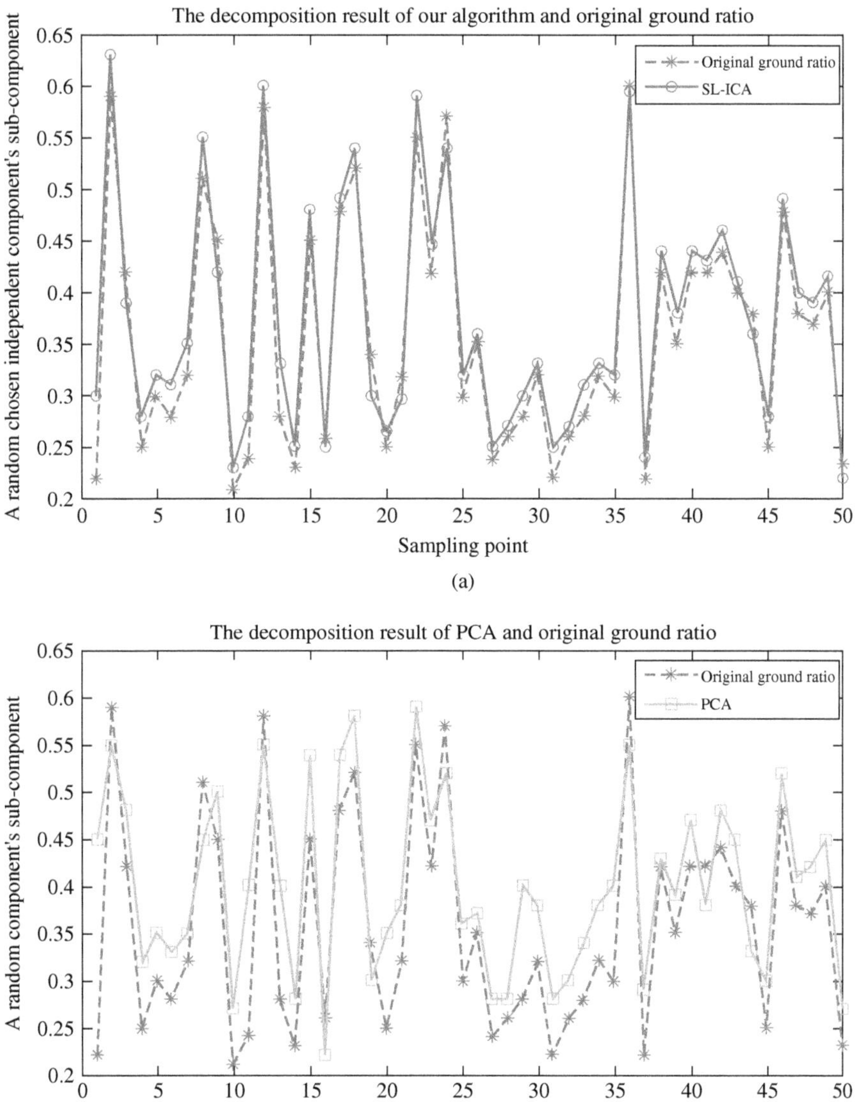

Figure 14.3 Comparison of the accuracy of the decomposition results for (a) SL-ICA, (b) PCA, and (c) ICA

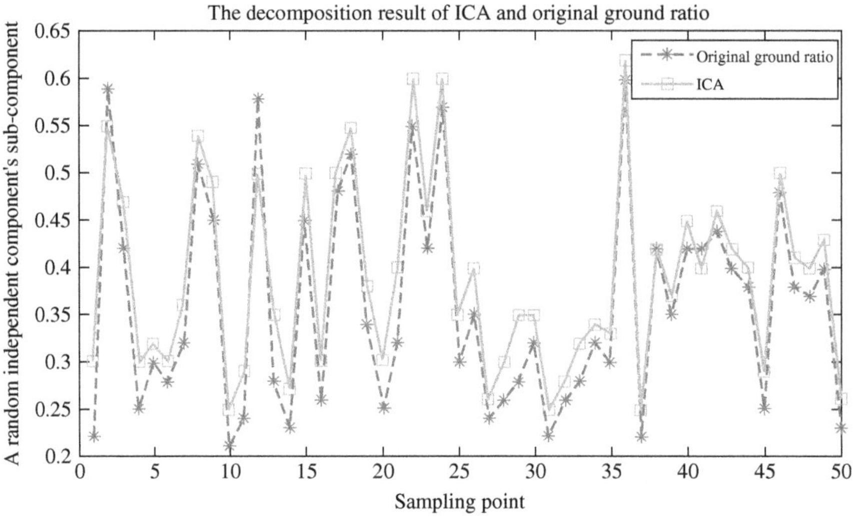

(c)

Figure 14.3 *Continued*

Figure 14.2a–c show the computer simulated SAR source images for three channels. Figure 14.2d–f present the mixed pixel images from three receivers. Next, we carry out the element decomposition operation using SL-ICA, PCA, and ICA methods.

From the comparison of the decomposition results shown in Figure 14.3 and the primitive independent components of 20 picture elements selected randomly from the results of SL-ICA, PCA, and ICA, we can see that the decomposition results for SL-ICA match the real terrain features more closely. According to the direct view on the chart, the small circles have a higher superposition rate with *, and the disparity of the non-superposing small circle with * is also smaller than that of the dices. Meanwhile, Figure 14.4 presents a better visual effect reflecting the same results as Figure 14.3. From Figure 14.4, we can see that the accuracy of the SL-ICA decomposition result is higher than that of PCA and ICA.

In the next step, we calculate the root-mean-square error (RMSE) and system error (SE) (Wu, 2004; Lu and Weng, 2006) for each of the three algorithms. The actual formulas are as follows:

$$\mathrm{RMSE}_i = \sqrt{\frac{\sum_{i=1,j=1}^{i=m,j=N} (s_i(j)^\wedge - s_i(j))^2}{N}}, \quad \mathrm{SE}_i = \frac{\sum_{i=1,j=1}^{i=m,j=N} |\, s_i(j)^\wedge - s_i(j)|}{N} \tag{14.12}$$

Table 14.1 gives two different indications of the SL-ICA, PCA, and ICA decomposition results, while Figure 14.4 illustrates the direct-view comparison. From left to right, Figure 14.5a shows the decomposition results for the three algorithms as RMSE_1, RMSE_2, RMSE_3, while Figure 14.5b gives SE_1, SE_2, SE_3.

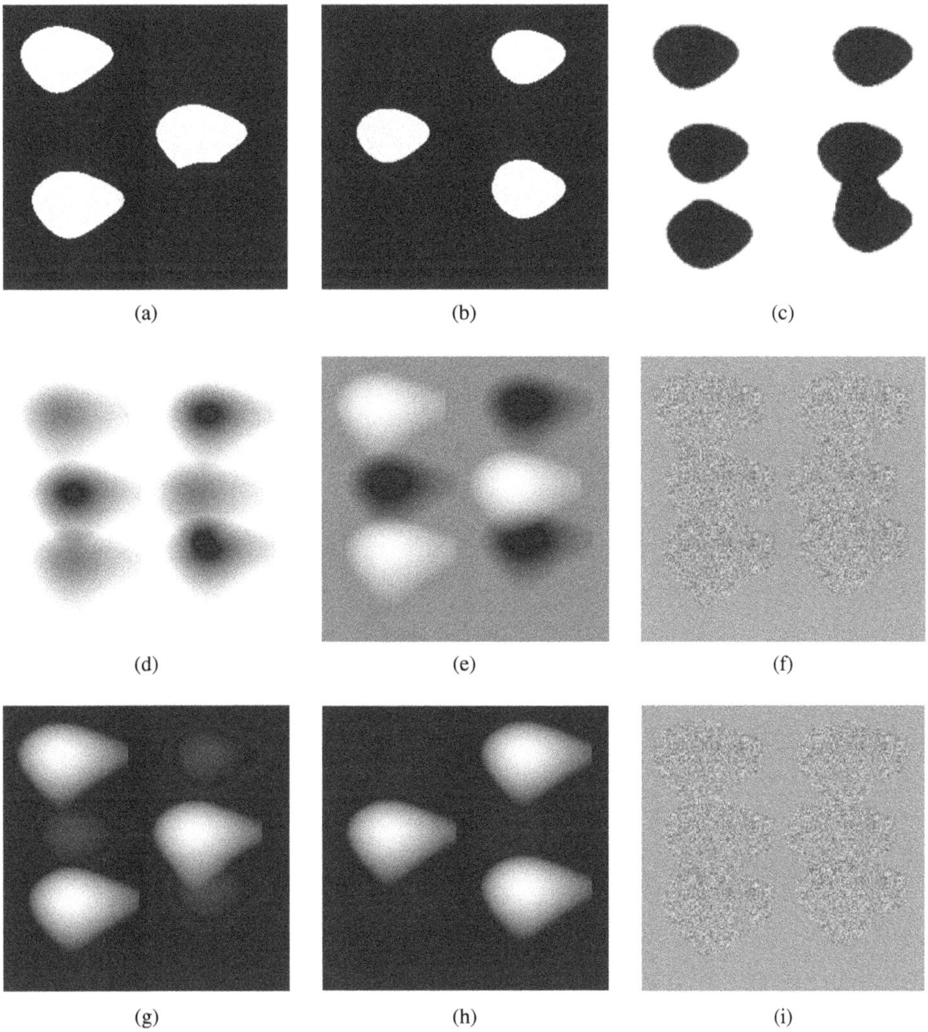

Figure 14.4 Comparison of the decomposition results for (a–c) SL-ICA, (d–f) PCA, and (g–i) ICA

Table 14.1 Accuracy indexes for the decomposition results of SL-ICA, PCA, and ICA algorithms

	$RMSE_1$	$RMSE_2$	$RMSE_3$	SE_1	SE_2	SE_3
SL-ICA	0.123	0.145	0.097	0.064	0.043	0.082
PCA	0.197	0.164	0.235	0.171	0.157	0.149
ICA	0.181	0.243	0.156	0.139	0.168	0.152

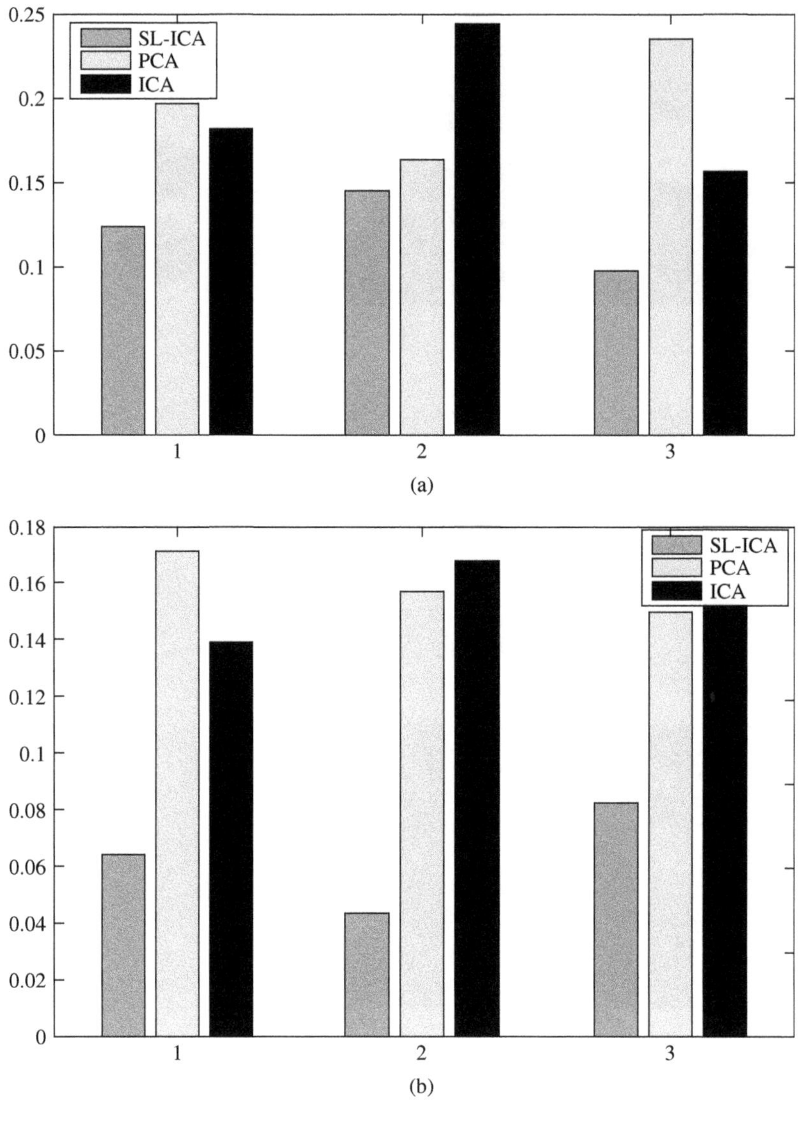

Figure 14.5 Comparison of the accuracy of the decomposition results for SL-ICA, PCA, and ICA: (a) comparison of $RMSE_1$, $RMSE_2$, $RMSE_3$ and (b) comparison of SE_1, SE_2, SE_3

14.5.2 Real Multi-polarization Channel SAR Images

Here, we use two polarization channel SAR images (ENVISAT SAR) of Beijing as the original images (see Appendix Figures 14A.1 and 14A.2, and the pixel size of the two intercepted images is 512×512, as shown in Figure 14.6. The experimental district is located in Liyuan, Tongzhou District of Beijing with simple terrain objects based on our personal field survey. Figure 14.6a,b illustrate the VV polarization image

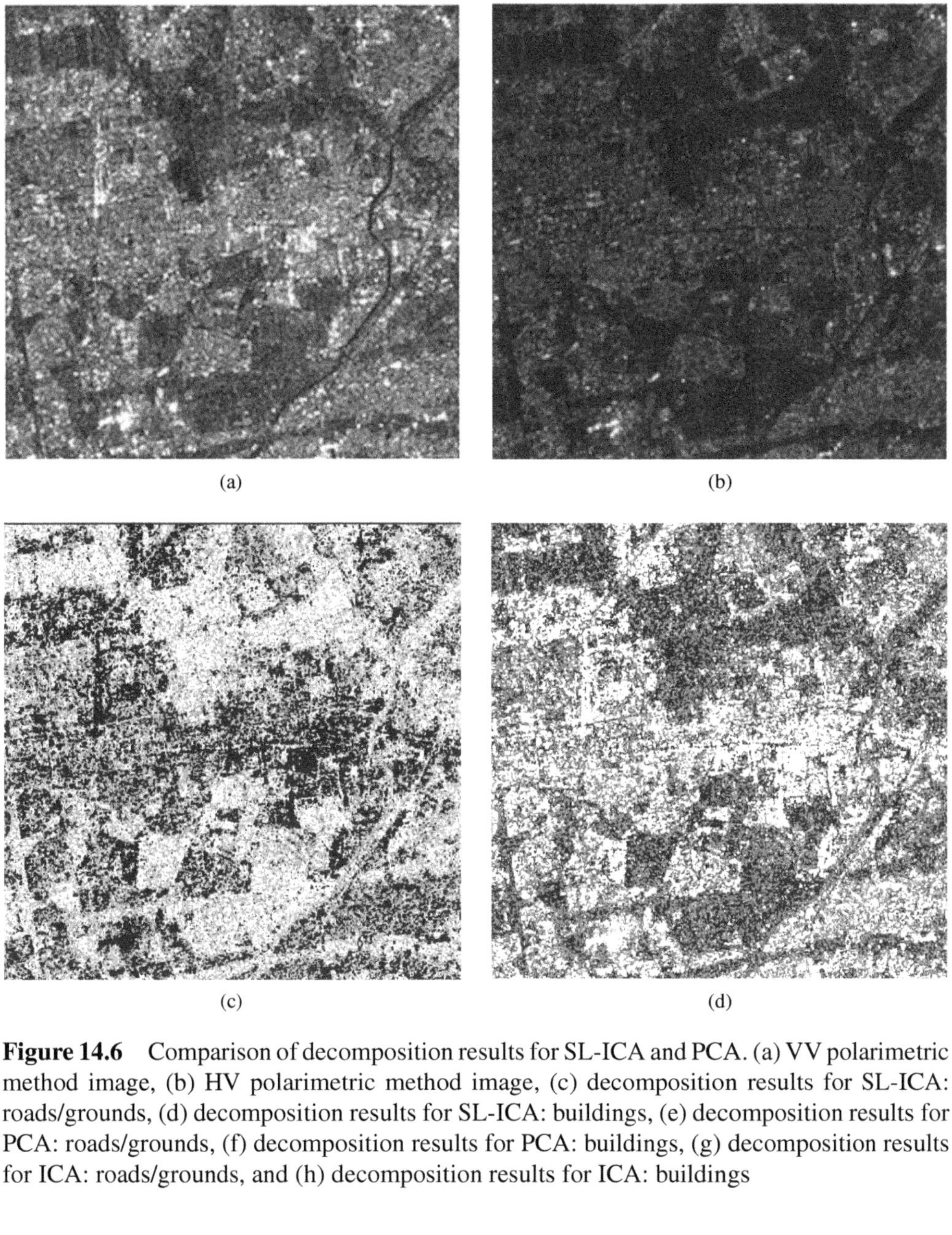

Figure 14.6 Comparison of decomposition results for SL-ICA and PCA. (a) VV polarimetric method image, (b) HV polarimetric method image, (c) decomposition results for SL-ICA: roads/grounds, (d) decomposition results for SL-ICA: buildings, (e) decomposition results for PCA: roads/grounds, (f) decomposition results for PCA: buildings, (g) decomposition results for ICA: roads/grounds, and (h) decomposition results for ICA: buildings

and HV polarization image, respectively. The images contain two types of features: roads/grounds and buildings.

Figure 14.6c,d illustrate the mixed pixel decomposition results for SL-ICA. Figure 14.6c shows a ground/road image, while Figure 14.6d depicts an artificial building image. In Figure 14.6c, the road outline is clearly visible, while in

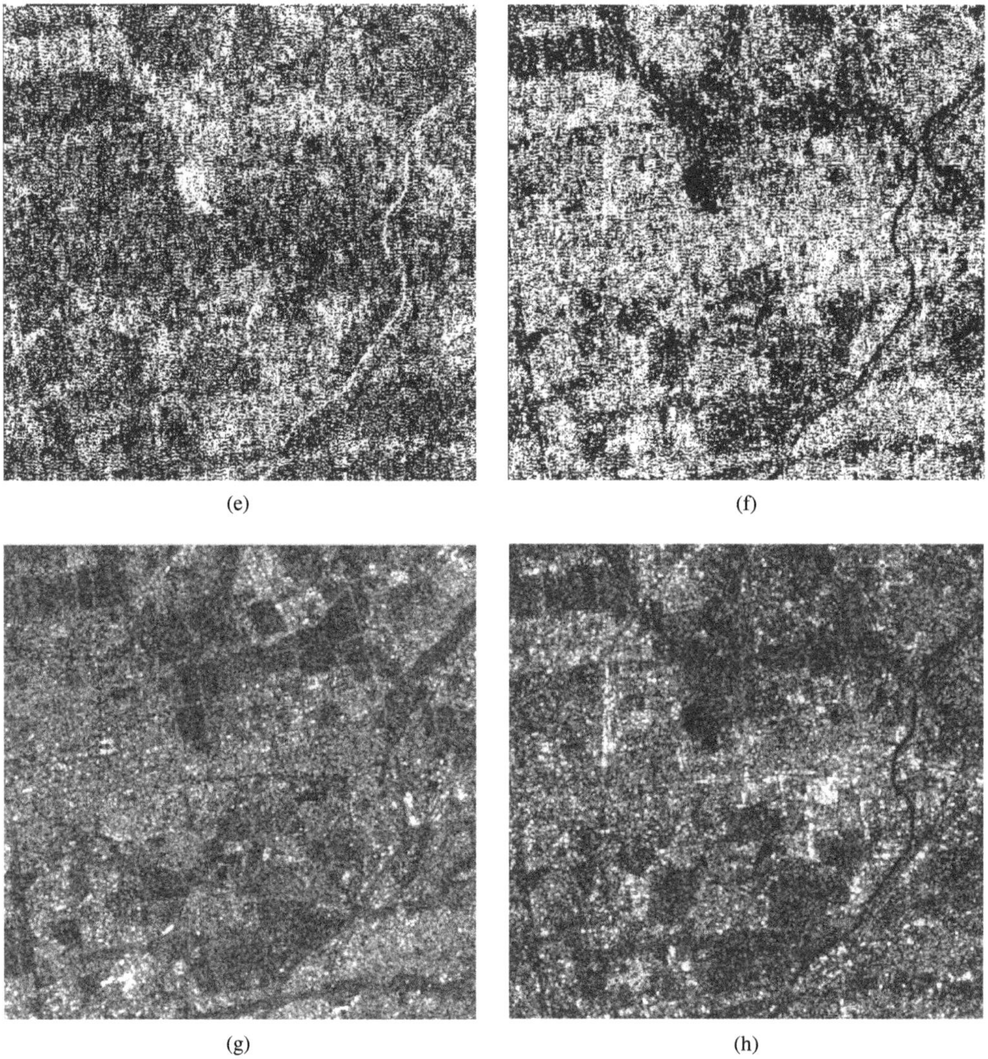

Figure 14.6 *Continued*

Figure 14.6d the building group is also neatly delineated. Thus, the results for SL-ICA yield perfect visual effects. Figure 14.6e,f depict the mixed pixel decomposition results for PCA. Figure 14.6g,h show the independent components decomposed by ICA. In Figure 14.6e,g, the road outline is clearly fuzzier than the outline in Figure 14.6c; while in Figure 14.6f,h, the subtle roads among the buildings are ambiguous and cannot be clearly identified. Comparing the results in Figure 14.6, we confirm that the decomposition results for SL-ICA are better than those for PCA and ICA. Figures 14A.3 and 14A.4 in the Appendix show the decomposition results of Figures 14A.1 and 14A.2. We can see that the SL-ICA also give satisfying results.

14.6 Chapter Summary

In this chapter, we proposed a new algorithm called SL-ICA, which is applied to decompose mixed pixels in SAR images, and presented the objective function and optimization algorithm, and provided a concrete implementation process for SL-ICA. We used both artificially simulated multi-polarization SAR images and real multi-polarization SAR images as experimental data. The results qualitatively and quantitatively show that SL-ICA has better decomposition precision and visual decomposition effects than PCA and ICA. The method has important theoretical and practical significance in solving high-precision ground classification, target detection, identification, and so on. At the same time, because we chose a mixed pixel decomposition model for optical remote sensing images as the mathematical model for our mixed pixel decomposition, SL-ICA can be applied to mixed pixel decomposition of optical remote sensing images.

Appendix 14.A Polarization Channel SAR Images of Beijing and the Decomposition Results using SL-ICA

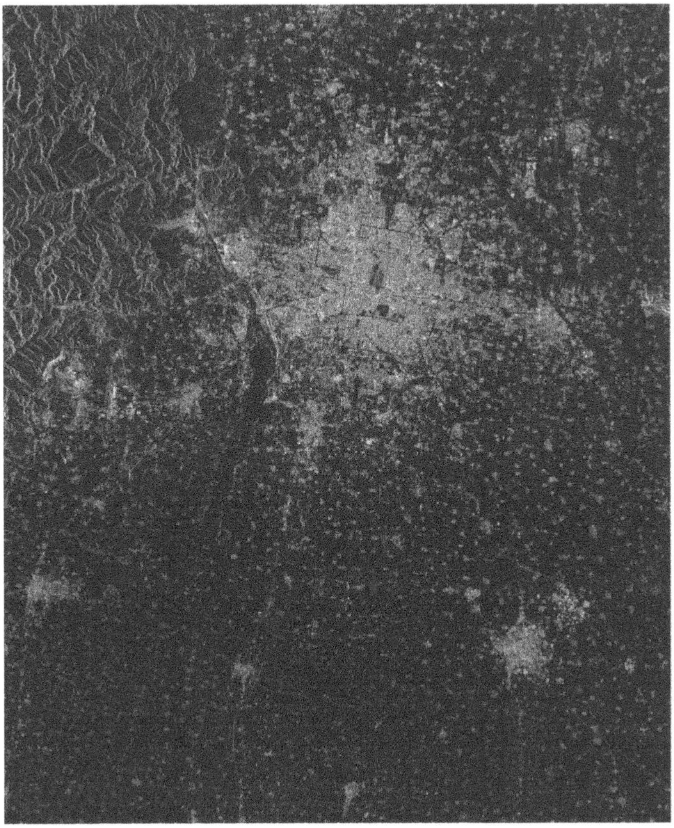

Figure 14A.1 VV polarimetric method panoramic image

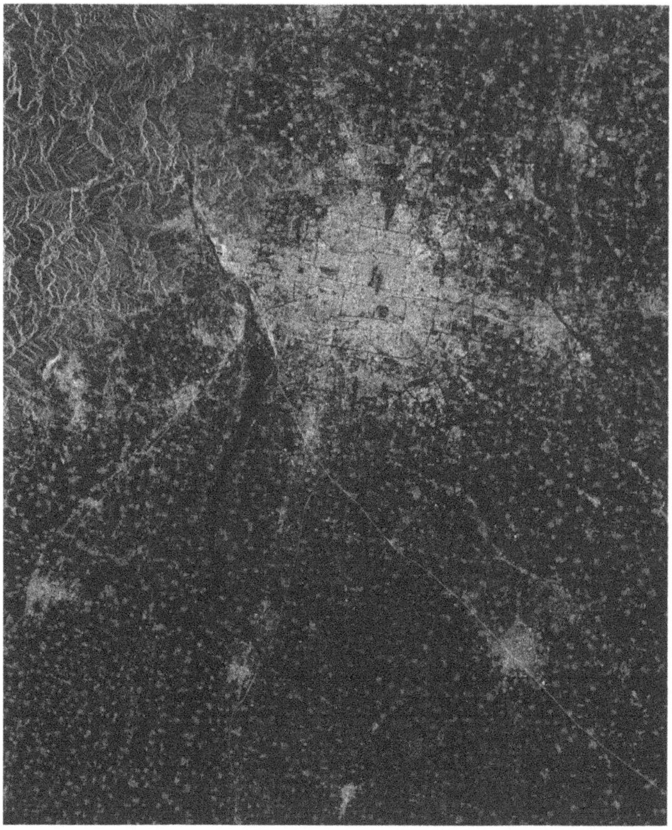

Figure 14A.2 HV polarimetric method panoramic image

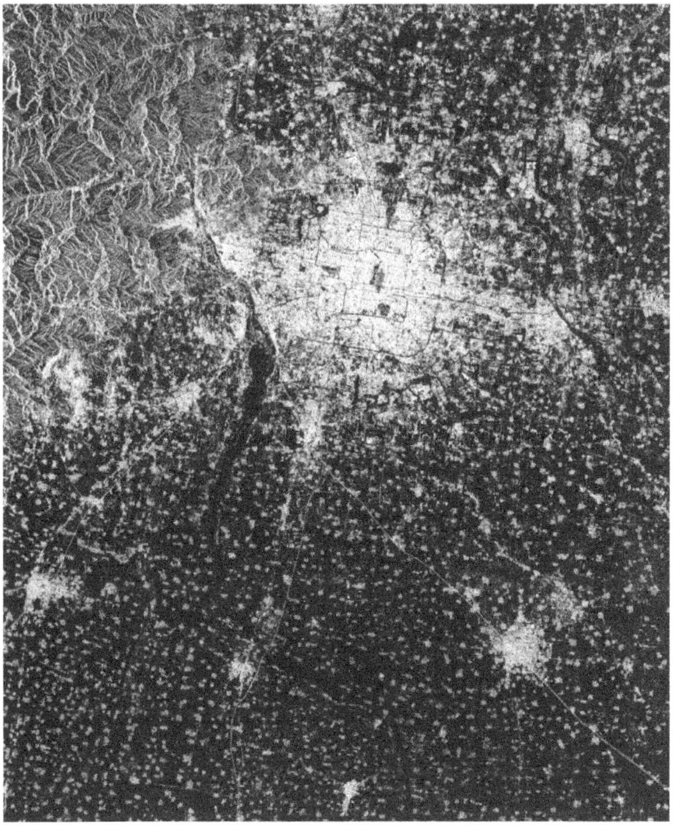

Figure 14A.3 SL-ICA based decomposition result for Figure 14A.1

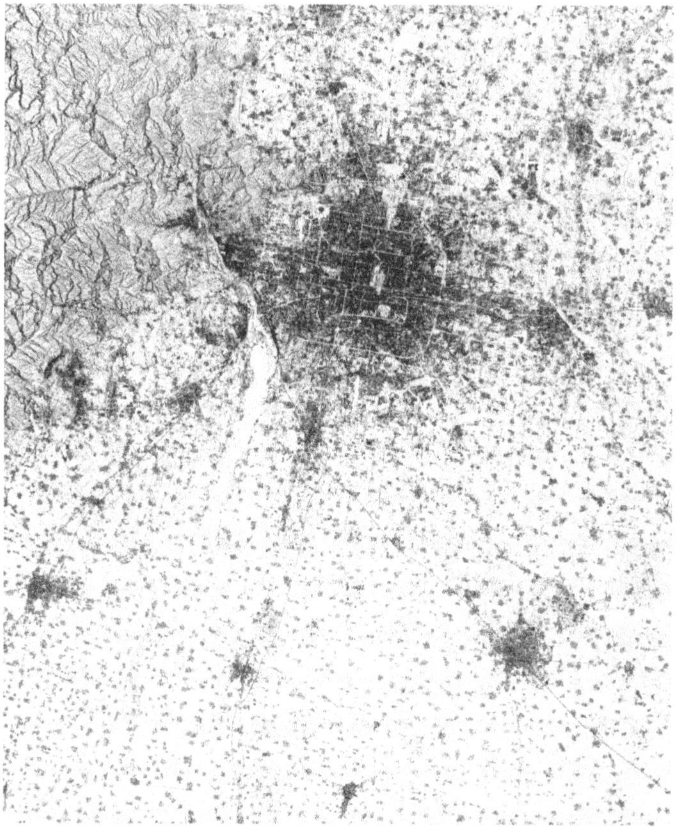

Figure 14A.4 SL-ICA based decomposition result for Figure 14A.2

References

Bastin, L. (1997) Comparison of fuzzy c-means classification, linear mixture modelling and MLC probabilities as tools for unmixing coarse pixels. *International Journal of Remote Sensing*, **18**(17): 3629–3648.

Bressan, M., Guillamet, D. and Vitria, J. (2003) Using an ICA representation of local color histograms for object recognition. *Pattern Recognition*, **36**(3): 691–701.

Chang, C.I. (2002) Target signature-constrained mixed pixel classification for hyperspectral imagery. *IEEE Transactions on Geoscience and Remote Sensing*, **40**(5): 1065–1081.

Chang, C.I., Ren, H., Chang, C.C. *et al.* (2004) Estimation of subpixel target size for remotely sensed imagery. *IEEE Transactions on Geoscience and Remote Sensing*, **42**(6): 1309–1320.

Chang, C.I., Zhao, X.L., Mark, L.G. and Pan, J.J. (1998) Least squares subspace projection approach to mixed pixel classification for hyperspectral images. *IEEE Transactions on Geoscience and Remote Sensing*, **36**(3): 898–912.

Giles, M.F. and Manoj, K.A. (1996) Incorporating mixed pixels in the training, allocation and testing stages of supervised classifications. *Pattern Recognition Letters*, **17**, 1389–1398.

Hyvdrinen, A., Karhunen, J. and Oja, E. (2001) *Independent Component Analysis*, New York, John Wiley & Sons, Inc.

Inglada, J. and Adragna, F. (2002) Blind source separation applied to multitemporal series of differential SAR interferograms. Geoscience and Remote Sensing Symposium (IGARSS), pp. 1240–1242.

Kanellopoulos, I., Varfis, A., Wilkinson, G.G., and Megier, J. (1991) Classification of remotely-sensed satellite images using multi-layer perceptron networks. Proceedings of the International Conference on Artificial Neural Networks, pp. 1067–1070.

Loghmari, M.A., Naceur, M.S., and Boussema, M.R. (2002) Mixed pixel decomposition of satellite images based on source separation method. Geoscience and Remote Sensing Symposium, IGARSS'02, pp. 914–916.

Lu, D.S. and Weng, Q.H. (2006) Use of impervious surface in urban land-use classification, *Remote Sensing of Environment*, **102**(1–2): 146–160.

Ristaniemi, T. and Joutsensalo, J. (1999) On the performance of blind source separation in CDMA downlink. Proceedings of the International Workshop on Independent Component Analysis and Signal Separation, pp. 437–441.

Shah, C.A., Varshney, P.K. and Arora, M.K. (2007) ICA mixture model algorithm for unsupervised classification of remote sensing imagery. *International Journal of Remote Sensing*, **28**(8): 1711–1731.

Shao, M.Z. and Deng, H.L. (1989) Mixed image cell decomposition based on principle component analysis. *Chinese Space Science Technology*, **2**(1): 51–56, (in Chinese).

Shaw, G. and Burke, H. (2003) Spectral imaging for remote sensing. *Lincoln Laboratory Journal*, **14**(1): 3–28.

Si, H.L. and Zhang, L.M. (2002) *Intelligent Visual Image Processing-Unsupervised Learning Methods and other Methods of Multi-channel Images*, Shanghai, Shanghai Science and Technology Education Press.

Szu, H. and Kopriva, I. (2001) Artificial neural networks for noisy image super-resolution. *Optics Communications*, **198**(1): 71–81.

Tsai, D.M., Tseng, Y.H., Chao, S.M., and Yen, C.H. (2006) Independent component analysis based filter design for defect detection in low-contrast textured images. 18th International Conference on Pattern Recognition, pp. 231–234.

Wu, C.S. (2004) Normalized spectral mixture analysis for monitoring urban composition using ETM+ imagery. *Remote Sensing of Environment*, **93**(4): 480–492.

Part III

Advances and Applications of BSS

15

Non-negative Matrix Factorization Algorithms and Applications

15.1 Introduction

15.1.1 The Origin of Non-negative Matrix Factorization Theory

Non-negative matrix factorization (NMF) is a recently proposed matrix factorization method in which the constraint condition of non-negative elements yields a completely different result. NMF theory has received extensive attention from researchers in various fields since being put forward: it provides a new method for processing large-scale data using a matrix factorization model and, compared with classical algorithms, the NMF algorithm has the advantages of easy implementation, good interpretability of the decomposition results, and a small storage space. Besides its continued theoretical development, NMF has gradually become one of the most popular multidimensional data processing tools in fields such as signal processing, biomedical engineering, pattern recognition, computer vision, and image engineering.

The theory of NMF can be traced back to a paper published by Paatero and Tapper (1994). As well as proposing NMF, they also conducted a factor analysis (FA) for actual environmental data in which each obtained factor is a non-negative linear combination of basic variables. From a computational perspective, the existence of non-negative values in the analytical results is correct. Indeed, from an application point of view, negative factors often have no practical meaning.

In a practical environment, the value of each factor has physical significance. If the factor has a positive value, it indicates that the factor performs some role, and a value of zero suggests that the factor has no physical relevance to the system. A negative value, however, can be difficult to explain. For such problems, it is important to ensure the non-negativity of factors in the transpose derivation and iterative operation. Therefore, Paatero and Tapper put forward an optimization model for such a problem, namely

$$\|\boldsymbol{Q} \cdot (\boldsymbol{V} - \boldsymbol{WH})\|^2, \ \boldsymbol{W} \geq 0, \ \boldsymbol{H} \geq 0 \tag{15.1}$$

Blind Source Separation: Theory and Applications, First Edition. Xianchuan Yu, Dan Hu and Jindong Xu.

where $\|\cdot\|^2$ indicates the L_2 paradigm, each column of the matrix $\boldsymbol{V}$ is an actual observed value, each column of the matrix $\boldsymbol{W}$ is a factor, and each row of the matrix $\boldsymbol{H}$ is the impact of each factor. The weight of each element, which represents the confidence level of each observed value, is given by the matrix $\boldsymbol{Q}$. The goal of the optimization model is to minimize the target function, for which Paatero and Tapper first calculated this model according to a constrained algorithm of least squares (ALS). Namely, we optimize $\boldsymbol{H}$ with a fixed $\boldsymbol{W}$, then switch the variable role and optimize $\boldsymbol{W}$ with a fixed $\boldsymbol{H}$, and continue this procedure until the optimal solution is reached.

Paatero subsequently designed a variety of methods to improve the above optimization algorithm, such as positive matrix factorization (PMF) (Paatero, 1997). However, PMF modifies the above algorithm too much, making the entire process more complicated. He also proposed a more general multilinear engine algorithm model (Paatero, 1999) to solve a multi-factor model by meeting the non-negative constraint. This algorithm makes use of an improved variable gradient algorithm to solve the optimization problem. In short, Paatero and other researchers were instrumental in pushing forward non-negative FA. However, from the current perspective, there are still some defects. First, the field of application of this research was too limited, being confined to some specific applications of NMF. No further research on the application and promotion of the algorithms in other areas was conducted. Second, we have no theory to prove the convergence and complexity of the algorithms. These properties are simply based on experience, with a lack of any theoretical foundation.

In 1997, Lee and Seung (1997) studied the following coding question in a paper about unsupervised learning: assuming each column of the matrix $\boldsymbol{W}$ is a fixed feature vector and $\boldsymbol{V}$ is an input vector using encoding processing, our goal is that the reconstruction error is minimized, that is, min $\|\boldsymbol{V}-\boldsymbol{WH}\|^2$. We can obtain different learning rules using different limits for h. For example, the principal component analysis (PCA) algorithm corresponds to unconstrained minimization, whereas vector quantization (VQ) requires that h is the unit vector. Lee and Seung proposed two techniques to generate the PCA and VQ algorithms: one is Convex Coding, which requires the elements of h to be non-negative and sum to 1, and the other is Conic Coding, which only requires that each element of the vector h is non-negative. Lee and Seung then studied how to determine the best feature vector set for their coding strategy. Hence, they studied the matrix approximation problem min $\|\boldsymbol{V}-\boldsymbol{WH}\|^2$ wherein each column of the matrix $\boldsymbol{V}$ is the observed data, the column number of $\boldsymbol{W}$ is far less than that of $\boldsymbol{V}$, each element of $\boldsymbol{W}$ and $\boldsymbol{H}$ is greater than zero at the same time, and both the sum of the column elements of $\boldsymbol{W}$ and the sum of the row elements of $\boldsymbol{H}$ are 1. To solve the above minimization question, they proposed a kind of alternating projection gradient algorithm. The realization of the algorithm is as follows. First, iterate the target function for $\boldsymbol{H}$ by fixing $\boldsymbol{W}$ using the gradient descent method, then set all negative elements of $\boldsymbol{H}$ to zero. Next, transform the roles of $\boldsymbol{W}$ and $\boldsymbol{H}$: iterate the target function for $\boldsymbol{W}$ using the gradient descent method by fixing $\boldsymbol{H}$. The sum of each column of $\boldsymbol{W}$ and each row of $\boldsymbol{H}$ is maintained at 1 by introducing a penalty function to the algorithm. This algorithm randomly selects the initial point and is iterated to reach the global optimum solution. It was found that Convex Coding could find the local linear relationship hidden in the data and Conic Coding could reveal the characteristics of the data. However, no proof of these algorithms was given.

In 1999, Lee and Seung (1999) proposed a definition of NMF that made all components non-negative after factorization and realized a reduction in nonlinear dimensionality. In 2001, the same authors (Lee and Seung, 2001) introduced two further simple matrix factorization algorithms. The target function of one algorithm is found by minimizing the square of the Euclidean distance, and that of the other is determined by minimizing the KL (Kullback–Leibler) divergence. Both algorithms adopt multiple iterative gradient descent rules and select many different initial points. Lee and Seung proved the convergence of the iterative rules of the algorithms, which theoretically guarantees the convergence of the algorithms themselves. The algorithms have been successfully applied in the fields of face recognition and syntax recognition, among others.

Subsequently, the above algorithms were generalized and improved by other researchers. NMF theory has become a new research hotspot in the field of information science. Some researchers have expanded the NMF formula by increasing the auxiliary limits of $\boldsymbol{W}$ and $\boldsymbol{H}$. For example, Guillamet, Bressan, and Vitrià (2001) multiplied both sides of the standard factorization by a diagonal matrix $\boldsymbol{Q}$ to give $\boldsymbol{VQ} = \boldsymbol{WHQ}$, eliminating the feature redundancy of the columns of the base matrix $\boldsymbol{W}$. This can be solved by the random column limitation of $\boldsymbol{H}$. Hoyer (2004) used a sparseness limitation to improve the local representation of data in $\boldsymbol{W}$ and $\boldsymbol{H}$. Catral *et al.* (2004) conducted a theoretical analysis of NMF, and found that the auxiliary limitation is dependent on the needs of the specific problem. Pauca, Piper, and Plemmons (2006) added a smoothness limitation to process the spectrum characteristic of remote sensing data based on standard NMF, and Xue *et al.* (2006) added Fisher's linear discriminant analysis based on the standard NMF formula to improve the local spatial geochemical characteristics. Rutkowski, Zdunek, and Cichocki (2007) used time smoothing and space restriction to improve the analysis of EEG (electroencephalography) data.

15.1.2 Characteristics of NMF Theory

Matrix factorization is the basis of many statistical analysis methods, such as PCA, FA, VQ, independent component analysis (ICA), and sparse component analysis (SCA). In these methods, the original matrix $\boldsymbol{V}$ is approximately decomposed into two low-rank matrices such that $\boldsymbol{V} = \boldsymbol{WH}$. The common characteristic of these methods is that elements of the factors $\boldsymbol{W}$ and $\boldsymbol{H}$ can be positive or negative. Even if the initial input matrix elements are all positive, the traditional rank cut algorithm cannot guarantee the non-negativity of the decomposition result. From a computational perspective, the existence of a negative value in the decomposition result is acceptable, but from the perspective of applications, negative elements often have no meaning. In geochemical and remote sensing data, negative values can be difficult to explain. NMF is a relatively recent multivariate data method that can represent the low-dimensional structure of patterns in high-dimensional space. We can obtain part of the linear expression of the data with non-negative constraints to give a smaller number of base vectors and an optimal linear estimation of the data. These base vectors use a pure additive description method that can reflect the local characteristics of data with definite physical meaning.

Unlike PCA, the base function space of NMF is not necessarily orthogonal, and all elements of vectors generated by the linear combination of basis vectors are non-negative, which makes NMF superior to multivariate analysis methods based on PCA. First, when data redundancy is relatively serious, NMF is still able to find a base function to show the data structure, but it is difficult to determine the relationship between variables using the principal component with orthogonal characteristics decomposed according to PCA. Second, the base function of NMF combines single components according to their linear weight, which is more suitable for image processing, and the analytical results of NMF are non-negative, which overcomes the existence of non-negative base vector values during the decomposition of PCA. The constraint of non-negativity conforms to the requirements of many practical problems in geochemistry and remote sensing. Current data on geology and mineral resources are the comprehensive results of long-term and multi-stage geological mineralization, and the scale of this data is very large. Therefore, the NMF algorithm for exploring matrices is a very significant research subject for mineral prediction.

ICA has two important limitations when it is applied to problems of blind source separation, namely the statistical independence and non-Gaussian distribution of the source. When we use NMF to conduct a statistical blind source separation, there are no limitations and we can separate the signals (Wei, 2004).

NMF can be seen as an optimization problem that provides an iterative method to obtain $\boldsymbol{W}$ and $\boldsymbol{H}$. This method has the characteristics of fast convergence, sparseness, non-negativity, and dimension reduction.

15.1.3 Application Fields of NMF Theory

Compared with ICA, NMF has no constraints on the statistical independence and non-Gaussian distribution of the source. We can achieve source separation (important in geochemistry and remote sensing data analysis) by NMF as long as no first-order statistical correlation exists between the source signals. In view of its characteristics, NMF has already become the main processing tool for multi-dimensional data in the fields of signal processing, biomedical engineering, pattern recognition, computer vision, and image engineering. Specifically, it has been applied in the study of text analysis and clustering, digital watermarks, face detection and recognition, image retrieval, image restoration, language modeling, speech classification, music signal analysis and instrument recognition, blind signal separation, network security, gene and cell analysis, and brain wave signal processing (Inderjitt *et al.*, 2001; Padilla *et al.*, 2010).

Guillamet, Bressan, and Vitrià (2001) developed a weighted non-negative matrix factorization (WNMF) algorithm. They applied this to image classification such that NMF was able to capture local positive features. Feng *et al.* (2002) introduced local non-negative matrix factorization (LNMF) and applied it in image recognition, and Hoyer (2002) put forward a non-negative sparse coding (NNSC) approach that was applied to non-negative variable analysis of the optic nerve. At present, image analysis and processing is one of the most successful applications of NMF (Miao and Wang, 2005). NMF improves the precision rate of image retrieval (Lu *et al.*, 2005), and exhibits better performance than the traditional wavelet transformation and Laplace tower method in image fusion applications (Yan, Xia, and Zheng, 2007). Perfect image classification and image compression results have been obtained using NMF (Liu, Zheng, and Liu, 2006). NMF can also be used for facial recognition, generating better results than the feature description of ICA and facial recognition of PCA due to its strong robustness (Lee and Seung, 1999). In medical information processing, NMF can be used to deal with the dynamic continuous images used in nuclear medicine, molecular cancer classification, and gene expression data analysis (Gao and Church, 2005). For text clustering, NMF can achieve the effect of expressing the local correlation effect of information, and has also shown good results in speech signal analysis and processing and language modeling (Plumbley *et al.*, 2002).

In addition, the NMF algorithm has a very good record in environmental data processing, signal analysis, and complex object recognition (Babji and Tangirala, 2010). In recent years, the application of NMF has expanded considerably. We believe that NMF will have more successful applications in every field, thus promoting the further research of NMF.

15.2 NMF Algorithms

15.2.1 The Mathematical Model of NMF

NMF theory is essentially a method of obtaining the representation of data according to non-negative constraints. It can be described as follows: for any given non-negative matrix $\boldsymbol{V}$, NMF can find a non-negative matrix $\boldsymbol{W}$ and a non-negative matrix $\boldsymbol{H}$ satisfying $\boldsymbol{V} \approx \boldsymbol{WH}$, thus decomposing one non-negative matrix into two non-negative matrices. As there are only non-negative elements in the decomposed matrix, the column vectors of the original matrix $\boldsymbol{V}$ can be thought of as the weighted sum of all column vectors (called base vectors) in the base matrix $\boldsymbol{W}$, and the weight coefficients are the elements corresponding to the column vectors of the coefficient matrix $\boldsymbol{H}$. This representation based on the combination of base vectors has a very intuitive explanation.

It is noteworthy that both the NMF and ICA algorithms are essentially matrix factorization, but NMF has increased non-negative constraints whereas ICA has an increased independence constraint. In the mathematical model of NMF, $\boldsymbol{W}$ corresponds to the matrix $\boldsymbol{A}$ in the ICA decomposition model, and $\boldsymbol{H}$ corresponds to

S. Interested readers can compare the two algorithms. The mathematical model of NMF theory is as follows:

Given the non-negative matrix $\boldsymbol{V}$, find the non-negative matrix factors $\boldsymbol{W}$ and $\boldsymbol{H}$ such that

$$\boldsymbol{V} = \boldsymbol{WH} + \boldsymbol{E}, \boldsymbol{V} \in \boldsymbol{R}^{n\times m}, \boldsymbol{W} \in \boldsymbol{R}^{n\times r}, \boldsymbol{H} \in \boldsymbol{R}^{r\times m} \tag{15.2}$$

That is, given an n-dimensional data vector set $\boldsymbol{V}$, wherein n is the number of variants, m is the number of the data samples in the set, and r is the number of principal components, the matrix can be decomposed as the sum of an error matrix $\boldsymbol{E}$ and the product of the base matrix $\boldsymbol{W}$ and the weight coefficient matrix $\boldsymbol{H}$. $\boldsymbol{V}$ is called the observation data matrix, and $\boldsymbol{W}$ and $\boldsymbol{H}$ are called the factor matrices. For simplicity, we do not generally consider the error factor. Hence, the model can be described as follows:

$$\boldsymbol{V} \approx \boldsymbol{WH} \tag{15.3}$$

or

$$\boldsymbol{V}_{ij} \approx (\boldsymbol{WH})_{ij} = \sum_{k=1}^{r} \boldsymbol{W}_{ik}\boldsymbol{H}_{kj} \tag{15.4}$$

In Equations 15.2 and 15.4, $\boldsymbol{V}_{ij}$ is an element of the non-negative matrix $\boldsymbol{V}$, and $\boldsymbol{W}_{ik}$ and $\boldsymbol{H}_{kj}$ are elements of the factor matrices $\boldsymbol{W}$ and $\boldsymbol{H}$. Usually, in Equations 15.2–15.4, the value of r is far smaller than n and m, and the condition $(n + m)r < nm$, which implies that a large quantity of data are described with few base vectors, should hold. Thus, we obtain a compressed model of the raw data matrix. If v_j and h_j are corresponding columns of $\boldsymbol{V}$ and $\boldsymbol{H}$, Equation 15.4 can be written in vector form: $v_j \approx \boldsymbol{W}h_j$. That is to say, each sample v_j can be considered as a non-negative linear combination of column vectors of the non-negative matrix $\boldsymbol{W}$, where the combination coefficients are the components of h_j. Therefore, the matrix $\boldsymbol{W}$ can be regarded as a set base for the linear approximation of the data matrix $\boldsymbol{V}$, and $\boldsymbol{H}$ is the non-negative projection coefficient of the sample set $\boldsymbol{V}$ on base $\boldsymbol{W}$. Usually, a small number of base vectors can represent a large amount of data vectors. This means that, if we can find the right base vector set and make it represent the potential structure of the relationship between the data, we will get a very good approximation.

15.2.2 Basis Algorithm of NMF

15.2.2.1 NMF Algorithm Based on the Maximum Likelihood

Lee and Seung (1999) were the first to put forward a definition of NMF. This supposes that each element of the matrix is statistically independent and $\boldsymbol{V}_{ij}$ obeys a Poisson distribution with $(\boldsymbol{WH})_{ij}$ as parameters. Based on this assumption, the NMF algorithm was designed according to the idea of maximum likelihood (ML), with the likelihood function as the objective function. This algorithm is the precursor of NMF.

The target function of the algorithm is defined as:

$$F = \sum_{i=1}^{n} \sum_{j=1}^{m} [V_{ij} \log (WH)_{ij} - (WH)_{ij}] \tag{15.5}$$

Algorithm NMF algorithm based on ML

Input: the sample set $V = (V_1, V_2, \cdots, V_m), V \in \mathbf{R}^{n \times m}$

Output: the feature matrix $W = (W_1, W_2, \cdots, W_r)$ and the projection coefficient matrix H

Steps:

1. Randomly initialize the matrices W^0 and H^0 with positive elements. Normalize the column vectors of W^0 and initialize $t = 0$.
2. Iterate until convergence:
   a. Iterate W, $W_{ia} \leftarrow W_{ia} \sum_j \frac{V_{ij}}{(WH)_{ij}} H_{aj}$
   b. Normalize the column vectors of W, $W_{ia} \leftarrow \frac{W_{ia}}{\sum_\mu W_{\mu a}}$
   c. Iterate H, $H_{aj} \leftarrow H_{aj} \sum_i W_{ia} \frac{V_{ij}}{(WH)_{ij}}$
   d. Update the iteration number $t = t + 1$.

The algorithm iterates according to the above rules, and converges when the target function attains a maximum. The proof of the convergence of the algorithm is similar to that of the EM (expectation–maximization) algorithm (Dempster, Laired, and Rubin, 1977). The iteration method ensures the non-negativity of W and H and that the sum of the elements of each column of W is 1. The process of the algorithm is shown in Figure 15.1.

Lee and Seung conducted a number of experiments using a face image database to compare NMF, PCA, and VQ. The experimental results (Figure 15.2) show that the obtained images of human brains are the linear combination of basis images, but the results differ according to which of the three methods is used. Using VQ, each feature is a whole face image and each column of the matrix H is a unit vector with just one element being 1 and the others being 0. The base feature of the PCA method is a twisted face image, and the column features of W and the line vectors of the matrix H are mutually perpendicular. The NMF method is a kind of representation in which each column feature of the matrix W contains local characteristics corresponding to each part of the face. The elements of W and H are all non-negative. The feature vectors are eyes, noses, mouths, and so on.

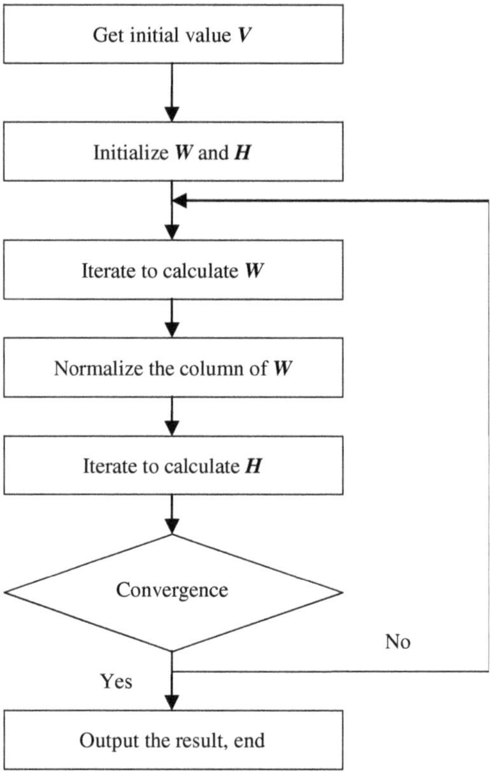

Figure 15.1 Flow chart of the non-negative matrix algorithm

The following sections describe two NMF algorithms based on the iteration of ***W*** and ***H***: one using the Euclidean distance and the other based on the divergence deviation. The two algorithms are simple and their convergence is guaranteed. Thus, they obtain good results in practical applications. NMF algorithms based on the Euclidean distance and the divergence deviation are widely regarded as reference NMF algorithms. In each iteration of the two algorithms, the values of ***W*** and ***H*** are obtained through the product of current values and some factors. Experimental results have proved that the effect of approximation can be guaranteed according to these iterative update rules, which means that the algorithms must be able to converge to a local optimum solution in accordance with the rules of iteration.

15.2.2.2 NMF Algorithm Based on Euclidean Distance

Early research methods adopted the traditional gradient descent technique and iterative rules for the NMF algorithm, with negative values being forced to 0. Lee and Seung (2001) proposed a reference algorithm that takes the square of the Euclidean

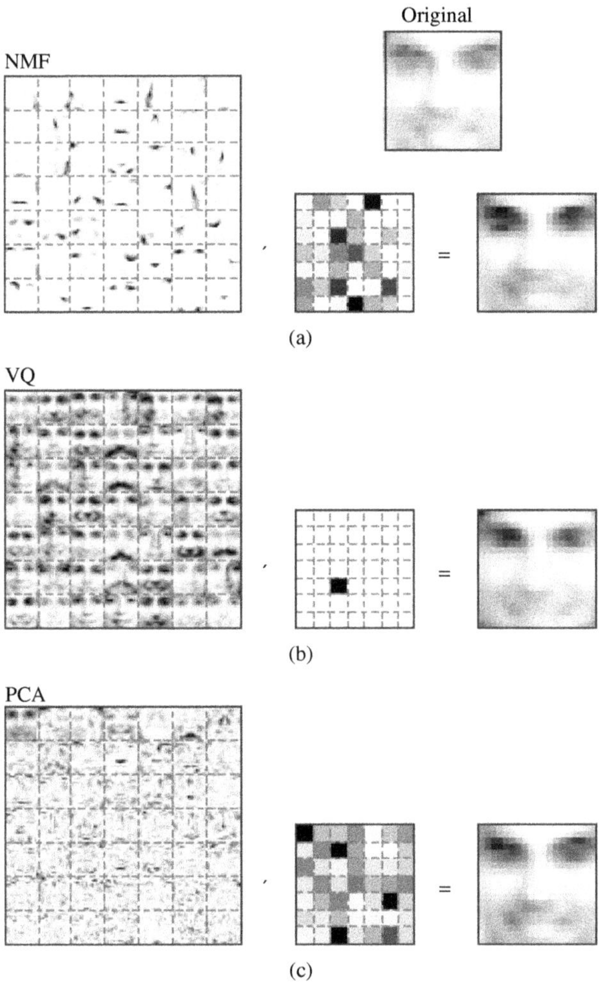

Figure 15.2 Comparison of (a) NMF, (b) VQ, and (c) PCA

distance between $\boldsymbol{W}$ and $\boldsymbol{H}$ as the target function. The algorithm is monotonically decreasing and uses a similar optimization strategy to that of the EM algorithm. Usually, the NMF algorithm uses a random initialization. Later, many researchers performed the initialization using K-means clustering or PCA. These methods improve the efficiency of the algorithm, but may converge to relatively bad solutions.

According to the mathematical model of NMF theory, we must find a decomposition $\boldsymbol{V} \approx \boldsymbol{WH}$ such that $\boldsymbol{WH}$ is close to $\boldsymbol{V}$, that is, the error should be as small as possible. Hence, we must define a corresponding measure; namely, we must define a target function to evaluate the effect of approximation. Usually, the target function is the Euclidean distance between $\boldsymbol{W}$ and $\boldsymbol{H}$.

We define the target function as:

$$\|V - WH\|^2 = \sum_{i,j} [V_{ij} - (WH)_{ij}]^2 \tag{15.6}$$

In the above equation, if $V = WH$, the target function is minimized and has a value of 0. The NMF algorithm based on the target function of Equation 15.6 can be defined as the following optimization problem: minimize $\|V - WH\|^2$, where the elements of W and H are non-negative.

Although the above target function is convex with respect to either W or H, it is not convex with respect to W and H at the same time. Thus, it is not realistic to find the global optimal solution. However, there are still many optimization techniques for finding a local optimal solution.

The gradient descent method is perhaps the simplest and most feasible method of finding an iterative local optimal solution that meets the above requirements, but its convergence speed is very slow. A convergent algorithm based on the gradient is very sensitive to the selection of step size, which is very inconvenient for practical applications. The above problem of optimization can use the method of alternately iterating W and H to obtain a local optimal solution.

Theorem 15.1 Under the iteration rule $H_{kj} \leftarrow H_{kj}\frac{(W^{\mathrm{T}}V)_{kj}}{(W^{\mathrm{T}}WH)_{kj}}$, $W_{ik} \leftarrow W_{ik}\frac{(VH^{\mathrm{T}})_{ik}}{(WHH^{\mathrm{T}})_{ik}}$, the Euclidean distance $\|V - WH\|^2$ is monotonic and does not increase. The algorithm converges when W and H reach some local optima.

Compared with the rules of the gradient descent method, under specific conditions, the incremental iterative rules that reduce H in $||V - WH||^2$ can be written as follows:

$$H_{au} \leftarrow H_{au} + \eta_{au}[(W^{\mathrm{T}}V)_{au} - (W^{\mathrm{T}}WH)_{au}] \tag{15.7}$$

When η_{au} is a positive minimum, Equation 15.6 is equivalent to the gradient descent method. As long as η_{au} is sufficiently small, the iteration rules will surely reduce $\|V - WH\|^2$. If we assume $\eta_{au} = \frac{H_{au}}{(W^{\mathrm{T}}WH)_{au}}$, we can find an iteration rule for H. The proof of the convergence of this algorithm is given after Lemma 15.3.

15.2.2.3 NMF Algorithm Based on Divergence Deviation

Lee and Seung (2001) proposed a reference algorithm based on divergence deviation. The algorithm is better than that based on the likelihood function. If and only if $V = WH$, the divergence deviation is equal to zero. We can determine the quality of each algorithm directly from the value of the target function corresponding to the result, simplifying the evaluation of the optimization. We obtain a monotonically decreasing target function by alternately optimizing the algorithm, a similar strategy to the EM algorithm. The most efficient and simple algorithm uses gradient descent in the self-learning iterative step.

The target function is:

$$D(V||WH) = \sum_{i,j}[V_{ij}\log\frac{V_{ij}}{(WH)_{ij}} - V_{ij} + (WH)_{ij}] \tag{15.8}$$

The target function of Equation 15.8 can be called KL divergence. Unlike the distance measure, it has no symmetry. If and only if $V = WH$, the formula attains its minimum value of 0.

Theorem 15.2 Under the iteration rule $H_{kj} \leftarrow H_{kj}\frac{\sum_i W_{ik}V_{ij}/(WH)_{ij}}{\sum_i W_{ik}}$, $W_{ik} \leftarrow W_{ik}\frac{\sum_j H_{kj}V_{ij}/(WH)_{ij}}{\sum_v H_{kv}}$, the divergence deviation $D(V||WH)$ is monotonic and does not increase. The algorithm converges when W and H reach some local optima.

Compared with the rules of the gradient descent method, under specific conditions, the incremental iterative rules that reduce H in $||V - WH||^2$ can be written as follows:

$$H_{au} \leftarrow H_{au} + \eta_{au}\left[\sum_i W_{ia}\frac{V_{iu}}{(WH)_{iu}} - \sum_i W_{ia}\right] \tag{15.9}$$

Similarly, when η_{au} is taken as a positive minimal value, the iteration rules will surely reduce $D(V||WH)$.

If we assume $\eta_{au} = \frac{H_{au}}{\sum_i W_{ia}}$, we can obtain an iteration rule for H.

It should be noted that if the value selected for η_{au} is not sufficiently small, we cannot ensure the decreasing characteristic of the target function. However, from the following proof, we can see that local convergence can be guaranteed. To prove Theorems 15.1 and 15.2, we introduce an auxiliary function (Definition 15.1) similar to that introduced in the EM algorithm (Dempster, Laired, and Rubin, 1977).

Definition 15.1 The condition that $G(h, h')$ becomes the auxiliary function of $F(h)$ is $G(h, h') \geq F(h), G(h, h) = F(h)$.

This auxiliary function is very useful, as shown in Lemma 15.1 to 15.3 and Figure 15.3.

Lemma 15.1 Assume that G is the auxiliary function. Under the iterative rule $h^{t+1} = \arg\min_h G(h, h^t)$, the function F is non-increasing.

Proof: Below, we give the proof of $F(h^{t+1}) \leq G(h^{t+1}, h^t) \leq G(h^t, h^t) = F(h^t)$.

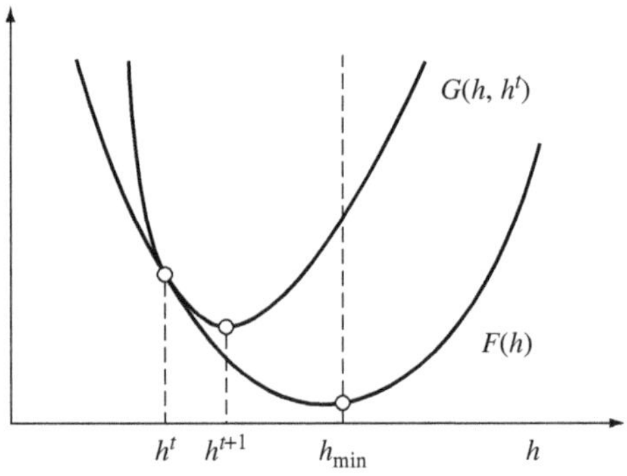

Figure 15.3 The auxiliary function

When h^t is the local minimum point of $G(h, h^t)$, $F(h^{t+1}) = F(h^t)$. If the derivative of the target function exists, then the differential $\nabla F(h^t) = 0$. Thus, each iteration according to $h^{t+1} = \arg\min_h G(h, h^t)$ will eventually converge to a local minimum $h_{\min} = \arg\min_h F(h)$:

$$F(h_{\min}) \leq \ldots F(h^{t+1}) \leq F(h^t) \ldots \leq F(h_2) \leq F(h_1) \leq F(h_0) \tag{15.10}$$

■

Therefore, we can see that the iterative rules of Theorems 15.1 and 15.2 can easily ensure $h^{t+1} = \arg\min_h G(h, h^t)$ for the target functions $||\boldsymbol{V} - \boldsymbol{WH}||^2$ and $D(\boldsymbol{V}||\boldsymbol{WH})$ with such an auxiliary function $G(h, h^t)$.

Lemma 15.2 Assume that $K(h^t)$ is a diagonal matrix $K_{ab}(h^t) = \delta_{ab}(\boldsymbol{W}^{\mathrm{T}}\boldsymbol{W}h^t)_a/h_a^t$. Then, $G(h, h^t) = F(h^t) + (h - h^t)\nabla F(h^t) + \frac{1}{2}(h - h^t)^{\mathrm{T}} K(h^t)(h - h^t)$ is the auxiliary function of $F(h) = \frac{1}{2}\sum_i \left(v_i - \sum_a \boldsymbol{W}_{ia} h_a\right)^2$.

Proof: Obviously, $G(h, h) = F(h)$, so we just need to prove $G(h, h') = F(h)$ by comparing $F(h) = F(h^t) + (h - h^t)^{\mathrm{T}}\nabla F(h^t) + \frac{1}{2}(h - h^t)^{\mathrm{T}}(\boldsymbol{W}^{\mathrm{T}}\boldsymbol{W})(h - h^t)$ and $G(h, h^t)$ (where ∇ represents the gradient). In other words, we must prove that $0 \leq (h - h^t)^{\mathrm{T}}[K(h^t) - \boldsymbol{W}^{\mathrm{T}}\boldsymbol{W}](h - h^t)$.

To prove the positive semi-definiteness, consider the matrix $\boldsymbol{M}_{ab}(h^t) = h_a^t(K(h^t) - \boldsymbol{W}^{\mathrm{T}}\boldsymbol{W})_{ab}h_b^t$, which is a partial scale of $\boldsymbol{K} - \boldsymbol{W}^{\mathrm{T}}\boldsymbol{W}$. $\boldsymbol{K} - \boldsymbol{W}^{\mathrm{T}}\boldsymbol{W}$ is positive semi-definite if and only if $\boldsymbol{M} = \boldsymbol{M}_{ab}(h^t)$, then

$$\begin{aligned} v^{\mathrm{T}}\boldsymbol{M}\mathbf{v} &= \sum_{ab} v_a \boldsymbol{M}_{ab} v_b \\ &= \sum_{ab} h_a^t(\mathbf{W}^{\mathrm{T}}\mathbf{W})_{ab}h_b^t v_a^2 - v_a h_a^t(\mathbf{W}^{\mathrm{T}}\mathbf{W})_{ab}h_b^t v_b \end{aligned}$$

$$
\begin{aligned}
&= \sum_{ab} (\boldsymbol{W}^{\mathrm{T}}\boldsymbol{W})_{ab} h_a^t h_b^t \left[\frac{1}{2}v_a^2 + \frac{1}{2}v_b^2 - v_a v_b\right] \\
&= \frac{1}{2}\sum_{ab} (\boldsymbol{W}^{\mathrm{T}}\boldsymbol{W})_{ab} h_a^t h_b^t (v_a - v_b)^2 \\
&\geq 0
\end{aligned}
$$

■

Lemma 15.3 $G(h, h') = \sum_i (v_i \log v_i - v_i) + \sum_{ia} w_{ia} h_a - \sum_{ia} v_i \frac{w_{ia} h_a^t}{\sum_b w_{ib} h_b^t} \left(\log w_{ia} h_a - \log \frac{w_{ia} h_a^t}{\sum_b w_{ib} h_b^t}\right)$ is the auxiliary function of $F(h) = \sum_i v_i \log\left(\frac{v_i}{\sum_a \boldsymbol{W}_{ib} h_a}\right) - v_i + \sum_a w_{ia} h_a$.

Proof: Obviously, $G(h, h) = F(h)$, so we will prove $G(h, h') \geq F(h)$.

According to the properties of the logarithm function, we have that:

$$
-\log \sum_a \boldsymbol{W}_{ia} h_a \leq -\sum_a a_a \log \frac{\boldsymbol{W}_{ia} h_a}{a_a} \tag{15.11}
$$

where $\sum_a a_a = 1$. Assuming $a_a = \frac{\boldsymbol{W}_{ia} h_a^t}{\sum_b \boldsymbol{W}_{ib} h_b^t}$, we can write:

$$
-\log \sum_a \boldsymbol{W}_{ia} h_a \leq -\sum_a \frac{w_{ia} h_a^t}{\sum_b w_{ib} h_b^t} \left(\log w_{ia} h_a - \log \frac{w_{ia} h_a^t}{\sum_b w_{ib} h_b^t}\right) \tag{15.12}
$$

Thus, we obtain $G(h, h') \geq F(h)$. ■

We will now prove the convergence of Theorem 15.1:

Proof: Substituting the above $G(h, h')$ into $h^{t+1} = \arg\min_h G(h, h^t)$, we get

$$
h^{t+1} = h^t - K(h^t)^{-1} \nabla F(h^t) \tag{15.13}
$$

$$
h_a^{t+1} = h_a^t \frac{(\boldsymbol{W}^{\mathrm{T}} v)_a}{(\boldsymbol{W}^{\mathrm{T}} \boldsymbol{W} h^t)_a} \tag{15.14}
$$

■

We now prove the convergence of Theorem 15.2:

Proof: The minimum of $G(h, h')$ can be obtained according to the gradient method

$$\frac{dG(h, h\prime)}{dh_a} = -\sum_i v_i \frac{\boldsymbol{W}_{ia} h_a^t}{\sum_b \boldsymbol{W}_{ib} h_b^t} + \sum_i \boldsymbol{W}_{ia} = 0 \tag{15.15}$$

Therefore, the corresponding iterative algorithm is:

$$h_a^{t+1} = \frac{h_a^t}{\sum_b \boldsymbol{W}_{ib}} \sum_i \frac{v_i \boldsymbol{W}_{ia}}{\sum_b \boldsymbol{W}_{ib} h_b^t} \tag{15.16}$$

Because $\boldsymbol{H}$ and $\boldsymbol{W}$ are symmetric, we can use the same method to prove the iterative rule for $\boldsymbol{W}$. ■

15.2.3 The Improved Non-negative Matrix Algorithm

The above basic NMF algorithms only impose the non-negative restriction on the results of decomposition. Researchers have put forward a series of improvements to NMF from different standpoints. Based on the basic algorithm, these enhancements adapt NMF to applications in different fields by adding new restrictions to the original basic algorithm. The new algorithms can be classified into three categories (see Figure 15.4):

1. NMF algorithms based on sparseness mainly include LNMF (Feng *et al.*, 2002), NNSC, sparse non-negative matrix factorization (SNMF) (Wei, 2004), non-negative matrix factorization with sparseness constraints (NMFSC) (Hoyer, 2004), and non-smooth non-negative matrix factorization (NSNMF) (Pascual-Montano *et al.*, 2006).
2. NMF algorithms based on weighting mainly include unbalanced weighted (Guillamet, Bressan, and Vitrià, 2001; Guillamet, Vitrià, and Schiele, 2003), uncertain weighted (Wang, Kossenkov, and Ochs, 2006), and important regional weighted (Blondel, Ho, and Van Dooren, 2007).
3. NMF algorithms based on classification include Fisher non-negative matrix factorization (FNMF) (Huang and Yu, 2008; Wang and Jia, 2004).

These methods improve the NMF algorithm in response to their application to some specific field.

15.2.3.1 NMF Algorithms Based on Sparseness

The basic NMF algorithm can obtain a relatively sparse description based on part of the observation signal. However, sometimes the degree of sparseness is not satisfactory. Hence, algorithms that can produce sufficiently sparse output have become an active research direction.

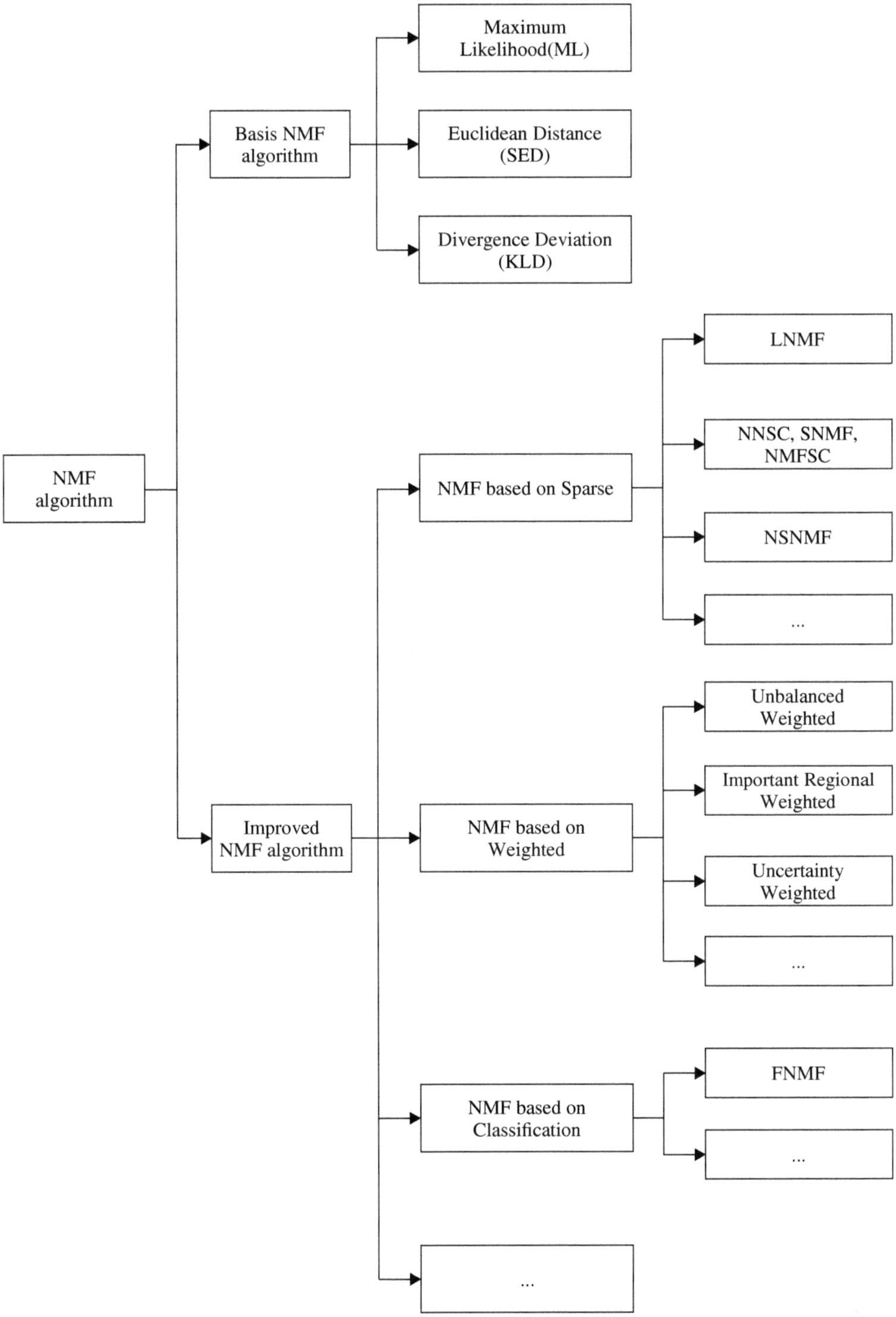

Figure 15.4 NMF and the improved algorithms

LNMF

In the study of basic images, the NMF algorithm adds non-negative constraints. This means that the values of basic image pixels and reconstruction weight coefficients are all non-negative. In this model, the "local integral" can be guaranteed with only linear superposition calculations. Thus, NMF is considered to be a method of extracting partial features, but the "part" from NMF is not always as expected, and the recognition of the basic NMF is occasionally unsatisfactory.

The research of Feng *et al.* (2002) showed that representation based in part by NMF proposed by Lee and Seung (2001) agrees with data in experiments. NMF was able to extract local features when selecting experimental data from CBC(http://www.xmarks.com/site/cbcl.mit.edu/software-datasets/FaceData2.html). In contrast, global features are obtained when we use data from the ORL face database (http://www.cl.cam.ac.uk/Research/DTG/attarchive/pub/data/att_faces.tar.Z), as shown in Figure 15.5.

Inspired by the basic NMF algorithms, Feng *et al.* (2002) introduced LNMF theory for the research of local space. By introducing a sparseness constraint, we can obtain the real local decomposition object of a coding vector (the matrix $\boldsymbol{H}$), from the basic component (the matrix $\boldsymbol{W}$), and strengthen the local features of the base component to make the algorithm suitable for the important application of local characteristics.

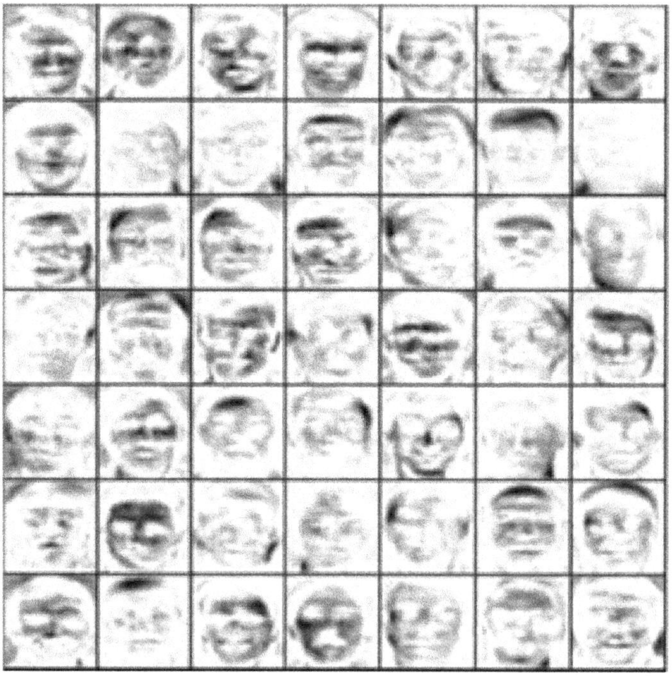

Figure 15.5 NMF processing results of ORL database

The basic NMF model does not reflect the constraint of characteristic localization. Thus, it sometimes cannot realize local feature decomposition of the data $\boldsymbol{V}$. We add the definition $\boldsymbol{U} = [u_{ij}] = \boldsymbol{W}^{\mathrm{T}}\boldsymbol{W}, \boldsymbol{V} = [v_{ij}] = \boldsymbol{H}\boldsymbol{H}^{\mathrm{T}}$; $\boldsymbol{U}, \boldsymbol{V} \in R^{r\times r}$ and three sparseness constraints to LNMF:

1. **Make *H* as sparse as possible.** $\boldsymbol{H}$ must contain as many zero elements as possible. A basis should be decomposed to its fullest extent, which ensures that the number of basis vectors needed by $\boldsymbol{V}$ is minimized. Make $w_{\mathbf{j}} = [w_{ij}]_{i=1}^{n}$ a basis vector. For the given constraint $\|w_{\mathbf{j}}\| = \sum_{i=1}^{n} w_{ij} = 1$, the target is to minimize $\sum_{i=1}^{n} {w_{ij}}^2 = u_{jj}$ for all j, making each $w_{\mathbf{i}}$ contain as many zeroes as possible to strengthen the partial capacity. When and only when $\sum_{i=1}^{n} w_{ij}^2 = u_{jj} = \min$, $\boldsymbol{H}$ attains its maximum sparseness.
2. **Maximize the local performance of *W*.** As mentioned in (1), the sparseness of $\boldsymbol{H}$ and the local performance ability of $\boldsymbol{W}$ are closely related. Hence, the described constraint strengthens the maximizing sparseness in (1) further. The idea is that only those compositions containing more information about the training sample can be reserved. For the sample $\boldsymbol{V}_j$, the information based on the basis component $\boldsymbol{W}_i$ is measured by the "activity" of the sample, defined as h_{ij}^2. The gross activity of all basis components is then $\sum_{i=1}^{m}\sum_{j=1}^{r} h_{ij}^2 = \sum_{i} v_{ii}$. When and only when $\sum_{i} v_{ii} = \max$, the local performance of $\boldsymbol{W}$ is at its strongest.
3. **Maximize the orthogonality of *W*.** Different bases should be orthogonal to minimize the redundancy of the base vectors. We can thus add the constraint $\sum_{i\neq j} u_{ij} = \min$, which can be modified as $\sum_{\forall i,j} u_{ij} = \min$.

Combining the above three constraints, we can obtain the divergence constraint as the target function of LNMF:

$$F(\boldsymbol{V}, \boldsymbol{WH}) = \sum_{i,j}\left(\boldsymbol{V}_{ij} \ln \frac{\boldsymbol{V}_{ij}}{(\boldsymbol{WH})_{ij}} - \boldsymbol{V}_{ij} + (\boldsymbol{WH})_{ij}\right) + \alpha \sum_{i,j} u_{ij} - \beta \sum_{i} v_{ii} \tag{15.17}$$

where α and β are constants greater than zero that can be eliminated through the minimization algorithm. LNMF aims to minimize the target function shown in Equation 15.17, which can be achieved through the following three updating rules:

The recursion formulas of LNMF:

$$h_{kl} = \sqrt{h_{kl} \sum_{i} \boldsymbol{V}_{il} \frac{\boldsymbol{W}_{ik}}{\sum_{k} \boldsymbol{W}_{ik} h_{kl}}},$$

$$W_{kl} = W_{kl}\frac{W_{kl}\sum_j V_{kj}\frac{h_{lj}}{\sum_k W_{kl}h_{lj}}}{\sum_j h_{lj}},$$

$$W_{kl} = \frac{W_{kl}}{\sum_k W_{kl}} \tag{15.18}$$

From the above rules, we know $F(\boldsymbol{X}, \boldsymbol{WH})$ is a non-increasing sequence value. It meets the constraint of the convergence of the local minimum point, as can be proved. Because LNMF is a spatial locality-oriented method, it can be used to solve some special circumstances, such as the image problems of local deformation, illumination change, and part missing. It can also improve the calculation efficiency, reduce image detection, and achieve system storage requirements.

NNSC

NMF and LNMF have become important tools in multivariate data analysis. Their most useful characteristic is that their decomposed results are sparse matrices. However, sometimes the sparseness of NMF and LNMF is not sufficient. In this case, the reasonable control of sparseness is very necessary and effective. Hoyer (2002) proposed a non-negative algorithm that combines sparse coding and NMF according to the information processing principle of visual neural science. The constraint problem to be solved is:

$$F = \frac{1}{2}\sum_{ij}(V_{ij} - (\boldsymbol{WH})_{ij})^2 + \lambda\sum_{kj} h_{kj} \tag{15.19}$$

where $\boldsymbol{W}_{ik} \geq 0, \boldsymbol{H}_{kj} \geq 0, \forall k, \sum_i \boldsymbol{W}_{ik} = 1, \lambda \geq 0$. In Equation 15.19, λ is a positive coefficient for regulating the control intensity.

NNSC uses different methods to iterate over $\boldsymbol{W}$ and $\boldsymbol{H}$. In the iteration process, we use the gradient method to solve $\boldsymbol{W}$ and a multiple iteration method to solve $\boldsymbol{H}$, similar to basic NMF (Lee *et al.*, 2001). The iterative rules of the algorithm are as follows:

$$\boldsymbol{W}_{ik} \leftarrow \boldsymbol{W}_{ik} - \lambda\sum_j^i((\boldsymbol{WH})_{ij} - \boldsymbol{V}_{ij})\boldsymbol{H}^{\mathrm{T}}_{kj},$$

$$\boldsymbol{W}_{ik} \leftarrow \frac{\boldsymbol{W}_{ik}}{\sqrt{\sum_i \boldsymbol{W}^2_{ik}}},$$

$$\boldsymbol{H}_{kj} \leftarrow \boldsymbol{H}_{kj}\sum_i \boldsymbol{W}_{ik}\frac{\boldsymbol{V}_{ij}}{(\boldsymbol{W}_{ik}(\boldsymbol{WH})_{ij} + \lambda)} \tag{15.20}$$

One advantage of the algorithm is that the base vector-learning is an additive iteration that cannot easily remain non-negative, so we must force the negative values to be zero.

Algorithm Non-negative sparse coding algorithm

Input: sample set $\boldsymbol{V} = (\boldsymbol{V}_1, \boldsymbol{V}_2, \cdots, \boldsymbol{V}_m)$, $\boldsymbol{V} \in R^{n\times m}$
Output: feature matrix $\boldsymbol{W} = (\boldsymbol{W}_1, \boldsymbol{W}_2, \cdots, \boldsymbol{W}_r)$ and projection sparse matrix $\boldsymbol{H}$
Steps:

1. Randomly initialize the matrices $\boldsymbol{W}^0$ and $\boldsymbol{H}^0$, ensuring the values of elements in the matrices are positive. Normalize the column vectors of the matrix $\boldsymbol{W}^0$ and initialize $t = 0$.
2. Iterate until convergence:
   a. $\boldsymbol{W}' \leftarrow \boldsymbol{W}^t - \lambda(\boldsymbol{W}^t\boldsymbol{H}^t - \boldsymbol{V})(\boldsymbol{H}^t)^{\mathrm{T}}$
   b. Set the negative values of the matrix $\boldsymbol{W}'$ to zero
   c. Normalize the column vectors of $\boldsymbol{W}'$, assign $\boldsymbol{W}^{t+1} = \boldsymbol{W}'$
   d. $\boldsymbol{H}^{t+1} \leftarrow \boldsymbol{H}^t .* ((\boldsymbol{W}^{t+1})^{\mathrm{T}}\boldsymbol{V})./((\boldsymbol{W}^{t+1})^{\mathrm{T}}(\boldsymbol{W}^{t+1})\boldsymbol{H}^t + \lambda)$
   e. Update the iteration number $t = t + 1$.

One defect of this algorithm is that we cannot guarantee the non-negative characteristic of the base vectors, so we must force the negative values to be zero. In practical applications, we can modify the form of the second term in the target function F so that it meets the specific requirements of sparseness, and so on. To realize the sparse coding, it is necessary to choose a suitable form for F. To add constraints on NMF to achieve the expected sparseness, we must consider whether to add sparseness constraints to the base vector $\boldsymbol{W}$ or the coefficient matrix $\boldsymbol{H}$. We cannot give a general answer to this question, and must instead decide through experiments according to the practical problem. Hence, Hoyer gave another form of F to control L_1 and L_2 of the base matrix and the coefficient matrix and form a project factor to reach the expected sparseness. This method not only adjusts the sparseness of $\boldsymbol{W}$, but also that of $\boldsymbol{H}$. The resulting NMFSC algorithm is introduced below.

SNMF

NMF and sparse coding are combined to form SNMF for the case when complex data learning cannot find the parts-based representation. This method not only obtains a sparser representation than NMF, but also finds the parts-based representation. SNMF decomposes a large number of data into positive sparse components based on linear sparse coding theory (Liu, Zheng, and Lin, 2003).

The target function of SNMF is defined as follows:

$$F(\boldsymbol{V}, \boldsymbol{W}\boldsymbol{H}) = \sum_{i=1}^{p}\sum_{j=1}^{n}\left(\boldsymbol{V}_{ij}\ln\frac{\boldsymbol{V}_{ij}}{(\boldsymbol{W}\boldsymbol{H})_{ij}} - \boldsymbol{V}_{ij} + (\boldsymbol{W}\boldsymbol{H})_{ij}\right) + \alpha\sum_{ii}\boldsymbol{H}_{jj}, \alpha \geq 0 \quad (15.21)$$

The iterative algorithm of SNMF is similar to that of LNMF:

$$
\begin{aligned}
H_{ab} &\leftarrow H_{ab} \frac{\sum_{i=1}^{p} (W_{ia} V_{ib}) / \sum_{k=1}^{q} W_{ik} H_{kb}}{1+\alpha}, \\
W_{ca} &\leftarrow W_{ca} \frac{\sum_{j=1}^{n} (H_{aj} V_{cj}) / \sum_{k=1}^{q} W_{ck} H_{kj}}{\sum_{j=1}^{n} H_{aj}}, \\
W_{ca} &\leftarrow \frac{W_{ca}}{\sum_{j=1}^{n} W_{ja}}
\end{aligned}
\tag{15.22}
$$

SNMF adds the sparseness constraint to the coefficient matrix $\boldsymbol{H}$. The algorithm has been applied to molecular cancer classification and obtained very good results.

NMFSC

NMFSC (Hoyer, 2004) is a kind of matrix factorization based on NMF. When factorizing the original matrix $\boldsymbol{V}$, we can control the sparseness of the feature matrix $\boldsymbol{W}$ and the coding matrix $\boldsymbol{H}$ to attain the specific application requirements. That is, the sparseness constraint may relate to either $\boldsymbol{W}$ or $\boldsymbol{H}$ according to the application. For example, when applied to feature extraction in facial recognition, we hope $\boldsymbol{W}$ is sparse so that its features stand out. A face that is related to as few characteristic faces as possible is easier to recognize. Hence, NMFSC can be effective in facial feature extraction.

The sparseness of NMFSC based on L_1 and L_2 is defined as follows:

$$
\text{sparseness}(\boldsymbol{X}) = \frac{\sqrt{n} - \left(\sum_i |X_i|\right) / \sqrt{\sum_i X_i^2}}{\sqrt{n} - 1} \tag{15.23}
$$

where n is the dimension of the non-negative vector $\boldsymbol{X}$. In simple terms, when $\boldsymbol{X}$ contains only one non-zero element, the sparseness is 1, and when all elements have equal values, the sparseness is 0. Generally speaking, when the sparseness is smaller, the numerical difference between the elements of $\boldsymbol{X}$ is also small. Different sparseness values are represented in Figure 15.6.

To obtain a better basic space or higher sparseness coefficient matrix, Hoyer proposed a kind of NMF with a sparseness constraint. The target function of NMFSC is defined as follows:

$$
F(\boldsymbol{W}, \boldsymbol{H}) = \|\boldsymbol{V} - \boldsymbol{W}\boldsymbol{H}\|^2 = \sum_{ij} (V_{ij} - (\boldsymbol{W}\boldsymbol{H})_{ij})^2 \tag{15.24}
$$

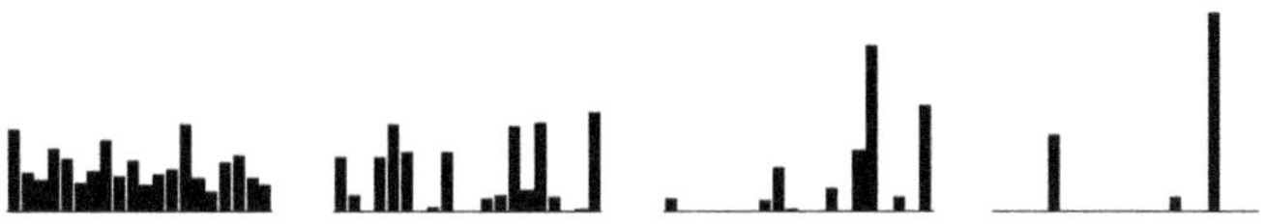

Figure 15.6 Different sparseness figures

NMFSC can be defined as the following optimization problem:

$$\min_{W,H} E(W,H) \quad W,H \geq 0, \sum_i h^2{}_{ij} = 1, \forall j \tag{15.25}$$

$$\text{and sparseness}(W_i) = S_w, \forall i \quad \text{or} \quad \text{sparseness}(H_i) = S_h, \forall i,$$

where W_i is the ith column of W and $h_i h_i$ is the ith row of H. S_w and S_h represent the size of the expected sparseness of W and H, respectively, which ranges from 0 to 1 according to the specific application.

NMFSC uses the projection operation many times. The algorithm controls the sparseness and non-negativity through constraints on the normal forms L_1 and L_2. NMFSC based on L_1 and L_2 is used to find the non-negative vector V nearest to the vector X.

Algorithm Projection algorithm

Input: vector X, k_1, and k_2 (k_1 and k_2 are the normal forms of L_1 and L_2 of the output vector V).

Output: the vector V, under some constraint on the normal form of L_1 and L_2. This vector is non-negative and has the smallest Euclidean distance to X.

Steps:

1. For any i, set $V_i = X_i + (L_1 - \sum X_i)/\dim(X), \forall i$ (dim represents the dimension of the vector).
2. Obtain all the indexes of the non-positive vector V and use them to form the array range = { }.
3. Iterative processing:

(a) Set $m_i = \begin{cases} L_1/(\dim(X) - size(range)) & i \in range \\ 0 & i \notin range \end{cases}$

(b) Set $V = m + \alpha(V - m), \alpha \geq 0$ to make the vector meet the constraint of the normal form L_2

(c) If all elements of V are non-negative, return V, otherwise continue

(d) Merge and update the range according to: $range = range \cup \{i; V_i < 0\}$

(e) Reset the element value of V according to: $V_i = 0, \forall i \in range$

(f) Recalculate and update the element value of V according to:

$$V_i = V_i - (\sum V_i - L_1)/(\dim(X) - size(range))$$

(g)Return to step i.

Algorithm Non-negative matrix factorization with sparseness constraints

Input: sample set $V = (V_1, V_2, \cdots, V_n)$
Output: feature matrix and project sparse matrix $W = (W_1, W_2, \cdots, W_r)$ and project sparse matrix H
Steps:

1. Select a certain initialization method (such as random initialization, PCA, fuzzy C-means initialization) to initialize W and H of the target function;
2. If W has the sparseness constraint, set each column of W to be non-negative according to the non-negative sparse projection algorithm, keeping the same normal form L_2 of W and setting L_1 to the specified sparseness;
3. If H has the sparseness constraint, set each line of H to be non-negative according to the non-negative sparse projection algorithm, keeping the same normal form L_2 of H and setting L_1 to the specified sparseness;
4. Iterate according to the following rules:
   a. If W has the sparseness constraint, first set $W = W - \mu_W(WH - V)H^T$, then set each column of W to be non-negative as above. If W does not have the sparseness constraint, according to the standard multiple iterative operation, then:
   b.

$$W = W \otimes (VH^T) \varnothing (WHH^T)$$

   c. If H has the sparseness constraint, first set $H = H - \mu_H(WH - V)$, then set each line of H to be non-negative as above. If H does not have the sparseness constraint, according to the standard multiplication, then:

$$H = H \otimes (W^T V) \varnothing (W^T WH)$$

In the above algorithm, the symbols $\otimes$ and $\varnothing$ represent matrix multiplication and division, respectively. μ_w and μ_H are very small positive numbers that control the iterative operation so as to guarantee convergence. The algorithm automatically adjusts the small step size.

NMFSC obtains a better base space by adding the sparseness constraint to the classical NMF algorithm, and exhibits a good performance in the field of facial recognition. NMFSC can control the sparseness of the feature matrix or the coefficient matrix according to the specific application. Later, Zhang and Mu improved NMFSC and applied it to ear recognition (Zhang and Mu, 2006).

NSNMF

To achieve the aim of global sparseness, NSNMF techniques have been proposed. The target function of NSNMF is:

$$V = WSH \tag{15.26}$$

where V, W, and H are consistent with the NMF algorithm, and the smooth positive-definite matrix $S \in \mathbf{R}^{q \times q}$ is defined as follows:

$$S = (1 - \theta)I + \frac{\theta}{q} 1 \cdot 1^{\mathrm{T}} \tag{15.27}$$

In the above equation, I is a feature matrix, 1 is the unit vector, and the range of θ is $0 \leq \theta \leq 1$. Let $Y = SX$. When $\theta = 0$, $Y = X$ (X has no smoothness constraint). When $\theta = 1$, each element of Y is the mean value of X, and the smoothness attains its maximum. Thus, θ controls the smoothness of S.

The target function of NSNMF can also be written as:

$$V = (WS)H = W(SH) \tag{15.28}$$

The iterative model of NSNMF is as follows, according to the conclusions from NMF:

$$H_{ab} \leftarrow H_{ab} \frac{\sum_{i=1}^{p} ((WS)_{ia} V_{ib}) / \sum_{k=1\}}^{q} (WS)_{ik} H_{kb}}{\sum_{i=1}^{p} (WS)_{ia}},$$

$$W_{ca} \leftarrow W_{ca} \frac{\sum_{j=1}^{n} ((SH)_{aj} V_{cj}) / \sum_{k=1\}}^{q} W_{ck} (SH)_{kj}}{\sum_{j=1}^{n} (SH)_{aj}}$$

$$W_{ca} \leftarrow \frac{W_{ca}}{\sum_{j=1}^{n} W_{ja}}. \tag{15.29}$$

NSNMF can retain the primitiveness of data and has found many applications in facial recognition and brain wave tomography imaging.

15.2.3.2 The Weighted NMF Algorithm

WNMF is another improved algorithm. The existing weighted algorithm can be classified into unbalanced weighted, important regional weighted, and uncertain weighted.

Unbalanced Weighted

The NMF algorithm aims to find the representation of non-negative data based on partial features. In the specific process of representation, we can obtain the local features of raw data when the NMF algorithm is applied to global data. However, the obtained features have redundant information when the algorithm is applied to the local data (Guillamet, Bressan, and Vitrià, 2001; Guillamet, Vitrià, and Schiele, 2003). Experiments have shown that, if training samples are not balanced (that is, the numbers in various classes are not balanced), the base vectors will be more reflective of features from classes having larger sample sizes after NMF factorization. WNMF was proposed to reduce this redundancy. WNMF overcomes the different sample proportions of various classes in the training set by assigning a higher weight to vectors of low frequency and a lower weight to vectors of high frequency. This can balance the training samples, reduce the repetition rate of training samples, and improve the local representation of NMF. WNMF introduces a diagonal matrix $\boldsymbol{Q}$ in which the elements along the diagonal define the degree of importance of each sample vector. WNMF defines a new factorization, and the weights of the training vectors are the diagonal elements q_μ of the matrix $\boldsymbol{Q}$, where $1 \le \mu \le m$ and the sum of the weights is 1. The weighting scheme counts the proportion of samples in various classes of the training set, and weights them according to the inverse of the proportion of the reconstruction error of each sample class in the process of optimization.

The target function of the algorithm is:

$$F_Q = \sum_{\mu=1}^{m} q_\mu \sum_{i=1}^{n} [V_{i\mu} \log((\boldsymbol{WH})_{i\mu} q_\mu) - (\boldsymbol{WH})_{i\mu}] \tag{15.30}$$

The iterative rule is defined as:

$$\begin{gathered} \boldsymbol{W}_{ia} \leftarrow \left(\boldsymbol{W}_{ia} / \sum_\mu q_\mu \boldsymbol{H}_{a\mu} \right) \times \sum_\mu (q_\mu V_{i\mu} / (\boldsymbol{WH})_{i\mu}) \boldsymbol{H}_{a\mu}, \\ \boldsymbol{W}_{ia} \leftarrow \boldsymbol{W}_{ia} / \sum_j \boldsymbol{W}_{ja}, \\ \boldsymbol{H}_{a\mu} \leftarrow \boldsymbol{H}_{a\mu} \sum_i \boldsymbol{W}_{ia} (V_{i\mu} / (\boldsymbol{WH})_{i\mu}) \end{gathered} \tag{15.31}$$

WNMF improves the classical NMF algorithm. It not only represents positive local data better, but is also very good at obtaining local features, meaning that the recognition effect of image classification has improved greatly.

Important Regional Weighted

Blondel, Ho, and Van Dooren (2007) proposed a new weighted algorithm with the aim of better describing important regions in the data. The weighting scheme is as follows: define the important regions in the data; define the weighted matrix according to the degree of importance of the data; give the reconstruction error of the important regions more weight in the process of optimization. Blondel *et al.* pointed out that the two algorithms developed by Lee and Seung can easily be adapted to incorporate the weight matrix. They also proved that the weighted target functions based on distance and the divergence deviation are essentially the same on the precondition of a reasonable choice of weight matrix.

1. The weighted Euclidean distance algorithm: The target function of the algorithm is defined as:

$$F = F_w(\boldsymbol{A}, \boldsymbol{UV}) = \frac{1}{2}(\boldsymbol{A} - \boldsymbol{UV})\mathbf{D}_w(\boldsymbol{A} - \boldsymbol{UV}) \tag{15.32}$$

   The iterative rule is:

$$\boldsymbol{V} \leftarrow \boldsymbol{V} \otimes \frac{\boldsymbol{U}^{\mathrm{T}}(\boldsymbol{W} \otimes \boldsymbol{A})}{\boldsymbol{U}^{\mathrm{T}}(\boldsymbol{W} \otimes (\boldsymbol{UV}))},$$
$$\boldsymbol{U} \leftarrow \boldsymbol{U} \otimes \frac{(\boldsymbol{W} \otimes \boldsymbol{A})\boldsymbol{V}^{\mathrm{T}}}{(\boldsymbol{W} \otimes (\boldsymbol{UV})\boldsymbol{V}^{\mathrm{T}})} \tag{15.33}$$

   We can prove that the target function is not increasing and convergent (Blondel, Ho, and Van Dooren, 2007).
2. The weighted KL divergence algorithm: The target function of the algorithm is defined as:

$$F = D_w(\boldsymbol{A}||\boldsymbol{UV}) = \sum_i w_i \left(a_i \log \frac{a_i}{\sum_j \boldsymbol{U}_{ij} v_j} - a_i + \sum_j \boldsymbol{U}_{ij} v_j \right) \tag{15.34}$$

   The iterative rule is:

$$\boldsymbol{V} \leftarrow \frac{\boldsymbol{V}}{\boldsymbol{U}^{\mathrm{T}}\boldsymbol{W}} \otimes \left(\boldsymbol{U}^{\mathrm{T}} \frac{\boldsymbol{W} \otimes \boldsymbol{A}}{\boldsymbol{UV}}\right),$$
$$\boldsymbol{U} \leftarrow \frac{\boldsymbol{U}}{\boldsymbol{W}\boldsymbol{V}^{\mathrm{T}}} \otimes \left(\frac{\boldsymbol{W} \otimes \boldsymbol{A}}{\boldsymbol{UV}}\right)\boldsymbol{V}^{\mathrm{T}} \tag{15.35}$$

   We can also prove that this target function is not increasing and convergent (Blondel, Ho, and Van Dooren, 2007).
3. The relation between the two weighting schemes is:
   The iterative rule for $\boldsymbol{V}$ in the weighted KL divergence algorithm can be transformed into:

$$\boldsymbol{V} \leftarrow \frac{\boldsymbol{V}}{\boldsymbol{U}^{\mathrm{T}}\boldsymbol{W}} \otimes \left(\boldsymbol{U}^{\mathrm{T}} \frac{\boldsymbol{W} \otimes \boldsymbol{A}}{\boldsymbol{UV}}\right) = \boldsymbol{V} \otimes \frac{\boldsymbol{U}^{\mathrm{T}} \frac{\boldsymbol{W} \otimes \boldsymbol{A}}{\boldsymbol{UV}}}{\boldsymbol{U}^{\mathrm{T}} \frac{\boldsymbol{W} \otimes (\boldsymbol{UV})}{\boldsymbol{UV}}}$$
$$= \boldsymbol{V} \otimes \frac{\boldsymbol{U}^{\mathrm{T}}(\boldsymbol{W}_{\boldsymbol{UV}} \otimes \boldsymbol{A})}{\boldsymbol{U}^{\mathrm{T}}[\boldsymbol{W}_{\boldsymbol{UV}} \otimes (\boldsymbol{UV})]}$$

Table 15.1 Comparison of the iterative rules of the weighted non-negative matrix factorization algorithms

	Euclidean distance (ED)	Kullback–Leibler divergence (KLD)
NMF	$V \leftarrow V \otimes \frac{U^{\mathrm{T}}(1 \otimes A)}{U^{\mathrm{T}}(1 \otimes (UV))}$	$V \leftarrow \frac{V}{U^{\mathrm{T}} 1} \otimes \left(U^{\mathrm{T}} \frac{1 \otimes A}{UV}\right)$
WNMF	$V \leftarrow V \otimes \frac{U^{\mathrm{T}}(W_1 \otimes A)}{U^{\mathrm{T}}(W_1 \otimes (UV))}$	$V \leftarrow \frac{V}{U^{\mathrm{T}} W_2} \otimes \left(U^{\mathrm{T}} \frac{W_2 \otimes A}{UV}\right)$
ED $\Longleftrightarrow$ *KLD*	$W_1 = \frac{W_2}{UV}$	

As we can see, the iterative rule based on the weighted KL divergence can be obtained by replacing W with W_{UV} in the weighted Euclidean distance iterative rule. Table 15.1 is about the comparison of the iterative rules of WNMF algorithms.

Uncertain Weighted

Wang *et al.* proposed a WNMF algorithm that aims to eliminate the influence of uncertainty on the result of NMF (Wang, Kossenkov, and Ochs, 2006). The weighting scheme is based on the uncertainty of each element in the training set. The reconstruction error of each element is weighted in descending order of its uncertainty in the optimization process.

To take advantage of the uncertainty information, we use the minimized error χ^2 in place of the minimized Euclidean distance in the classical methods, that is:

$$\chi^2 = \sum_{ij} \left(\frac{V_{ij} - M_{ij}}{\sigma_{ij}} \right)^2 \tag{15.36}$$

The iterative rule is:

$$H_{a\mu} \leftarrow H_{a\mu} \frac{\sum_i W_{ia} \frac{V_{i\mu}}{\sigma_{i\mu}}}{\sum_i W_{ia} \frac{M_{i\mu}}{\sigma_{i\mu}}},$$

$$W_{\delta a} \leftarrow W_{\delta a} \frac{\sum_j \frac{V_{\delta j}}{\sigma_{\delta j}} H_{aj}}{\sum_j \frac{M_{\delta j}}{\sigma_{\delta j}} H_{aj}},$$

$$M_{ij} = \sum_k W_{ik} H_{kj} \tag{15.37}$$

where σ_{ij} is the uncertainty metric of V_{ij}, and each element of U is the uncertainty metric of the corresponding element of V_{ij}. We set the certainty to 0 for data of low accuracy and to 1 for data of high accuracy. Thus, the algorithm can reduce the error introduced by uncertainties in the data. We can prove that the error function is non-increasing under the iterative rule.

15.2.3.3 NMF Algorithm Based on Classification

The classical NMF algorithm does not have any class information, so the addition of discriminative information is very important for classification (Wang and Jia, 2004). FNMF adds identification information (the within-class scatter and the between-class scatter) to the classical KL target function in scatter form, and minimizes the within-class scatter and between-class scatter so that the decomposition results contain more class information. Experimental results show that the FNMF algorithm outperforms the LNMF algorithm.

One of the target functions of FNMF can be defined as:

$$F(V||\boldsymbol{WH}) = \sum_{i,j}\left(v_{ij}\log\frac{v_{ij}}{y_{ij}} - v_{ij} + y_{ij}\right) + \alpha\boldsymbol{S}_w - \alpha\boldsymbol{S}_B \tag{15.38}$$

where $\alpha > 0$ is a constant, and $\boldsymbol{S}_w$ and $\boldsymbol{S}_B$ are the within-class and between-class scatter matrices of $\boldsymbol{H}$, respectively.

The corresponding iterative rule of FNMF is:

$$h_{kl} \leftarrow -b + \sqrt{b^2 + 4\left(\sum_i v_{il}\frac{w_{ik}h'_{kl}}{\sum_k w_{ik}h'_{kl}}\right)\left(\frac{2}{n_iC} - \frac{4}{n_i^2(C-1)}\right)},$$

$$w_{kl} \leftarrow \frac{w_{kl}\sum_j v_{kj}\frac{h_{lj}}{\sum_k w_{kl}h_{lj}}}{\sum_j h_{lj}},$$

$$w_{kl} \leftarrow \frac{w_{kl}}{\sum_k w_{kl}} \tag{15.39}$$

where C is the number of classes:

$$u_i = \frac{1}{n_i}\sum_{j=1}^{n_i} h_j,$$

$$b = \frac{4}{n_iC(C-1)}\sum_j\left(u_{kj} - \left(u_{ki} - \frac{h'_{kl}}{n_i}\right)\right) - \frac{2}{n_iC}u_{ki} + 1 \tag{15.40}$$

FNMF introduces Fisher constraint conditions based on classical NMF algorithms, and obtains the local base of a space from the training set. FNMF can attain a higher recognition rate than classical NMF for facial recognition.

15.2.4 The Initialization Problem of the Target Function

The final processing result of NMF is often a local optimum. Therefore, for $\boldsymbol{W}$ in the target function, the initialization of $\boldsymbol{H}$ will directly affect the whole solution, that is, different initial choices of $\boldsymbol{W}$ and $\boldsymbol{H}$ often lead to different results. The classical NMF algorithm generally uses a random initialization method. However, this method is not recommended, because it may lead to the non-repeatability of experiments. We can use improved initialization methods to get good convergence results from an exact initial value. The main initialization methods are based on PCA and fuzzy C-means.

15.2.4.1 Initialization Based on PCA

From the perspective of mean square error (MSE), PCA is the optimal representative, and the reconstruction error is:

$$E_{mse} = E||\overline{\boldsymbol{A}} - \boldsymbol{WH}||^2 \tag{15.41}$$

where $\overline{\boldsymbol{A}}$ is the set of samples that are subtracted from the total sample mean, $\boldsymbol{W}$ is the transformation matrix, and $\boldsymbol{H}$ is the projection coefficient. This function is very similar to the target function of NMF. PCA can find the projection direction in which the energy is most concentrated. According to this characteristic, we can use PCA to initialize the non-negative matrices $\boldsymbol{W}$ and $\boldsymbol{H}$. First, we use PCA to find the transformation matrices, then we perform the non-negative processing on the transformation matrices $\boldsymbol{W}$ and $\boldsymbol{H}$, that is:

$$w_{ij} = \max(0, w_{i,j}) \tag{15.42}$$

$$h_{ij} = \max(0, h_{i,j}) \tag{15.43}$$

This gives the non-negative matrices $\boldsymbol{W}'$ and $\boldsymbol{H}'$, which we take as the initial transformation matrices of NMF. The method of initializing non-negative matrices $\boldsymbol{W}$ and $\boldsymbol{H}$ according to PCA has the following two advantages: first, the subspace of PCA that contains most information about the sample space can extract large amounts of raw data, allowing us to minimize the error between the extracted data and the raw data; second, to some extent, the sparseness increases during the non-negative processing of the transformation matrices $\boldsymbol{W}$ and $\boldsymbol{H}$. These two points are difficult to achieve using general random initialization methods.

15.2.4.2 Initialization Based on Fuzzy C-Means

The basic principles of cluster analysis are to measure the degree of intimacy between things and to realize a classification. The essence of fuzzy clustering analysis is the construction of a fuzzy matrix according to object attributes and the determination of a relationship according to certain membership rules. Fuzzy clustering is applicable to the processing of imprecise or noisy data sets. The specific implementation procedure is as follows: first, obtain the cluster centers using a fuzzy cluster algorithm; second, initialize the matrix $\boldsymbol{W}$ in the NMF algorithm according to the obtained cluster centers; finally, select a method to initialize the matrix $\boldsymbol{H}$, such as the least-squares method, and set the negative elements of the matrix to 0 to obtain the initialization matrix of $\boldsymbol{H}$, or use the cluster center corresponding to each sample and then obtain the initialization matrix of $\boldsymbol{H}$ according to the membership matrix.

15.2.5 Advantages and Disadvantages of the NMF Algorithm

The NMF algorithm has three main advantages:

1. It has the internal mechanism of a non-negative restriction algorithm, that is, the algorithm itself imposes a non-negative constraint. The restriction of "non-negativity" is finding more extensive applications in various fields. Generally speaking, non-negativity is a kind of "external restriction" in traditional metrology methods, meaning that the algorithm itself cannot guarantee the non-negativity of the analytic results. For example, when a negative value appears, we can set it to zero or limit it according to the principle of non-negativity least-squares. The constraint for non-negativity is entirely different in NMF. Because of the application of a multiplicative update rule, NMF guarantees that negative values cannot appear in the decomposition result of $\boldsymbol{W}$ and $\boldsymbol{H}$ if the raw data are non-negative.
2. The NMF algorithm is a relatively recent multivariate statistical analysis method that can represent raw data through a few variables (base variables or basis functions). Compared with PCA, the basis functions of NMF are not necessarily orthogonal. Raw data can be regarded as the linear weight of the decomposition result (basis function). We can obtain the relationship between variables from NMF decomposition results, and do not need complex analysis techniques such as projection or rotation to perform the FA. The basic variables in NMF cannot be negative, which immediately overcomes the problem of negative elements in the principal components of PCA. At the same time, the decomposed results given by NMF have more direct physical significance and are easier to explain than those from PCA. Furthermore, when the interaction of information between variables is more serious, NMF can still find the right basis function to represent the data. In terms of practical variables, because of the orthogonality constraint of PCA (or singular value decomposition, SVD), its principal component space is worse than

that of NMF. In short, NMF is a new kind of multivariate analysis method. The base variables that are decomposed according to NMF do not contain negative values, and these base variables constitute a non-orthogonal basis vector space.

3. Finally, the theory of NMF is fundamental and its algorithms are brief and simple to understand and implement. NMF is a point-to-point calculation, which means that each element of data in the matrix is calculated, rather than each vector in the PCA matrix. Thus, the decomposition results of NMF are better able to represent local features of data.

However, the NMF algorithm also has the following deficiencies:

1. The NMF algorithm has the problem of zero values. As stated above, according to the multiplicative update formula in the iterative NMF procedure, if there are zero elements in the matrices $\boldsymbol{W}$ and $\boldsymbol{H}$, then the corresponding elements in the reconstruction data matrix must be zero. In such cases, the iteration may not converge, or cannot reach a local minimum value, and so the solution will not be optimal.
2. The NMF algorithm has a slow rate of convergence and needs thousands or even tens of thousands of iterative calculations to achieve ideal convergence results. A well-known deficiency of gradient-based methods is that the step size has a great impact on the convergence rate. NMF is based on the gradient descent method, and the multiplicative update formula is converted from the additive update formula of the gradient descent method. If and only if the step size is a given value, the additive update formula can be converted to a multiplicative update formula. Because the learning rate of the NMF algorithm does not correspond to this value, the convergence rate is understandably slow.
3. Finally, the NMF algorithm has the problem of multiple solutions, that is, the analytical results of NMF are not unique, which is similar to the rotational uncertainty of classical FA. The characteristic that the iterative calculation of the NMF algorithm is based on individual elements, instead of a single vector containing multiple elements, makes the analytical results reflect local data characteristics better. Because it only has the weaker constraint condition of non-negativity, the NMF algorithm can obtain different analytical results for the same raw data on successive runs. The results are thus uncertain, and will be distributed in a solution domain that depends on the NMF constraint condition. When NMF is used for image analysis processing, the problem of multiple solutions has a finite influence on the analysis results but a great influence for other specific areas.

15.3 Applications and Analysis

15.3.1 Applications of the NMF Algorithm in Remote Sensing Image Preprocessing

High-spatial-resolution remote sensing images mainly refer to those for which the ground resolution is higher than 5 m. When the image resolution is of the order of meters, most individual trees, cars, roads, and houses can be seen in the images. Making full use of the advantages of this high spatial resolution and improving the accuracy of classification and the reliability of object extraction is of great significance. Traditional methods mainly use multispectral image grayscale features in classification. In the past 30 years, more attention has been focused on the role of the image space features (such as texture, shape, and geological data) rather than spectral characteristics in information extraction. High-spatial-resolution image classification by computers, which use rich subtle information and favorable visual effects to enhance the features of geometry structure and texture information, is prone to error. Large changes in the angular field of view of high-spatial-resolution images and complex illumination conditions cause some problems, such as changeable reflections from the same artificial surface, the displacement of significant images, and obscured scenes. These also increase the complexity of technology, such as automatic target identification, and present challenges for the use of traditional mature methods in low-resolution image processing.

To extract effective information from multispectral images, we apply NMF to high-spatial-resolution remote sensing image processing. First, we preprocess the multispectral images before using automatic target recognition on the high-resolution images. We select QuickBird remote sensing images as experimental data. These are 1024×1024 images with a spatial resolution of 1 m. The image, taken near Beijing Normal University on 31 May 2005, has three bands. n bands of a multispectral image make up a data vector set $\boldsymbol{V}$. We use NMF to process $\boldsymbol{V} \approx \boldsymbol{WH}$ with $r = n$ to obtain n feature bases. A new multispectral image is given by recovering the n feature bases.

15.3.1.1 Preprocessing

We translate the multispectral image data into a form of the matrix $\boldsymbol{V}$ ($\boldsymbol{V} \in \mathrm{R}^{m \times n}$, where m represents the number of pixels in the multispectral image and n denotes the number of bands in the image). To reduce the computational complexity and avoid an overflow in the calculation, the matrix $\boldsymbol{V}$ is normalized. After preprocessing, the elements of $\boldsymbol{V}$ are in the range 0–1. $\boldsymbol{V}$ is then processed using the NMF algorithm based on the traditional Euclidean distance. Part of the dataset is shown in Table 15.2.

In this experiment, the algorithm converges after 877 iterations to give the matrix $\boldsymbol{W}$ (part of the dataset in matrix $\boldsymbol{W}$ is shown in Table 15.3).

As a result of the data preprocessing and $\boldsymbol{WH} = (\boldsymbol{W}\lambda)/(\boldsymbol{H}\lambda)$, there is a discrepancy between the scale of the matrix $\boldsymbol{W}$ and that of the original matrix after decomposition. Therefore, we perform a stretch transform on $\boldsymbol{W}$. Following this, all of the data in $\boldsymbol{W}$ are in the range 0–255 (some of the data in $\boldsymbol{W}$ after recovery are shown in Table 15.4).

15.3.1.2 Experimental Results and Analysis

NMF is represented by linear, non-negative, and approximate data. It obtains a data representation using a non-negativity restriction, which results in a representation

Table 15.2 Part of the data after preprocessing

ID of pixels	Band 1	Band 2	Band 3
1	0.0285	0.0000	0.0285
2	0.1179	0.0163	0.0854
3	0.1341	0.0610	0.1016
4	0.1098	0.0244	0.0813
5	0.0691	0.0203	0.0772
6	0.1138	0.0610	0.0935
7	0.0407	0.0041	0.0569
8	0.0285	0.0000	0.0285
9	0.1138	0.0366	0.0813
10	0.1260	0.0366	0.0691

Table 15.3 Part of the data in matrix $\boldsymbol{W}$

ID of pixels	Band 1	Band 2	Band 3
1	0.0000	0.0137	0.0301
2	0.0000	0.0767	0.0893
3	0.0217	0.0820	0.1091
4	0.0000	0.0715	0.0867
5	0.0000	0.0313	0.0856
6	0.0275	0.0634	0.1007
7	0.0000	0.0111	0.0625
8	0.0000	0.0137	0.0301
9	0.0000	0.0772	0.0881
10	0.0000	0.0965	0.0727

Table 15.4 Part of the data in matrix $\boldsymbol{W}$ after data recovery

ID of pixels	Band 1	Band 2	Band 3
1	0.0000	4.7030	6.9367
2	0.0000	26.2952	20.5962
3	5.2614	28.1115	25.1712
4	0.0000	24.5162	19.9929
5	0.0000	10.7178	19.7478
6	6.6810	21.7176	23.2224
7	0.0000	3.8102	14.4142
8	0.0000	4.7030	6.9367
9	0.0000	26.4555	20.3272
10	0.0000	33.0791	16.7612

Figure 15.7 Original multispectral image

based on partial image information. Local information can be obtained by applying NMF in the image field. Tables 15.2 and 15.3 show that $\boldsymbol{W}$ is sparser than $\boldsymbol{V}$, which means that the NMF algorithm can extract the underlying data structure from $\boldsymbol{V}$.

The original multispectral image (QuickBird multispectral image of Beijing Normal University) is as shown in Figure 15.7. The three bands of the original multispectral image are shown in Figures 15.8–15.10 and after decomposition are shown in Figures 15.11–15.13. The multispectral image after processing is shown in Figure 15.14. These results show that NMF can extract information effectively. Compared with the original image, that after processing has richer local information, a clearer data structure, and can more easily be used for target recognition.

Currently, high-spatial-resolution images generally use simple, traditional classification methods. Thus, the degree of automatic identification is low and the identification accuracy needs to be improved. In addition, storage space is not used effectively. The original identification method cannot meet the needs of high-spatial-resolution images, so information extraction techniques such as NMF are of great significance in improving the classification accuracy and reliability of target extraction.

15.3.2 The Application of NMF in Remote Sensing Image Fusion

Remote sensing images are an expression of surface information. Multi-temporal, polymorphic, and multi-band images have certain limitations in terms of space,

Figure 15.8 Band 1 of the original multispectral image

Figure 15.9 Band 2 of the original multispectral image

Figure 15.10 Band 3 of the original multispectral image

Figure 15.11 Component 1 after decomposition

Figure 15.12 Component 2 after decomposition

Figure 15.13 Component 3 after decomposition

Figure 15.14 False color image after processing

spectrum, and geometric resolution, and cannot provide rich, reliable, and accurate information. Image fusion technology can combine the advantages of different remote sensing images into a single image containing accurate, clear, and comprehensive information. This technology is more suitable for the processing of remote sensing data, and can improve the accuracy of data processed by computer.

Compared with high-spatial-resolution panchromatic images, multispectral remote sensing images have better spectral characteristics and lower spatial resolution. We can obtain high-quality fused images using complementary fusion algorithms that not only maintain the spectral characteristics of images, but also obtain high spatial resolution. Current fusion methods include principal component fusion and wavelet transform fusion.

As it is based on the combination of base vectors, NMF has a very intuitive interpretation. The algorithm expresses the partial raw data as a "local composition of the whole". The data obtained is an additive, non-negative combination of raw data. The non-negative vector set $\boldsymbol{W}$ obtained by the algorithm, which can express the features and structure of raw data effectively, has a certain linear independence and sparsity. Features can be elicited from raw data by choosing an appropriate r value in NMF, whereby the non-negative constraint is in line with the conditions of non-negative grayscale values for an image pixel. Thus, NMF has been widely used in imaging fields.

Based on the above theory, we have researched NMF and applied it to remote sensing image fusion. According to the characteristics of the algorithm, we combine it with other fusion algorithms to obtain an improved technique (Huang *et al.*, 2008; Wang, Yu, and Zhang, 2008a,b). We compare and analyze the basic NMF algorithm, a fusion of NMF and the principal component method, and a fusion of NMF and the wavelet transform algorithm. We also evaluate the performance of the improved fusion algorithms in accordance with the experimental results (Huang *et al.*, 2008).

15.3.2.1 Experimental Data

The image used in the experiment is from Zhuhai, in the south of Guangdong province. On the west bank of the Pearl River, Zhuhai is situated at 21°48′–22°27′N latitude and 113°03′–114°19′E longitude, close to the South China Sea. Guangzhou is 140 km to the north, Shenzhen and Hong Kong lie to the east, and Macao is situated to the south. Zhuhai has a beautiful natural environment, picturesque scenery, and a coastal stretch. There are 146 islands within the city boundary, which encloses a sea area of 1653 km^2 and a land area of 6000 km^2.

The surface features of the experimental image of the Doumen District in Zhuhai are of five types: town land, water, woodland, agricultural land, and unused land. Using the remote sensing image processing software Erdas 8.5, we perform image registration and other image preprocessing for the original remote sensing image, and select an image block of 1192 × 1205 pixels.

1. The first set of image data was captured by the ETM+ (enhanced thematic mapper) sensor in 1999. The multispectral image display is a synthetic false color image of Band 4, Band 3, and Band 2 (see Figure 15.15).
2. The second set of image data is from the ETM+ sensor and SPOT sensor. The multispectral image display is a synthetic false color image of Band 4, Band 2, and Band 1 (see Figure 15.16).
3. The third set of image data was captured by the ETM+ sensor and the CBERS (China-Brazil Earth Resource Satellite) satellite sensor in 1999. The multispectral image display is a synthetic false color image of Band 3, Band 2, and Band 1 (see Figure 15.17).

15.3.2.2 Fusion Algorithm Based on NMF

We form the multispectral image $\boldsymbol{M}$ (with k bands) and a panchromatic image from the data vector set $\boldsymbol{V}$ containing k column vectors, $\boldsymbol{V} \in \mathrm{R}^{n \times m}$, where $m = k + 1$ and n is the number of pixels in each image. $\boldsymbol{V}$ can be expressed as:

$$\boldsymbol{V} = [v_1, v_2 \ \cdots \ v_m] = \begin{bmatrix} v_{11} & v_{12} & \vdots & v_{1m} \\ v_{21} & v_{22} & \vdots & v_{2m} \\ \vdots & \vdots & \vdots & \vdots \\ v_{n1} & v_{n2} & \vdots & v_{nm} \end{bmatrix} \tag{15.44}$$

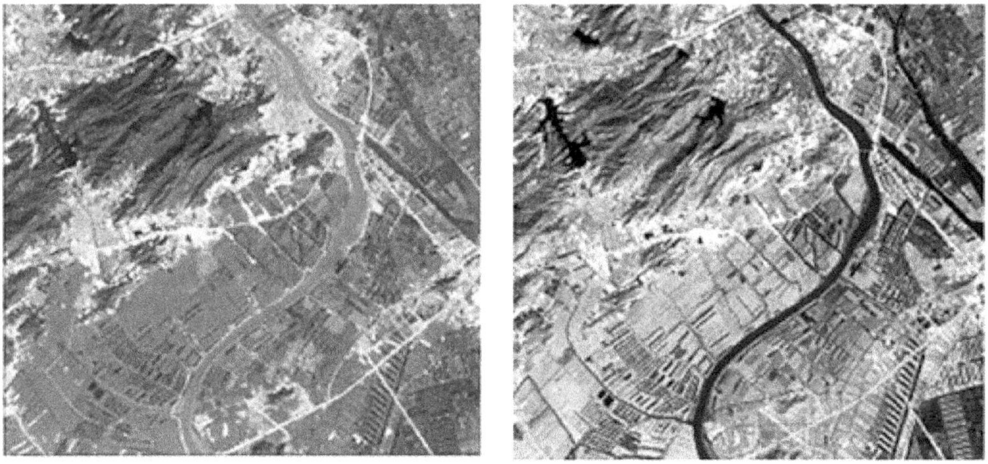

(a) (b)

Figure 15.15 Images in the first group (a) ETM+ multispectral image and (b) ETM+ panchromatic image

(a) (b)

Figure 15.16 Images in the second group (see color Figure 15.16). (a) SPOT satellite multispectral image and (b) ETM+ panchromatic image

The raw data in $\boldsymbol{V}$ can be decomposed as the approximate product of the matrix $\boldsymbol{W}(\boldsymbol{W} \in \mathrm{R}^{n\times r})$ and the matrix $\boldsymbol{H}(\boldsymbol{H} \in \mathrm{R}^{r\times m})$. Supposing that w and h are the corresponding column vectors of $\boldsymbol{W}$ and $\boldsymbol{H}$, it follows that $\boldsymbol{V} = \boldsymbol{WH}$, that is, each image can be seen as a non-negative linear combination of the column vectors of a non-negative matrix W. Therefore, the non-negative matrix $\boldsymbol{W}$ can be seen as a linear approximation of a set of bases of the raw data $\boldsymbol{V}$, and $\boldsymbol{H}$ is the non-negative projection coefficient of

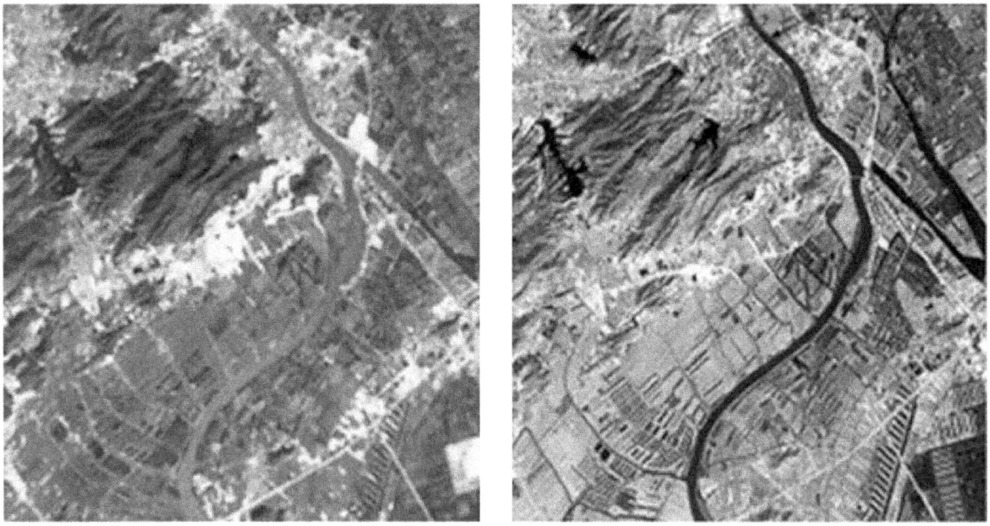

(a) (b)

Figure 15.17 Images in the third group. (a) CBERS multispectral images and (b) ETM+ panchromatic image

raw data $\boldsymbol{V}$ about the base $\boldsymbol{W}$. Normally, r is less than m or n, implying that $\boldsymbol{W}$ and $\boldsymbol{H}$ will be less than the raw data $\boldsymbol{V}$. Thus, we have a compression model of the raw data $\boldsymbol{V}$.

Algorithm Multispectral image and panchromatic image fusion algorithm based on NMF

Input: a three-band multispectral image $\boldsymbol{M}$ and the panchromatic image $\boldsymbol{P}$
Output: the multispectral fused image
Steps:

1. The multispectral image and the panchromatic image compose a data vector set $\boldsymbol{V}$ (each band image is a column vector of the matrix $\boldsymbol{V}$).
2. Take $r = 3$ for three feature bases. These generally contain most of the information in the raw data. Ensure a non-negative $\boldsymbol{V} = \boldsymbol{WH}$ and obtain three feature bases: $\boldsymbol{W} = [w_1, w_2, w_3]$
3. The base vector set contains approximate complete information of the four bands of the images involved in fusion. Thus, this set can be used to find an approximate representation of the original image and reduce noise.
4. Arrange the three feature bases of $\boldsymbol{W}$ in descending order according to the size of the variance. The feature base with the greatest variance contains most of the energy and information of the raw data $\boldsymbol{V}$.
5. Perform histogram matching with the panchromatic image $\boldsymbol{P}$ and the first feature base of $\boldsymbol{W}$ so that the panchromatic image $\boldsymbol{P}$ has a similar mean and variance as the first feature base of $\boldsymbol{W}$.

6. Replace the first feature base with the panchromatic image after histogram matching, and apply an inverse transform to restore the feature base to the pixel level of the original image, thus obtaining the multispectral fused image with rich information details.

15.3.2.3 Fusion Algorithm Based on NMF and PCA

The PCA fusion algorithm is a commonly used approach. The multispectral images to be fused should be preprocessed according to the NMF algorithm prior to the PCA fusion algorithm to obtain higher image quality.

Algorithm Fusion algorithm based on non-negative factorization and principal component algorithm (N-PCA)

Input: a three-band multispectral image $\boldsymbol{M}$ and the panchromatic image $\boldsymbol{P}$

Output: the multispectral fused image

Steps:

1. Input the data matrix $\boldsymbol{V}$. The multispectral image $\boldsymbol{M}$ contains three bands. Take the three bands as the column vectors of the input matrix $\boldsymbol{V}$.
2. Perform decomposition according to NMF to obtain the reference image $\boldsymbol{W}$ and the weight matrix $\boldsymbol{H}$.
3. Input the reference image $\boldsymbol{W}$ as the multispectral image and apply principal component transform fusion with the panchromatic image $\boldsymbol{P}$ to obtain the fused image $\boldsymbol{G}$.

15.3.2.4 Fusion Algorithm Based on NMF and Wavelet Transform

Processing images according to the wavelet transform fusion algorithm gives good time–frequency localized features and multi-resolution analysis features. This method also retains complete low-frequency information about the multispectral images and high-frequency information about the panchromatic images. However, it suffers from spectral distortion and blocking. A fusion of the wavelet transform and NMF algorithm can ameliorate these disadvantages. Other limitations of the traditional wavelet transformation include complex computation, low efficiency, high memory consumption, and strict requirements that the size of processed images is a multiple of integer powers of 2, but these cannot be overcome by fusion with NMF. Fortunately, the lifting wavelet transform, which is based on maintaining local features of the traditional wavelet transform, overcomes these disadvantages, and so we choose to combine NMF with the lifting wavelet transform algorithm.

The basic principle of the lifting wavelet transform is to obtain better performance in accordance with the basic wavelet, obtaining high-frequency components using polynomial interpolation and low-frequency components by building a scale function.

There are three main steps: division, updating, and prediction. Division is the process whereby similar images and the detail in images are divided from the original signals $\boldsymbol{R}_i$ into two mutually disjoint subsets $\boldsymbol{R}_{i-1}$ and $\boldsymbol{D}_{i-1}$. Updating refers to the substitution of approximate images with information from the detailed images to produce a better data subset $\boldsymbol{R}_{i-1}$. The prediction process involves predicting detailed information in the approximate images after the updating procedure, and then substituting the detailed images with prediction error into the similar images. This is essentially using $\boldsymbol{R}_{i-1}$ data to predict $\boldsymbol{D}_{i-1}$ according to the relationship between the data, that is, selecting a prediction operator $\boldsymbol{P}$ such that $\boldsymbol{D}_{i-1} = \boldsymbol{P}(\boldsymbol{R}_{i-1})$. The inverse lifting wavelet transform is the reverse process of the decomposition, that is, anti-prediction, anti-updating, and merging. Anti-prediction obtains detailed signals from approximate image data, and then substitutes the detailed signals with filtering error in images according to $d_{i-1} = d_{i-1} + \boldsymbol{P}(\boldsymbol{R}_{i-1})$. The anti-updating process is described by $\boldsymbol{R}_{i-1} = \boldsymbol{R}_{i-1} + \boldsymbol{U}(d_{i-1})$, and refers to updating approximate images with information from the detailed images after anti-prediction. The merging process can be expressed as $\boldsymbol{R}_i = \boldsymbol{M}(\boldsymbol{R}_{i-1}, d_{i-1})$, where $\boldsymbol{M}$ refers to merging $\boldsymbol{R}_{i-1}$ and $\boldsymbol{D}_{i-1}$ into the original signal $\boldsymbol{R}_i$.

Algorithm NMF and lifting wavelet fusion algorithm

Input: a three band multispectral image $\boldsymbol{M}$ and the panchromatic image $\boldsymbol{P}$
Output: the multispectral fused image
Steps:

1. Input the data matrix $\boldsymbol{V}$. The multispectral image $\boldsymbol{M}$ contains three bands. Take the three bands as the column vectors of the input matrix $\boldsymbol{V}$.
2. Perform decomposition according to NMF to obtain the reference image $\boldsymbol{W}$ and the weight matrix $\boldsymbol{H}$.
3. Obtain the largest component of variance of the reference image, that is, the first feature base $\boldsymbol{K}$. Use this to perform histogram matching with the panchromatic image $\boldsymbol{P}$ to ensure it has a similar mean and variance to the first feature base $\boldsymbol{K}$.
4. Perform the lifting wavelet transform to the first feature base $\boldsymbol{K}$ and the panchromatic image $\boldsymbol{P}$. Select relevant fusion rules and conduct fusion processing on the high- and low-frequency coefficients. Apply the lifting wavelet transform to obtain the first feature base $\boldsymbol{K}$ after fusion.
5. Substitute the original feature base $\boldsymbol{K}$ with the original first feature base $\boldsymbol{K}$ and apply the transformation according to NMF to obtain the fusion image $\boldsymbol{G}$.

The fusion rule used in the experiment is as follows:

$$d_{i,j} = \max\{\text{abs}(c_{i,j}), \text{abs}(s_{i,j})\} \tag{15.45}$$

where $c_{i,j}$, $s_{i,j}$ are the high-frequency coefficients of the first feature base $\boldsymbol{K}$ and the panchromatic image $\boldsymbol{P}$, respectively, after the lifting wavelet transform.

The low-frequency part, according to the mean variance weighting principle, is given by:

$$m_1 = \frac{1}{l}\sum_{i=1}^{M}\sum_{j=1}^{N}(a_{i,j} - \overline{a})^2, m_2 = \frac{1}{l}\sum_{i=1}^{M}\sum_{j=1}^{N}(b_{i,j} - \overline{b})^2 \tag{15.46}$$

In accordance with the size of m_1m_1 and m_2m_2, select weight coefficients to make the larger MSE value correspond to the larger weight value. The fused low-frequency coefficients are:

$$t_{i,j} = \alpha a_{i,j} + (1 - \alpha)b_{i,j} \tag{15.47}$$

15.3.2.5 Analysis of Experimental Results

In this experiment, we analyzed and compared the effect of the above fusion algorithms. The experimental data are the image data from the third set of ETM+ (1999) sensor and CBERS satellite sensor images. The size of the image is 256 × 256, and the multispectral images, the panchromatic image, and the fusion images given by the algorithms are shown in Figure 15.18.

The fusion results can be evaluated visually and by calculating some statistical parameters. The fusion of multispectral images and panchromatic images is a complementary fusion of spectral characteristics and spatial resolution. We can give an objective evaluation of these parameters using statistical parameters. The parameters reflecting spatial details are the variance, information entropy, and clarity, and the parameter reflecting spectral information is the torsion resistance. We introduce these as follows:

1. **The standard variance.** Variance reflects the range of gray relative to the average gray value. The larger the standard deviation, the more dispersed is the distribution of gray levels. At this time, the probability of all gray levels in images tends to be equal, so more information tends to be contained within.
2. **Information entropy.** The information entropy of an image histogram with a gray scale range of $\{0, 1, \cdots, L - 1\}$ is as shown in Equation 15.47, where $\boldsymbol{P}_i$ is the ratio of pixels with a value of I to the total number of pixels in the image:

$$H = \sum_{i=0}^{l-1} \boldsymbol{P}_i \ln \boldsymbol{P}_{ii} \tag{15.48}$$

3. **Clarity.** The sharpness of an image is measured using the gradient method. The gradient is calculated as follows:

$$\overline{q} = \frac{1}{n}\sqrt{(\Delta l_x^2 + \Delta l_y^2)/2_i} \tag{15.49}$$

where Δl_x and Δl_y are the differences in the x and y directions, respectively, and n is the size of the image.

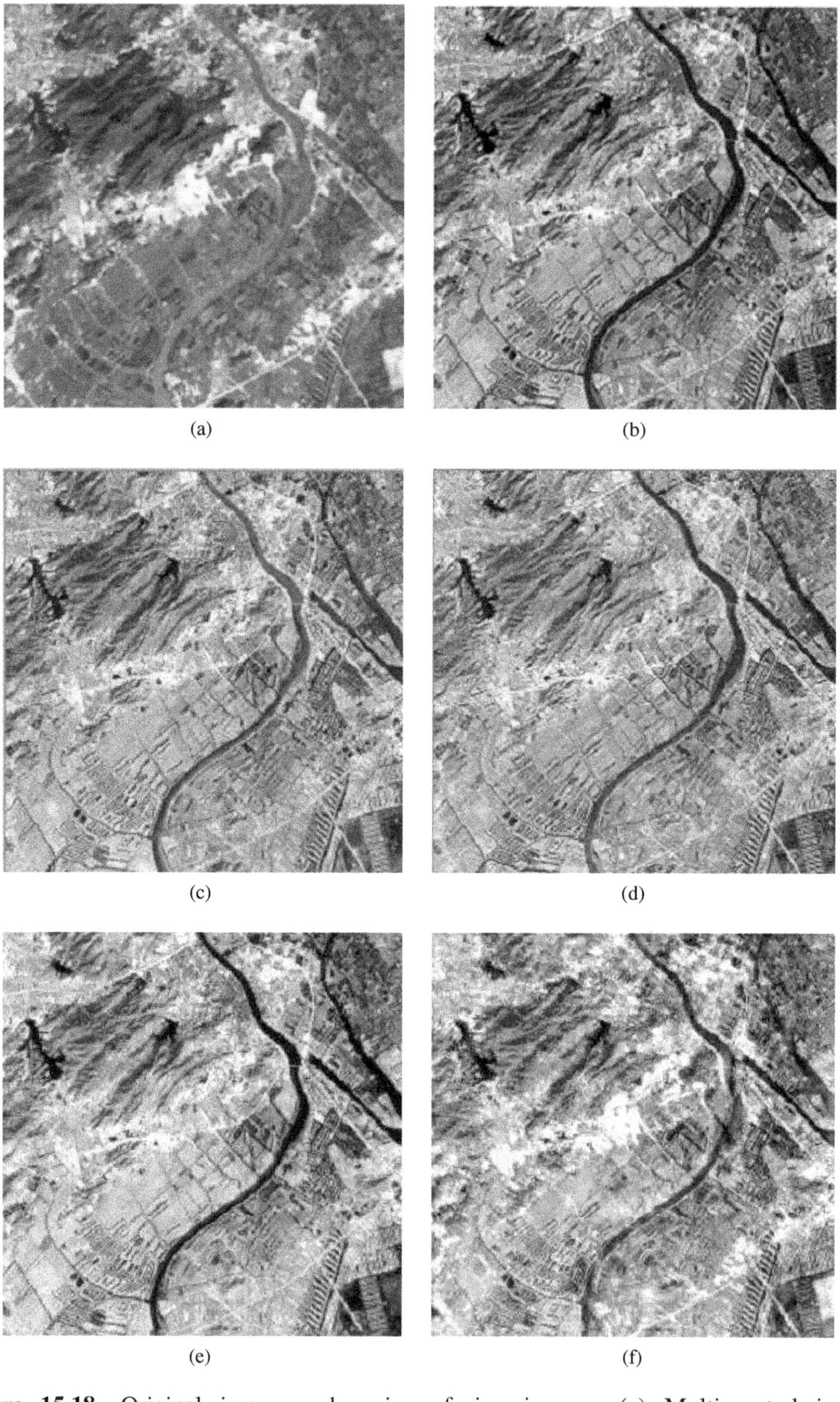

Figure 15.18 Original image and various fusion images. (a) Multispectral image, (b) panchromatic image, (c) PCA fusion image, (d) N-PCA fusion image, (e) NMF fusion image, and (f) NMF lifting wavelet fusion image

Table 15.5 Comparison of the parameters of the fusion methods

Fusion method	Variance	Information entropy	Clarity	Torsion resistance
The multispectral image	51.186	7.6224	15.7177	–
The panchromatic image	46.598	7.5246	24.18	–
The PCA fusion image	50.501	7.60307	25.9667	18.050
The N_PCA fusion image	50.514	7.6644	26.4213	18.088
The NMF fusion image	57.845	7.7633	27.376	18.433
The NMF lifting wavelet fusion image	58.268	7.8010	27.833	18.106

4. **Torsion resistance.** The torsion resistance of an image reflects the degree of spectral distortion in multispectral images. This is defined as:

$$D = \frac{1}{n}\sum_{i}\sum_{j}\|V'_{ij} - V_{ij}\| \tag{15.50}$$

where V'_{ij} and V_{ij} are the gray values of the fused image and the original image, respectively, and n is the size of the image.

The statistical parameters for the evaluation of multispectral images, the panchromatic image, and the fusion images processed according to the various fusion algorithms are shown in Table 15.5.

A comprehensive comparison of the above statistics shows that the fusion algorithm based on NMF and the lifting wavelet transform gave the best results, and the other NMF fusion algorithms outperformed the PCA fusion algorithm. Therefore, better fusion results can be obtained by applying NMF to remote sensing image fusion.

In short, the application of NMF to image fusion not only improves spatial resolution, but also increases the quality of detailed features, maintains spectral characteristics as far as possible, and combines the advantages of multispectral images and panchromatic images. The performance of the NMF fusion algorithms is significantly better than that of the PCA fusion algorithm.

15.3.3 Design of a Subspace Classifier Based on the NMF Algorithm

In designing a classifier, we should perform NMF preprocessing on the data to obtain the basis matrix of data to be processed. This basis matrix is then classified according to the C-means clustering algorithm.

15.3.3.1 Discriminant Function of Subspace Classification

During the process of C-means clustering, we compare the distance of each basis matrix and cluster in which the Euclidean distance and the Mahalanobis distance are common distance functions. There are many distance functions for subspace classifiers. The distance based on the non-negative vector similarity coefficient (NVSC) is:

$$d(\boldsymbol{X},\boldsymbol{Y}) = 1 - \gamma^2(\boldsymbol{X},\boldsymbol{Y}),$$

$$\gamma(\boldsymbol{X},\boldsymbol{Y}) = \frac{\sum_{i=1}^{n} \min(x_i, y_i)}{\sum_{i=1}^{n} \max(x_i, y_i)} \tag{15.51}$$

where ***XX*** and ***YY*** adopting the above distance function are the two feature matrices. Experimental results show that the adoption of the two distance functions gives very similar output. However, NVSC-based methods require fewer iterations and converge faster than Euclidean distance approaches.

15.3.3.2 Design of Subspace Classifier

Suppose there are m class data $\boldsymbol{V}_1, \boldsymbol{V}_2, \cdots, \boldsymbol{V}_m$. Obtaining the non-negative subspace $\boldsymbol{K}_i$ of the first class training data $\boldsymbol{V}_i (i = 1, 2, 3, \ldots, m)$ involves finding r non-negative base vectors of the sample space to minimize the projection error of $\boldsymbol{V}_i$ in $\boldsymbol{K}_i$. For test samples $\boldsymbol{V}$ in n-dimensional space, the appropriate classification criterion is (Pu *et al.*, 2005):

$$\textit{If } l = \arg\min g(\boldsymbol{V}, \boldsymbol{K}_i), \quad \textit{then } C(\boldsymbol{V}) = l; \tag{15.52}$$

where $g(\boldsymbol{V}, \boldsymbol{K}_i)$ is the discriminant function of the pattern class that represents the proximity of the vector $\boldsymbol{V}$ and the subspace $\boldsymbol{K}_i$, and $C(\boldsymbol{V})$ is the class of test samples. Obviously, classification aims to minimize the distance between the vector $\boldsymbol{V}$ and its corresponding generic subspace to obtain the smallest classification error. The discriminant function is obtained by calculating the reconstruction error of the test sample $\boldsymbol{V}$ in the corresponding subspace $\boldsymbol{K}_i$.

For sub-spatial representation based on NMF or the PCA algorithm, if we have found a non-negative base/orthogonal feature base $\boldsymbol{W}$ of a dataset $\boldsymbol{V}$ and the coefficients h representing the data $\boldsymbol{V}$ in the corresponding subspace $\boldsymbol{K}_i$, the reconstruction error given by this representation is the same. Both of them can be written as:

$$E(\boldsymbol{V}; \boldsymbol{W}) = \frac{1}{2}\|\boldsymbol{V} - \boldsymbol{W}h\|^2 \tag{15.53}$$

However, the meaning of the expression in Equation 15.52 for NMF and PCA is very different. In general, the base vector data in a subspace classifier based on the NMF algorithm is more compact and effective than that based on the PCA algorithm for the representation of data. Because the data are just full of part of the plane, the subspace based on the PCA algorithm contains considerable redundancy, resulting in a great waste in the representation of data. Furthermore, for pattern classification problems, it is entirely possible for several models of data to exist in the same subspace due to the hyper-redundant subspace given by the PCA algorithm. Hence, this subspace is apparently unable to realize the classification of these patterns. On the contrary, even if different models of data exist in the subspace based on the PCA algorithm, we can still achieve the correct classification because of the squeezing property that subspaces based on the NMF algorithm have on data.

In the NMF subspace classification algorithm, for the test sample $\boldsymbol{V}$, we structure the following discriminant function for identification:

$$g(\boldsymbol{V}, \boldsymbol{K}_i) = k_1 \|\boldsymbol{V} - \boldsymbol{W}^i h_i\|_2^2 + k_2 \|h_i - \overline{h}_i\|_2^2 \tag{15.54}$$

where $\boldsymbol{W}^i = \{\boldsymbol{W}_1^i, \boldsymbol{W}_2^i, \cdots, \boldsymbol{W}_r^i\}$ is the base vector L_i of the ith class training samples, h_i is the projection coefficient of the test sample $\boldsymbol{V}$ on r-dimensional subspace, $\overline{h}_i$ is the mean vector of the projection of the ith training sample set on $\boldsymbol{W}^i$, and $k_1, k_2 > 0$ are the weight coefficients, whose values are determined according to application conditions.

The first part of the discriminant function is the reconstruction error of the test sample $\boldsymbol{V}$ on subspace L_i, and the second part guarantees the correct classification of samples whose subspace base overlaps. This latter term is the Euclidean distance between the reconstruction error of test sample $\boldsymbol{V}$ on subspace L_i and the center vector of the ith class training sample.

Algorithm Subspace design algorithm based on the NMF algorithm

Input: m training sample sets $\boldsymbol{V}_1, \boldsymbol{V}_2, \cdots, \boldsymbol{V}_m$ and test sample set $\boldsymbol{V}$
Output: the class of the training sample $c(\boldsymbol{V})$

1. Obtain the non-negative base of feature subspaces of the various kinds of samples $\boldsymbol{W}^1, \boldsymbol{W}^2, \boldsymbol{W}^3, \cdots, \boldsymbol{W}^m$ according to the NMF algorithm, where $\boldsymbol{W}^i$ consists of r non-negative bases $\boldsymbol{W}^i = \{\boldsymbol{W}_1^i, \boldsymbol{W}_2^i, \cdots, \boldsymbol{W}_r^i\}$, and calculate the mean vector $\overline{h}_1, \overline{h}_2, \cdots, \overline{h}_m$ of the various samples on their own feature subspace.
2. Calculate the projection coefficients $h_i (i = 1, \cdots, m)$ of the test sample $\boldsymbol{V}$ on their own subspaces $\boldsymbol{W}^1, \boldsymbol{W}^2, \boldsymbol{W}^3, \cdots, \boldsymbol{W}^m$ according to the non-negative matrix iterative rule, taking $\boldsymbol{W}^i = \{\boldsymbol{W}_1^i, \boldsymbol{W}_2^i, \cdots, \boldsymbol{W}_r^i\}$ $(i = 1, \ldots, m)$ as the non-negative feature base of the ith class data.
3. Calculate the discriminant function $g(\boldsymbol{V}, \boldsymbol{K}_i), i = 1, 2, \ldots, m$ of $\boldsymbol{V}$ for the various class subspaces.
4. Determine the class of $c(\boldsymbol{V})$.

15.3.3.3 Selection and Optimization of Subspace Number and Classifier

In the NMF algorithm, the radix r is a very important parameter that not only determines the dimension of the local feature subspace extracted from the raw data space by the NMF algorithm, but also affects the performance of the NMF algorithm itself.

Generally, the closer r is to the dimension of the subspace of raw data, the slower the convergence speed of the algorithm. If r is too large, we cannot obtain good dimension reduction, whereas if r is too small, it is difficult to extract most of the information from the raw data and the fast convergence of the algorithm cannot be guaranteed. Thus, the choice of r is a key decision. For a particular data set, the dimension of the feature space hidden within is known. That is to say, when the selected value of r is in accordance with the dimension of the feature space of the actual data set, the squeeze characteristic of the obtained subspace of NMF and the non-negative base of the subspace feature are the most effective at handling the redundancy of data representation. However, there is no good way to determine the value of r. In this section, we select different values of r in the NMF applications to obtain the feature subspace that best reflects the characteristics of the internal structure of the data set. In this way, we are able to maximize the information extracted from the raw data and avoid data redundancy as far as possible.

The PCA algorithm can find the projection direction in which the energy is most concentrated. According to the characteristics of PCA, we can adopt the algorithm to determine the number of NMF subspaces and initialize NMF, and we can perform a non-negative conversion for negative values obtained in the PCA process.

After NMF preprocessing, the data need to undergo a certain classification. We can adopt a C-means algorithm for this purpose. The C-means algorithm is a kind of unsupervised method based on a similarity measure between samples. As this algorithm is simple, fast, and can handle large databases effectively, it is chosen as the classification algorithm. However, most remote sensing images have different characteristic surface textures, which usually have large or small distinctions. Therefore, we can conduct a classification of remote sensing images by combining the macro characteristics of the image region and the regional texture digital characteristics. This improves the selection method of the original cluster centers given by the C-means algorithm, resulting in a better representation. In addition, we can perform the subspace classification by combining fuzzy set theory with the fuzzy C-means clustering algorithm.

15.3.4 Applications of NMF in Mineral Prediction

15.3.4.1 Introduction of the Experimental Data

The copper polymetallic deposits in Gaoyao City, Guangdong province, are mainly found in low mountain landforms at elevations of 240–350 m, with a highest altitude of 387.71 m and a relative elevation of 150 m. The overall terrain is higher in the east, and has steep gradients of up to 40°. There are two large rocks, namely Stone Oxhead rock and Wu Cun rock, in the deposit. In addition, at the Shuangren mining area to the northwest of Stone Oxhead, there is a roughly northeast parallel distribution of granite dikes (bodies). This exposed range is not big, generally about 10 km^2 with some being greater than 20 km^2. Copper-gold polymetallic mine sites are mainly located to the west side of the north contact zone of Stone Oxhead rock, being controlled by the structure of the northeast fracture.

According to a 1/50 000 stream sediment survey report of the working area, the area of mineralized anomaly contains elements including Cu, Ag, Zn, Bi, Au, and As in the Sinian and lower Paleozoic sedimentary and metamorphic rocks. The fault zone and the mylonite zone appear mostly in this area. Cu, Ag, and Zn all have concentration zones of Level 3 (where a higher concentration zone and mineralization zoning of Cu is better), and Cu is the main ore-forming element in the area. The correlative analytical results of the R-factor in the trace elements show that the relation between Au, Ag, and Mo and that between Cu, Zn, and Mo is close, which indicates that the Au and Ag are spatially close, but the relation between Cu and Au, Ag is not close. The above suggests that the migration and enrichment of Cu and Au, Ag occurred during different chronological periods.

In Figure 15.19, the box denotes the survey area. We take 10 geochemical lines from 330°NW to 150°SE, each having 51 sampling points numbered 1/X, 2/X, … ,51/X. The length of each geochemical line is 500 m, with a between-line distance of 200 m and a point distance of 20 m. The tested elements include Au, Ag, Cu, Pb, Zn, Sn, As, Mo, Bi, and W. The reconnaissance elements include Au, Ag, Cu, Pb, and Zn. In the prophase work, the geological unit responsible for the exploration delineated

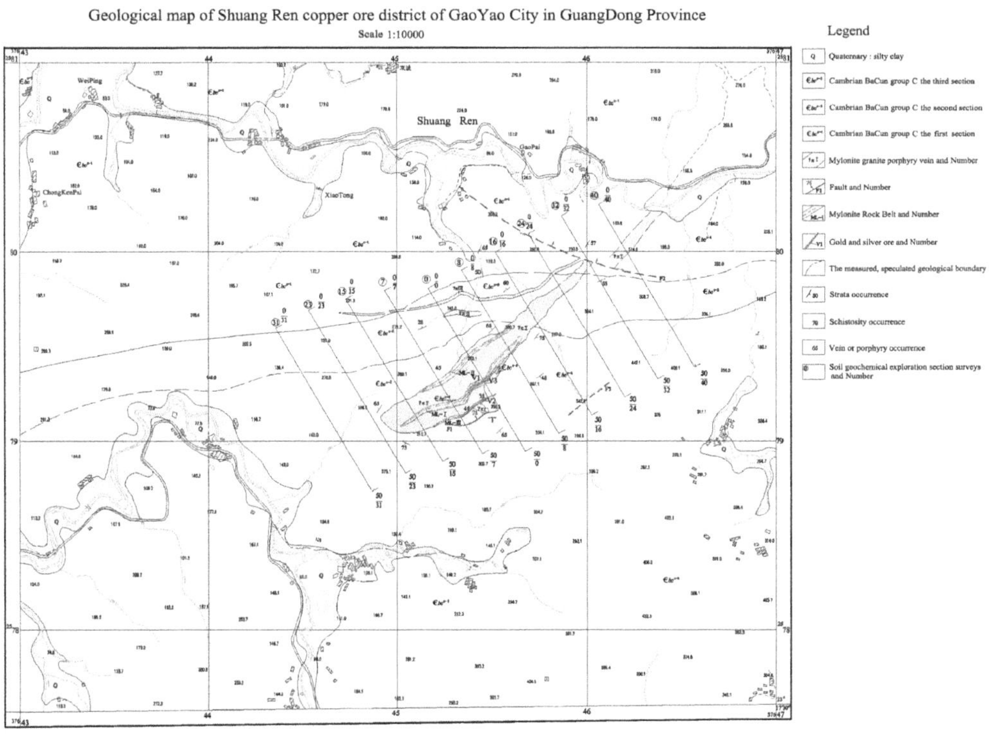

Figure 15.19 Geological map of Shuangren copper ore in Gaoyao city in Guangdong province (see color Figure 15.19)

Table 15.6 Part of Shuangren copper mining data in Gaoyao, Guangdong (Au, Ag, Cu, Pb, Zn)

Sample number	x-coordinates	y-coordinates	Au	Ag	Cu	Pb	Zn
1	1 732.0508	1 866.0254	1	0.21	18.8	61	78.8
2	1 742.0508	1 848.7049	1	0.07	14.3	47.2	31.3
3	1 752.0508	1 831.3844	1	0.35	26.3	54.9	31
4	1 762.0508	1 814.0639	1.5	0.03	15.1	46.9	52.3
5	1 772.0508	1 796.7434	1	0.04	16.9	54.1	68.4
6	1 782.0508	1 779.4229	2	0.03	17.1	52.7	56.8
7	1 792.0508	1 962.1024	2	0.17	24.1	60.6	53.5
8	1 802.0508	1 744.7818	1	0.27	18.7	59.2	30.8
9	1 812.0508	1 727.4613	1	0.57	30.4	66.1	47.9
10	1 822.0508	1 710.1408	1.5	0.12	27.9	66.6	57.5

the geochemical anomalies according to the lower limit (when the element content of a point is greater than $E + 2\sigma$, it is known as a mineralized anomaly). There is a significant deviation between the prediction results and the ore body distribution based on asurvey using this method of mineral prediction (the actual ore body does not have mineralized anomalies and the ore body was found accidentally).

The source data obtained in the experiment is in .txt files, which can be transformed into other formats. The data in the files include x- and y-coordinates and the measured value of other elements, as shown in Table 15.6.

15.3.4.2 Predictions Based on the Classical NMF Algorithm

The mineral prediction steps based on NMF are as follows:

1. Perform NMF processing for the element content of the samples in the survey area.
2. Take the survey data and the data after NMF processing as new observed variables and conduct a FA for the data.
3. Determine the element corresponding to each component after decomposition according to the FA results.
4. Select the elements of interest to construct a mineral resources prediction map, which is an isoline map superimposed on a geological base map. If we find an area that has an obvious concentration center consistent with the actual distribution from the figure, we can outline the mineral prediction area according to the concept of analogy forecasting.

In this experiment, we use the classical NMF algorithm and obtain an oscillograph of 10 mineral elements, as shown in Figure 15.20 (S1–S10 denote Au, Ag, Cu, Pb, Zn, Sn, As, Mo, Bi, and W, respectively).

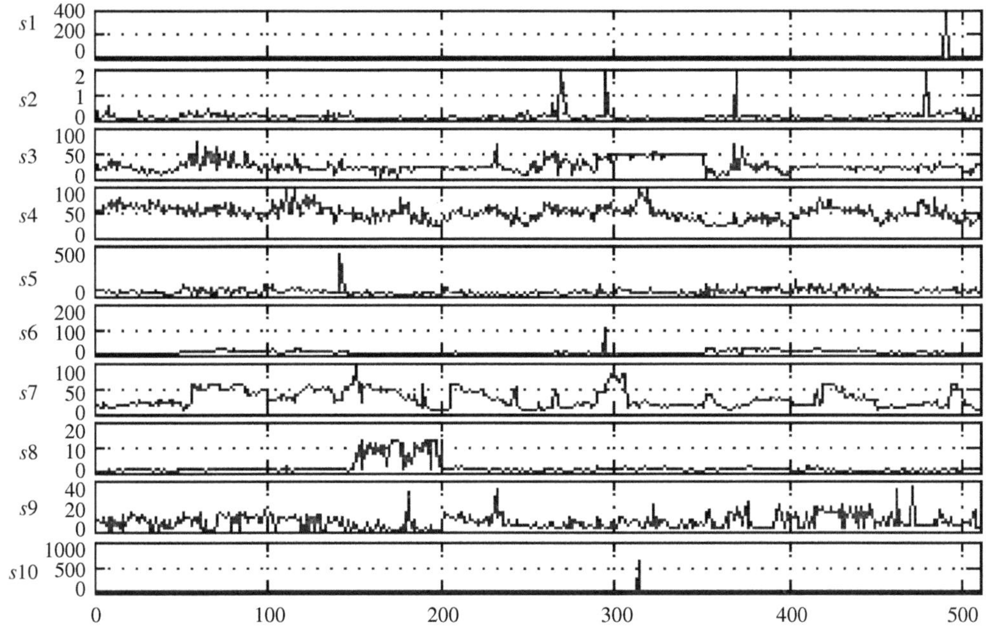

Figure 15.20 Oscillograph of 10 original mineral data elements. The *x*-axis denotes the sample point and the *y*-axis denotes the sample content of each element

We process the mineral data using the classical NMF algorithm and obtain the oscillograph of the 10 mineral elements after processing, which are denoted by X1–X10. The oscillograph is shown in Figure 15.21.

Because the order of the 10 elements is uncertain after the NMF algorithm processing, we conduct further processing to determine the element corresponding to each component using a FA method. The FA result is shown in Table 15.7.

We can determine the order of the 10 elements after NMF processing to be W, Au, Ag, Zn, Sn, Cu, As, Mo, Pb, and Bi based on the FA results. In the geological map (see Figures 15.22–15.24), we can find an Au–Ag–Cu mine belt in the actual exploration survey area, and so we choose a component that is independent of these three elements to construct our mineral prediction map. The square areas represent the scope according to the practical exploration results, and the circular regions denote the range of prediction in these figures.

We can see from Figure 15.24 that the Cu distribution is obviously near the geochemical prospecting lines, and the distribution exhibits the same trend as the lines. This is because the point distribution is too dense along a given direction, as the distance between sampling points and the distance between lines are not of the same order of magnitude. From the experimental results, we can conclude that the geological data conforms to the actual distribution of mineral reserves, and the area that has an obvious Cu concentration center indicates a larger probability of the existence of minerals. Thus, the area is more specific than the previous exploration results.

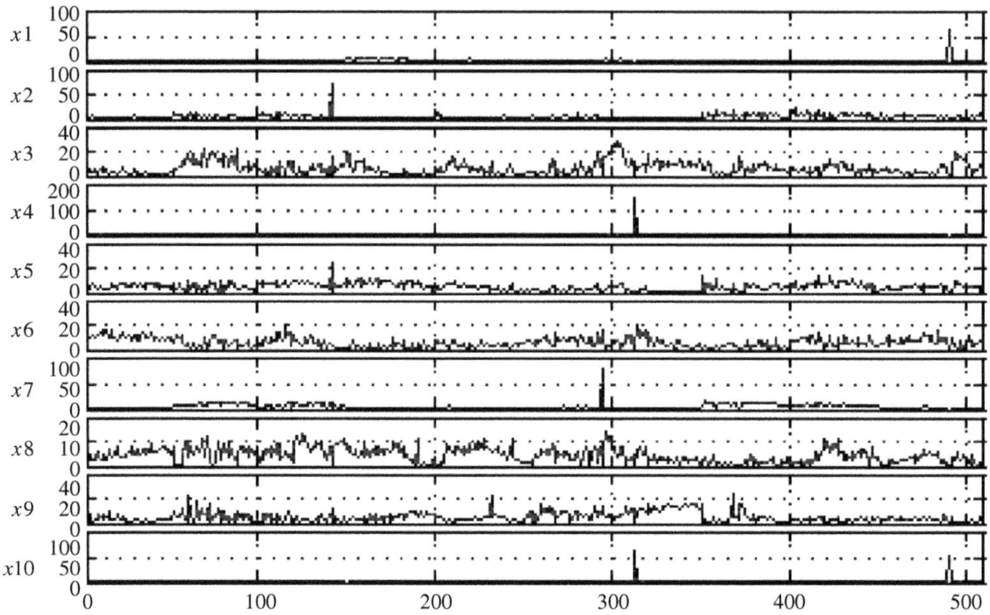

Figure 15.21 Oscillograph of 10 mineral data elements after classical NMF algorithm processing

Table 15.7 Factor loading matrix

Variable	Component									
	1	2	3	4	5	6	7	8	9	10
V1	−0.013	−0.018	−0.008	0.999	0.001	−0.006	0.005	0.013	−0.020	0.031
V2	0.007	0.043	−0.010	0.023	−0.012	0.018	0.995	−0.067	0.020	0.039
V3	0.003	0.052	0.036	0.007	0.051	0.008	0.044	−0.057	0.977	0.185
V4	0.178	0.038	0.065	0.037	0.083	0.033	0.034	−0.77	0.152	0.961
V5	−0.016	0.124	0.126	0.001	0.008	0.977	0.013	−0.081	0.015	0.080
V6	0.104	0.964	0.102	-2.2×10^{-5}	0.009	0.189	0.041	−0.075	−0.009	0.009
V7	0.989	0.096	0.033	0.006	−0.018	-6.3×10^{-5}	0.013	0.099	0.026	−0.011
V8	0.092	−0.087	−0.127	0.016	−0.033	−0.074	−0.061	0.976	−0.034	−0.063
V9	0.021	0.083	0.977	0.018	0.021	0.120	−0.016	−0.129	−0.007	0.072
V10	−0.021	0.011	0.011	−0.003	0.995	0.006	0.002	−0.040	0.049	0.070
V11	−0.350	−0.119	0.036	0.029	0.085	0.033	0.030	−0.092	0.130	0.909
V12	−0.013	−0.016	−0.008	0.999	−0.007	−0.012	0.006	0.001	−0.023	0.023
V13	0.039	0.093	0.977	−0.035	−0.007	0.120	−0.016	−0.123	−0.057	0.018
V14	0.991	0.066	0.025	−0.032	−0.022	−0.042	−0.012	0.066	0.022	−0.068
V15	0.071	0.984	0.076	−0.036	0.006	0.079	0.007	−0..095	−0.016	−0.065
V16	−0.029	0.136	0.113	−0.020	−0.005	0.980	0.008	−0.073	−0.027	−0.021
V17	−0.022	0.003	0.002	−0.003	0.995	−0.003	−0.025	−0.038	0.041	0.069
V18	−0.007	0.001	−0.020	−0.011	−0.011	0.003	0.998	−0.062	0.006	0.015
V19	0.085	0.085	−0.128	−0.001	−0.051	−0.082	−0.078	0.972	−0.038	−0.086
V20	0.040	−0.074	−0.097	−0.051	0.042	−0.020	−0.015	−0.018	0.987	0.063

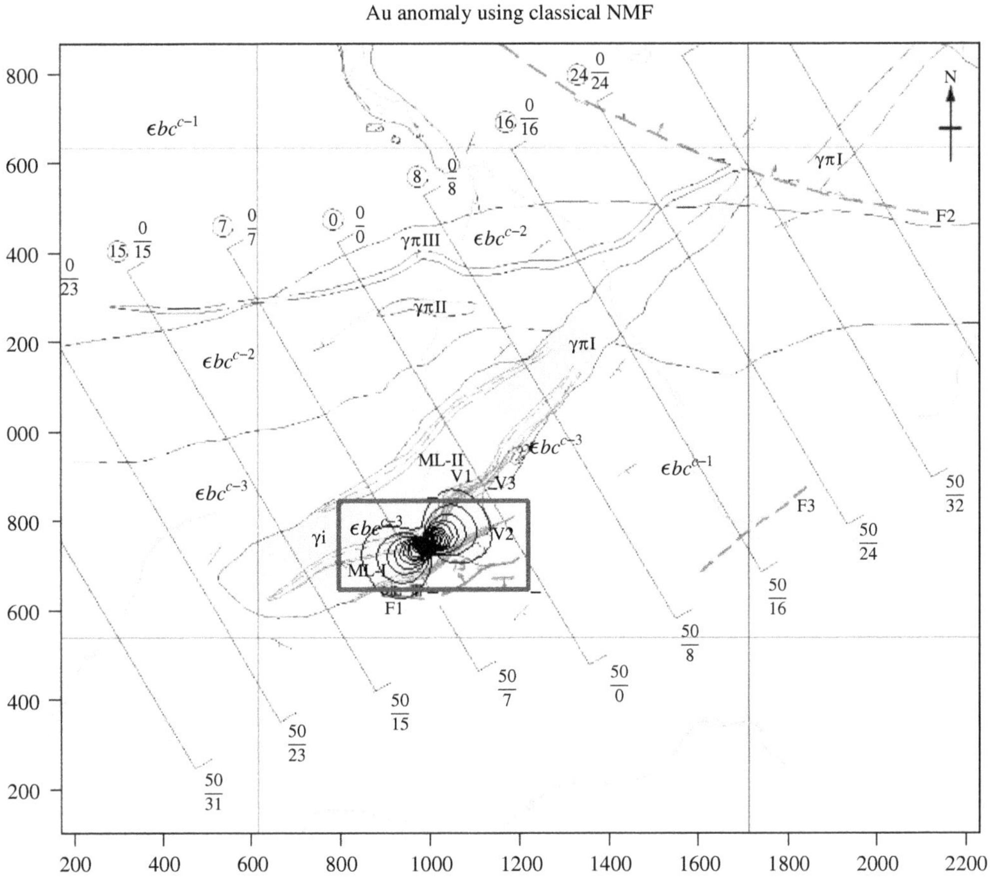

Figure 15.22 Au anomaly using classical NMF processing

15.3.4.3 Determination of the Sparseness Factor

In the process of improving the NMF algorithm, the selection of a sparseness factor is related not only to the degree of sparseness of mineral data, but also to the convergence speed of the algorithm and the accuracy of mineral prediction. The greater the sparseness factor, the larger the degree of sparseness. However, if the sparseness factor is too great, we may lose important information from the raw data. Therefore, we must select a suitable sparseness factor to realize the optimum sparseness and prediction accuracy. We choose several different sparseness factor values and obtain different degrees of sparseness when processing the mineral data. We then conduct an analysis and prediction for the data with different degrees of sparseness, and determine the value of the sparseness factor according to the prediction effect.

When the sparseness factor increases, it will affect the execution speed and the error size of the algorithm. Thus, we must increase the number of iterations to attain the

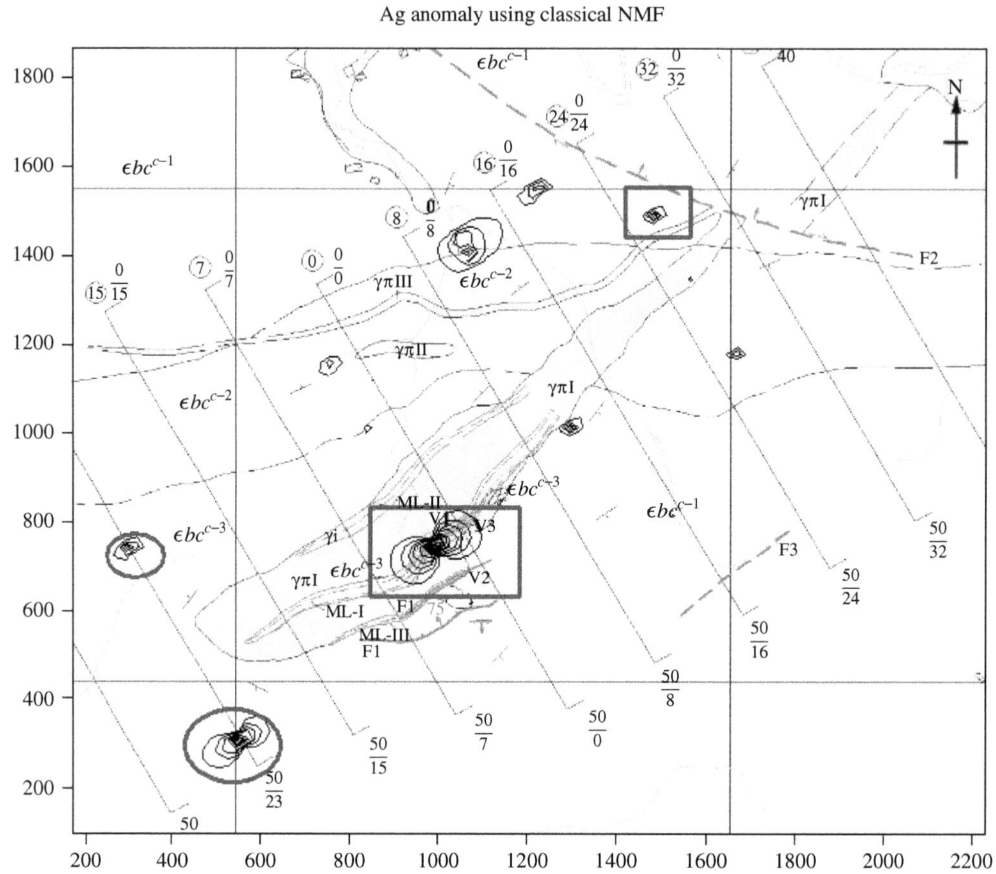

Figure 15.23 Ag anomaly using classical NMF

same target error, that is, the larger the sparseness factor, the larger the target error for the same data and number of iterations, as shown in Figure 15.25.

In Figure 15.25, the x-axis represents the size of the sparseness factor and the y-axis represents the corresponding error. In the experiment, the error increases with the sparseness factor for the same data, fixed number of iterations, and different sparseness factor values. Hence, we must choose an appropriate sparseness factor to ensure we can acquire a good degree of sparseness and a reasonable target error.

15.3.4.4 Mineral Prediction Based on the Improved NMF Algorithm and Factor Analysis

In this experiment, we use the improved NMF algorithm to process the mineral data and obtain the oscillograph of the 10 elements represented by X1–X10. The oscillograph is shown in Figure 15.26.

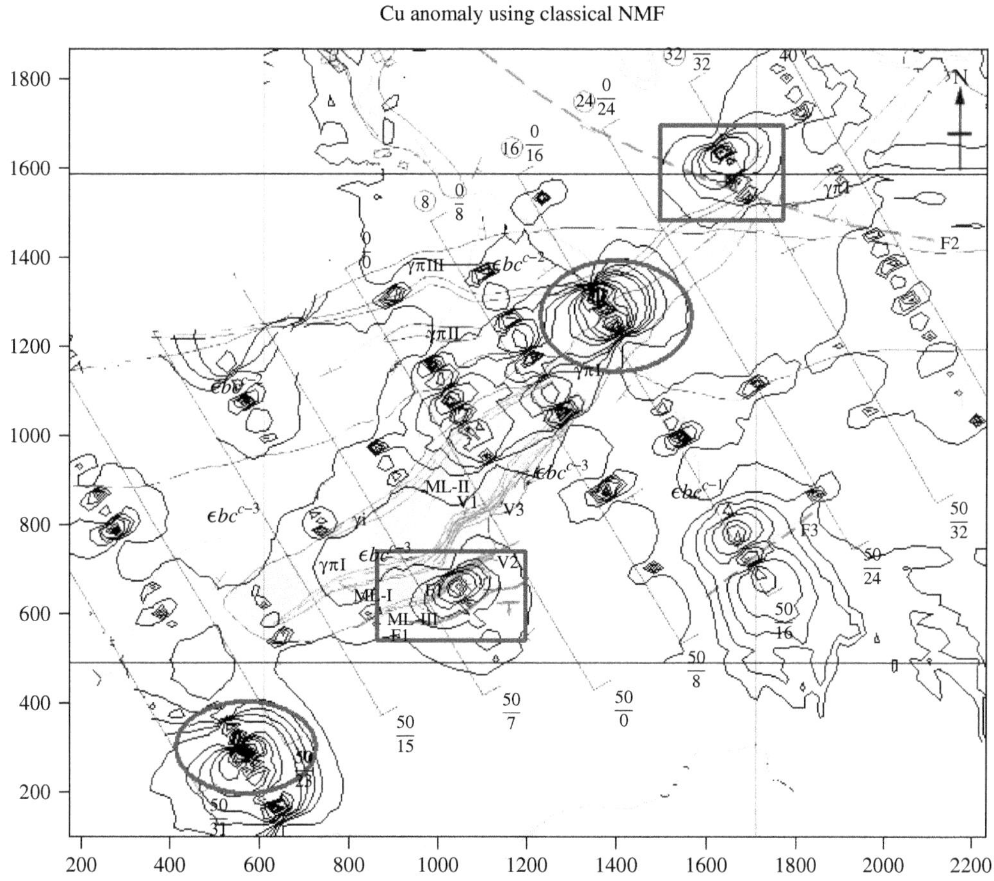

Figure 15.24 Cu anomaly using classical NMF

We use the SNMF algorithm in this experiment to allow for a kriging space estimation (interpolation) with highly sparse mineral data (Table 15.8) obtained by adjusting the sparseness factor.

Because the order of elements after processing is uncertain, we process the results further to determine the element corresponding to each component. The results of the FA are shown in Table 15.9.

We can determine the order of the 10 elements after NMF processing to be W, Au, Ag, Zn, Sn, Cu, As, Mo, Pb, and Bi, according to the FA results. Because we can find an Au–Ag–Cu ore belt in the practical survey area from the geological topographic map (see Figures 15.27–15.29), we form the mineral prediction map based on the decomposition results (Au, Ag, Cu). In these figures, the square areas (the obvious concentration center) accord with the practical results and the circular regions are the predicted range.

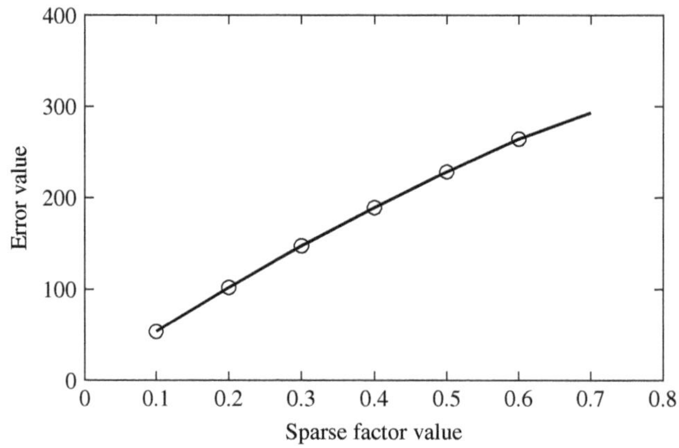

Figure 15.25 Mapping of sparse factor and error values

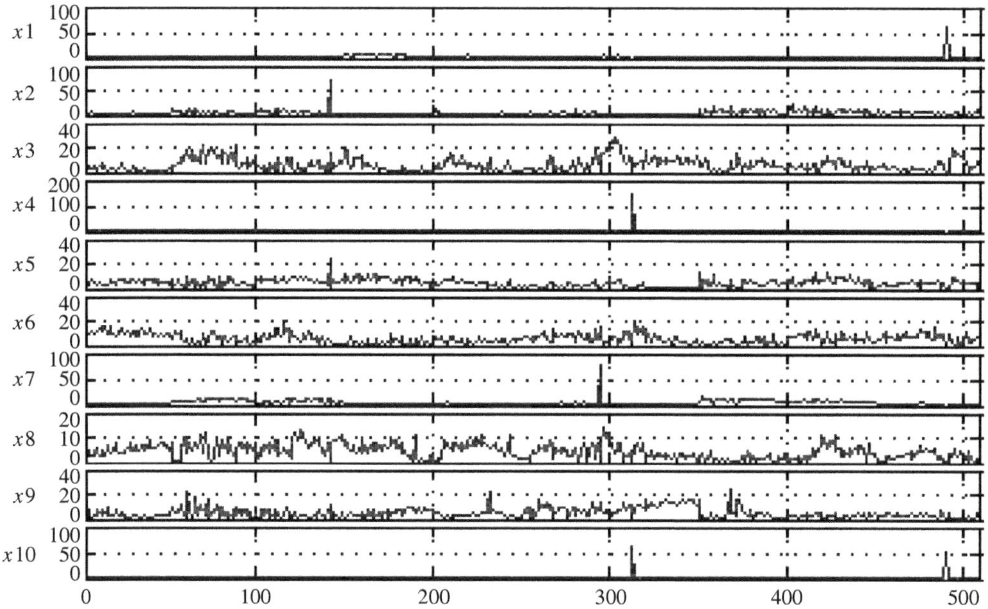

Figure 15.26 Oscillograph of 10 elements after the improved NMF processing

After processing the geological data with SNMF, we can conclude that sparse processing is effective for the raw data and gives a prediction effect that agrees with the actual distribution of mineral reserves more accurately than that given by the classical NMF algorithm.

In addition, the obvious concentration centers represent a greater probability of mineral resources. Compared with previous exploration results, SNMF delineates the range more clearly.

Table 15.8 Part of data based on NMF sparse processing

Sample number	*x*-coordinates	*y*-coordinates	Au	Ag	Cu	Pb	Zn
1	1 732.0508	1 866.0254	0	0	3 296.6	0	0
2	1 742.0508	1 848.7049	0	0	0	0	0
3	1 752.0508	1 831.3844	0	0	0	0	0
4	1 762.0508	1 814.0639	0	0	0	0	0
5	1 772.0508	1 796.7434	0	0	0	0	0
6	1 782.0508	1 779.4229	0	0	0	1 411.7	0
7	1 792.0508	1 962.1024	0	0	0	0	0
8	1 802.0508	1 744.7818	0	0	1 659.4	0	0
9	1 812.0508	1 727.4613	0	0	0	0	0
10	1 822.0508	1 710.1408	0	0	0	0	0

Table 15.9 Factor loading matrix table

Variable	Component									
	1	2	3	4	5	6	7	8	9	10
V1	−0.013	−0.018	−0.008	0.999	0.001	−0.006	0.005	0.013	−0.020	0.031
V2	0.007	0.043	0.010	0.023	0.012	0.018	0.9995	0.067	0.020	0.030
V3	0.003	0.052	0.036	0.007	0.051	0.008	0.044	0.057	0.077	0.185
V4	0.178	0.038	0.055	0.037	0.033	0.033	0.034	0.077	0.152	0.951
V5	0.016	0.124	0.125	0.001	0.008	0.977	0.013	0.081	0.015	0.080
V6	0.105	0.957	0.102	-2.2×10^{-5}	0.009	0.189	0.041	0.075	0.009	0.009
V7	0.989	0.096	0.033	0.006	−0.018	-6.3×10^{-5}	0.013	0.099	0.028	−0.011
V8	0.092	−0.087	−0.127	0.016	−0.033	−0.074	−0.061	0.970	−0.034	−0.053
V9	0.021	0.083	0.977	0.018	0.021	0.120	−0.018	−0.129	−0.007	0.072
V10	−0.021	0.011	0.011	−0.003	0.995	0.006	0.002	−0.040	0.049	0.070
V11	−0.350	−0.119	0.036	0.029	0.085	0.033	0.030	−0.092	0.130	0.909
V12	−0.350	−0.119	0.036	0.029	0.085	0.033	0.030	−0.092	0.130	0.909
V13	0.039	0.093	0.977	−0.035	−0.007	0.120	−0.015	−0.123	−0.057	0.018
V14	0.991	0.066	0.025	−0.032	−0.022	−0.042	−0.012	−0.066	0.022	−0.063
V15	0.070	0.984	0.076	−0.036	0.006	0.070	0.007	0.095	0.016	0.065
V16	0.020	0.136	0.113	−0.020	−0.005	0.080	0.008	0.073	−0.027	−0.021
V17	0.022	0.003	0.002	0.003	0.995	0.003	0.025	0.038	0.041	0.069
V18	0.007	0.001	0.020	0.011	0.011	0.003	0.999	0.082	0.006	0.015
V19	0.085	−0.085	0.128	−0.001	−0.051	−0.082	−0.078	0.972	0.038	0.086
V20	0.040	−0.074	0.097	0.051	0.042	0.020	0.015	0.013	0.987	0.063

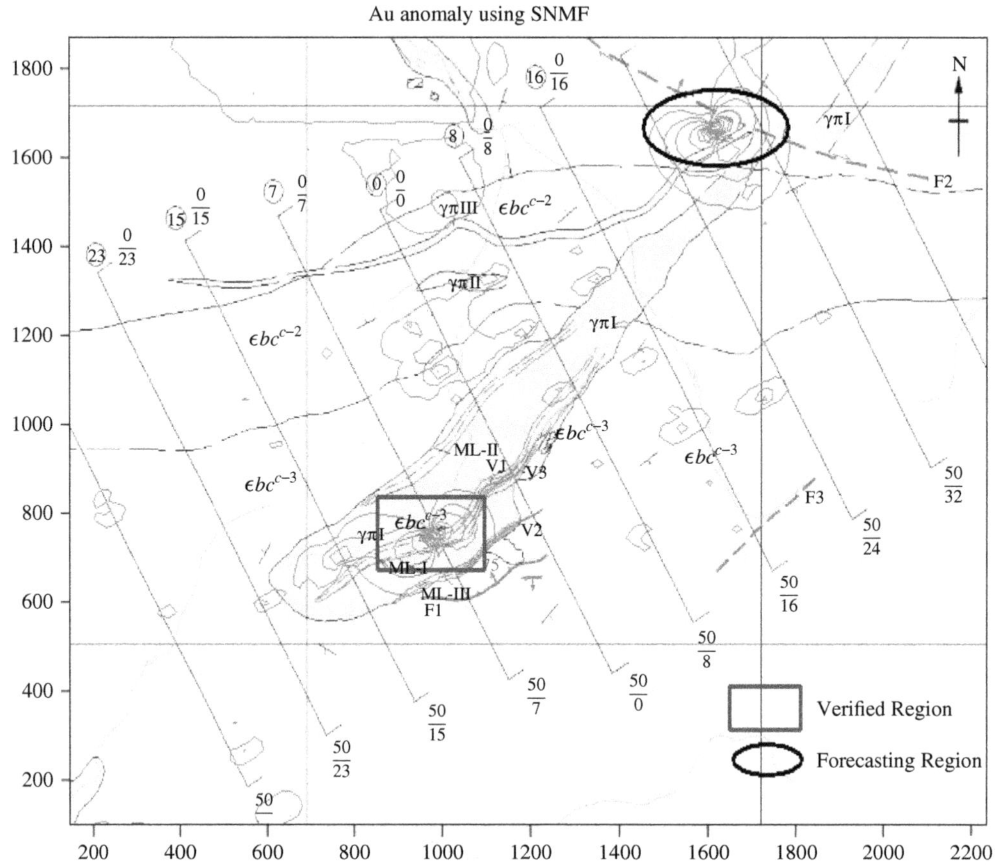

Figure 15.27 Au anomaly using SNMF

15.3.4.5 Mineral Prediction Based on the Improved NMF Algorithm and Fuzzy C-Means Classifier

We can construct a classifier to realize digital mineral prediction. The results will have two classes, one to denote the area in which there is a mineralized anomaly and the other for the region of no anomaly. Thus, we can construct a classifier for the mineral data to identify the mineral distribution area.

We extract the hidden structure of the geochemical data using NMF, and combine this with other characteristics, such as the geological structure, magmatic rocks, and strata to obtain new sample points for classification. We then classify the points according to fuzzy C-means, analyze the results of classification, calculate the accuracy of classification, and verify the correctness of the prediction. Part of the data to be classified in the experiment is given in Table 15.10.

If the characteristic value of the geological structure, magmatic rocks, and strata is 1, it is said that this attribute of the sample point is consistent with that in the mineralized anomaly area. Conversely, 0 means this attribute is inconsistent.

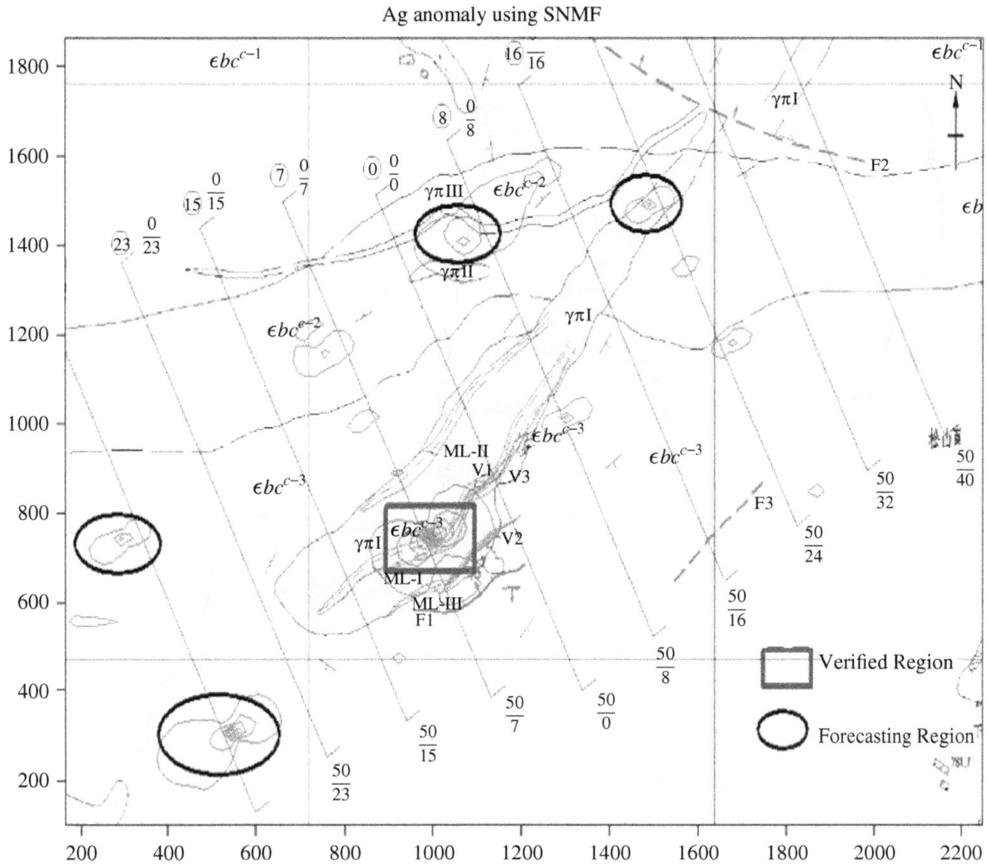

Figure 15.28 Ag anomaly using SNMF

The main steps of mineral prediction with the improved NMF and fuzzy C-means classifier are as follows:

1. Input raw data matrix V and get the basis matrix W and the weight coefficient matrix H using the improved NMF algorithm.
2. Combine the corresponding elements in W with characteristics of the data.
3. Classify the new sample points according to the fuzzy C-means algorithm with a classification number of 2.
4. Take the class that contains more mineralized anomaly points as the mineral area and the other class as the common area. Calculate the classification accuracy and determine the prediction range according to a comparison of the classification result and the geological map.

Part of the reassembled data to be classified is shown in Table 15.11.

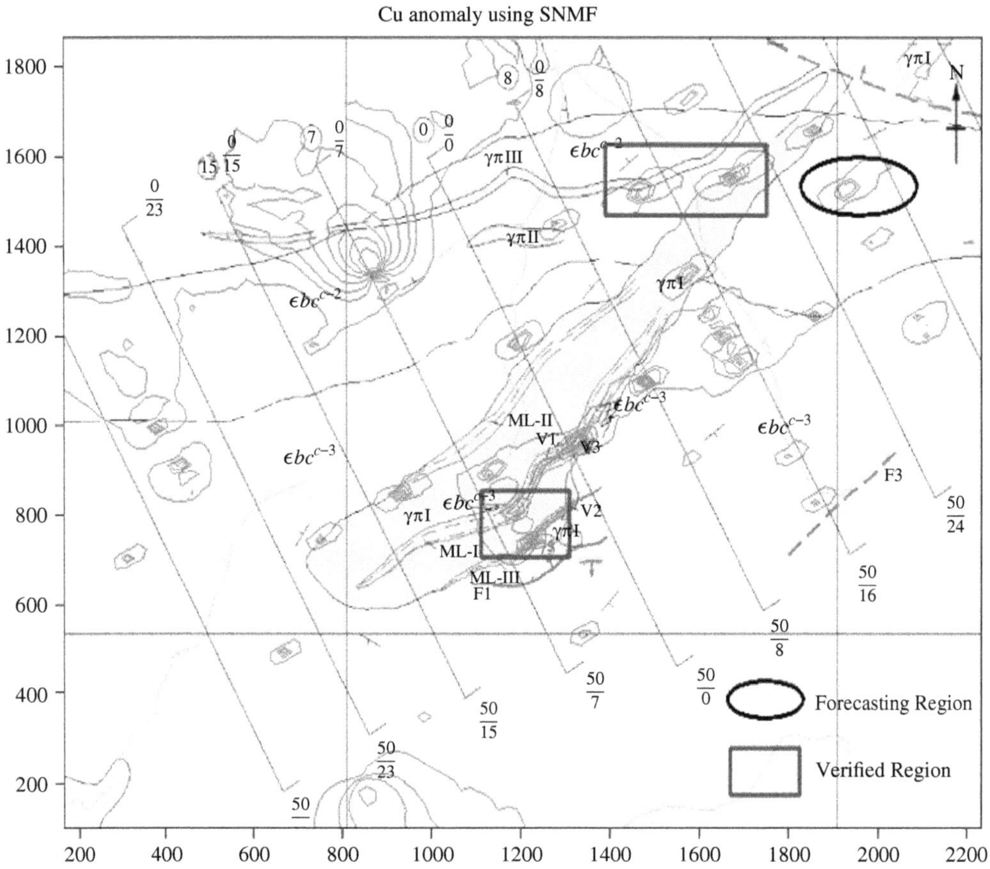

Figure 15.29 Cu anomaly using SNMF

We process the resulting sample data with the fuzzy C-means classifier to obtain two sample classes according to their membership values. The membership values of some of the samples are shown in Table 15.12.

The classification result shows that sample points in the proven mineral area are mostly distributed in the second class. Thus, we take the second class as the mineralized anomaly class and the first class as the common area. We can then determine the scope of mineral resources by comparing the geological map and the area of the second class of sample points. The predicted distribution of mineral resources is shown in Figure 15.30.

We compare the predicted area of mineral resources and the actual survey results. If they correspond to one another, we can apply the improved NMF algorithm and fuzzy C-means classifier to mineral prediction to good effect. Because we determine the prediction area according to the specific sample points after classification, the prediction range is more accurate.

Table 15.10 Geological structure, magmatic rocks, part of element data of Shuangren copper ore district in Gaoyao, Guangdong (Au, Ag, Cu, Pb)

Sample number	Geological structure	Magmatic rocks	Strata	x-coordinates	y-coordinates	Au	Ag	Cu	Pb
1	0	0	0	1 732.0508	1 866.0254	1	0.21	18.8	61
2	0	0	0	1 742.0508	1 848.7049	1	0.07	14.3	47.2
3	0	0	0	1 752.0508	1 831.3844	1	0.35	26.3	54.9
4	0	0	0	1 762.0508	1 814.0639	1.5	0.03	15.1	46.9
5	0	0	0	1 772.0508	1 796.7434	1	0.04	16.9	54.1
6	0	0	1	1 782.0508	1 779.4229	2	0.03	17.1	52.7
7	1	0	0	1 792.0508	1 962.1024	2	0.17	24.1	60.6
8	0	0	0	1 802.0508	1 744.7818	1	0.27	18.7	59.2
9	0	0	0	1 812.0508	1 727.4613	1	0.57	30.4	66.1
10	0	1	0	1 822.0508	1 710.1408	1.5	0.12	27.9	66.6

Table 15.11 Geological structure, magmatic rocks, part of element data of Shuangren copper ore district in Gaoyao, Guangdong after processing according to NMF (Au, Ag, Cu, Pb)

Sample number	Geological structure	Magmatic rocks	Strata	x-coordinates	y-coordinates	Au	Ag	Cu	Pb
1	0	0	0	1 732.0508	1 866.0254	0	0	3 296.6	0
2	0	0	0	1 742.0508	1 848.7049	0	0	917.1	0
3	0	0	0	1 752.0508	1 831.3844	0	0	0	0
4	0	0	0	1 762.0508	1 814.0639	0	0	0	0
5	0	0	0	1 772.0508	1 796.7434	0	0	0	0
6	0	0	1	1 782.0508	1 779.4229	0	0	0	1 411.7
7	1	0	0	1 792.0508	1 962.1024	0	0	0	0
8	0	0	0	1 802.0508	1 744.7818	0	0	1 659.4	0
9	0	0	0	1 812.0508	1 727.4613	0	0	0	0
10	0	1	0	1 822.0508	1 710.1408	0	0	0	2 814.8

Table 15.12 Membership of selected samples from Shuangren copper ore district in Gaoyao, Guangdong after processing according to NMF and FCM (fuzzy C-means) classifier

Sample number	Membership		Category
437	0.45951	0.54059	2
438	0.42679	0.57321	2
442	0.31174	0.68826	2
495	0.37743	0.62257	2
496	0.43991	0.56009	2
499	0.22524	0.77476	2

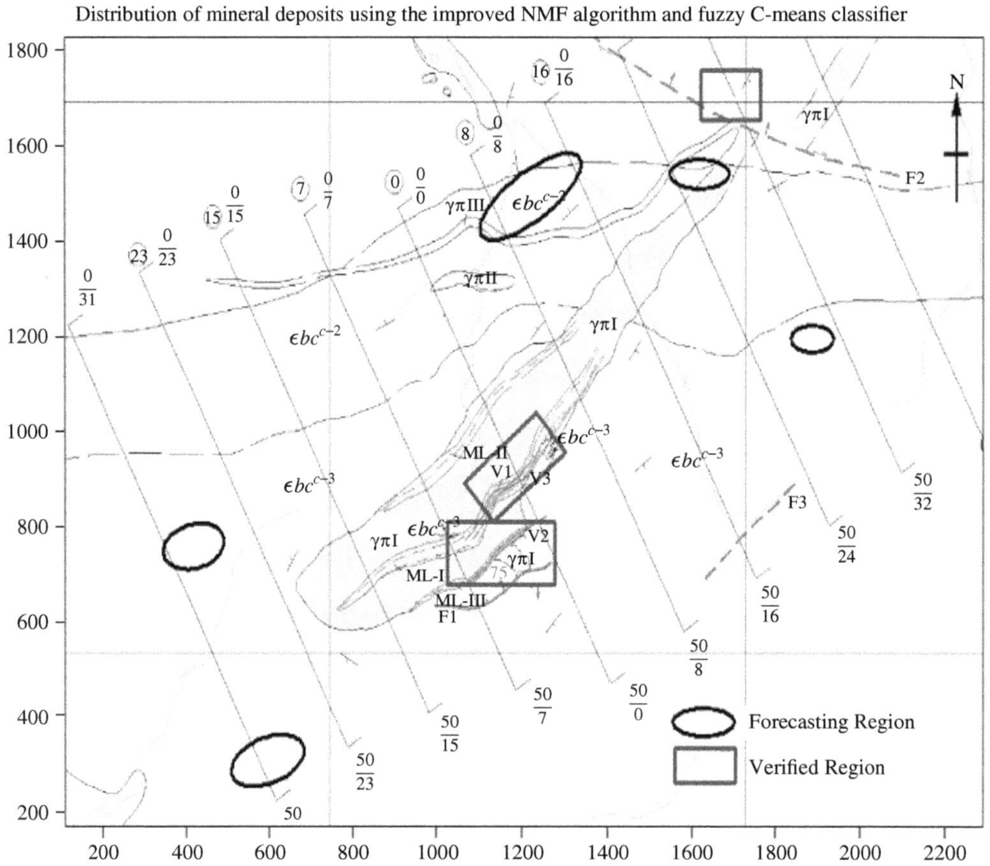

Figure 15.30 Distribution of mineral deposits using the improved NMF algorithm and fuzzy C-means classifier

15.3.4.6 Analysis of Experimental Results

There are three characteristics when applying the NMF algorithm to mineral prediction:

1. The NMF algorithm can ensure the non-negativity of the decomposed matrix. The decomposed results have better interpretability.
2. The solving method of the NMF algorithm has the characteristics of fast convergence, sparseness, and reduced dimensionality, meaning that the NMF algorithm can achieve fast and accurate processing of mineral data.
3. We processed copper polymetallic deposit data from Gaoyao, Guangdong province. The experimental results show that the geological data processed by the NMF algorithm reflected the actual distribution of the mineral reserves. Compared with mineral prediction using the classical NMF method, the information extracted

by the improved NMF algorithm better reflected the hidden mineralogical structure characteristics in the data. In the mineral prediction based on the improved NMF algorithm and FA, we predicted the distribution area of each element. These prediction results conformed to the actual distribution of mineral reserves, and the areas with obvious concentration centers denote a greater probability. Using the improved NMF and fuzzy C-means classifier, we divided the sample points into two classes, one representing the mineralized anomaly area and the other representing common areas. We then compared the classification results with a geological map to determine the mineral distribution range. The delineated area was more precise than previous exploration results, which has significance for the further prediction of mineral resources.

15.4 Chapter Summary

The recently proposed matrix decomposition method of NMF operates under the constraint condition that all elements in the matrices are non-negative. The method has attracted extensive attention in related research fields. Image fusion, proposed in the late 1970s, is a very important branch of data fusion involving subjects such as image processing, signal processing, and artificial intelligence. It can integrate multi-source information, eliminate potential redundancy and contradictions, and improve interpretation accuracy, reliability, and utilization rate. Data fusion technology offers the potential for the comprehensive utilization of multi-source remote sensing data, and improving the data quality has been a hot research topic in the international remote sensing community in recent years. In this chapter, we fused multispectral and panchromatic images to obtain high-quality remote sensing images using NMF combined with other algorithms.

This chapter mainly covered the following three areas:

1. We introduced the origins, characteristics, and application fields of NMF, including image analysis, text clustering/data mining, biomedical engineering, data processing in chemical engineering, signal analysis, and complex object recognition.
2. We introduced the object function in detail, including its selection, the initialization problem, the selection of the cardinal number r, the iterative rule and its convergence, the basic NMF algorithm, and several improved algorithms (LNMF, WNMF, FNMF, NSNMF, NMFSC). We also discussed the advantages and disadvantages of the various NMF algorithms.
3. We introduced specific applications of NMF theory. We introduced NMF to the field of remote sensing image processing, and demonstrated how to preprocess multispectral images prior to high-spatial-resolution image processing. We proposed several fusion algorithms based on NMF and obtained better, high-quality remote sensing fused images according to these algorithms. We then discussed the application of NMF to subspace classifiers. We applied the NMF and SNMF algorithms to the processing of geology mineral resources data. The analysis results

given by the different NMF algorithms for metal mineral deposits in Gaoyao, Guangdong province, enabled better mineral forecasting maps to be produced.

Readers can continue their research, according to the requirements of specific application fields, in the following topics: improved NMF algorithms; improved NMF algorithms for more professional fields; the initialization problem of the target function; selection of the base value r, and the convergence speed of the algorithms.

References

Babji, S., and Tangirala, A.K. (2010) Source separation in systems with correlated sources using NMF. *Digital Signal Processing*, **20**(2), 417–432.

Blondel, V., Ho, N.D., and Van Dooren, P. (2007) Algorithms for Weighted Nonnegative Matrix Factorizations, http://www.inma.ucl.ac.be/publi/303209.pdf (accessed 10 March 2013).

Catral, M., Han, L., Neumann, M. *et al.* (2004) On reduced rank non-negative factorization for symmetric non-negative matrices. *Linear Algebra Application*, **393**, 107–126.

Dempster, A.P., Laired, N.M., and Rubin, D.B. (1977) Maximum likehood from incomplete data via the EM algorithm. *The Royal Statistical Society*, **39**, 1–38.

Feng, T., Li, S.Z., Shum, H.Y. *et al.* (2002) Local non-negative matrix factorization as a visual representation. Proceedings of the 2nd International Conference on Development and Learning, 2002, pp. 178–183.

Gao, Y., and Church, G. (2005) Improving molecular cancer class discovery through sparse non-negative matrix factorization. *Bioinformatics*, **21**(21), 3970–3975.

Guillamet, D., Bressan, M., and Vitrià, J. (2001) A weighted non-negative matrix factorization for local representations. Proceedings of the 2001 IEEE Computer Society Conference Computer Vision and Pattern Recognition, Vol. 1, pp. 942–947.

Guillamet, D., Vitrià, J., and Schiele, B. (2003) Introducing a weighted non-negative matrix factorization for image classification. *Pattern Recognition Letters*, **24**(14), 2447–2454.

Hoyer, P.O. (2002) Non-negative sparse coding. Proceedings of the 2002 12th IEEE Workshop on Neural Networks for Signal Processing, 2002, pp. 557–565.

Hoyer, P.O. (2004) Non-negative matrix factorization with sparseness constraints. *Journal of Machine Learning Research*, **5**(9), 1457–1469.

Huang, Z., and Yu, X.C. (2008) Application of several non-negative matrix factorization-based methods in remote sensing image fusion. *Proceedings of FSKD*, **4**, 29–33.

Huang, Z., Yu, X.C., Wang, G. *et al.* (2008) Non negative matrix fusion algorithm in remote sensing image fusion application. *Journal of Beijing Normal University (Natural Science Edition)*, **44**(6), 599–602, (in Chinese).

Inderjitt, S., and Dharmendra, S. (2001) Concept decompositions for large sparse text using clustering. *Machine Learning*, **42**(1), 143–175.

Lee, D., and Seung, H.S. (1997) Unsupervised learning by convex and conic coding. *Advances in Neural Information Processing Systems*, **9**, 515–522.

Lee, D., and Seung, H.S. (1999) Learning the parts of objects by non-negative matrix factorization. *Nature*, **401**(10), 788–791.

Lee, D., and Seung, H.S. (2001) Algorithms for non-negative matrix factorization. *Advances in Neural Information Processing Systems*, **13**, 556–562.

Liu, W.X., Zheng, N.N., and Lin, X.F. (2003) Non-negative matrix factorization for visual coding. *Proceedings IEEE International Conference on Acoustics, Speech, and Signal Processing*, **3**, 293–296.

Liu, W.X., Zheng, N.N., and Liu, Q.B. (2006) Non negative matrix factorization and its applications in pattern recognition. *Chinese Science Bulletin*, **51**(3), 241–250, (in Chinese).

Lu, J.J., Yang, J., Liang, D. *et al.* (2005) Non negative matrix factorization based on relevance feedback algorithm in image retrieval. *Journal of Shanghai Jiao Tong University*, **39**(4), 578–581, (in Chinese).

Miao, Q.G., and Wang, B.S. (2005) Gathering image fusion research based on non-negative matrix factorization. *Journal of Optics*, **25**(6), 755–759.

Paatero, P. (1997) Least squares formulation of robust non-negativity factor analysis. *Chemometrics and Intelligent Laboratory Systems*, **37**, 23–35.

Paatero, P. (1999) The multilinear engine-a table-driven, least squares program for solving multilinear problems, including the n-way parallel factor analysis model. *Journal of Computational and Graphical Statistics*, **8**(4), 854–888.

Paatero, P., and Tapper, U. (1994) Positive matrix factorization: a non-negative factor model with optimal utilization of error estimates of data values. *Environmetrics*, **5**(2), 111–126.

Padilla, P., Górriz, J.M., Ramírez, J. *et al.* (2010) Analysis of SPECT brain images for the diagnosis of Alzheimer's disease based on NMF for feature extraction. *Neuroscience Letters*, **479**(3), 192–196.

Pascual-Montano, A., Carazo, J.M., Kochi, K. *et al.* (2006) Nonsmooth nonnegative matrix factorization (nsNMF). *IEEE Transactions on Pattern Analysis and Machine Intelligence*, **28**(3), 403–415.

Pauca, V.P., Piper, J., and Plemmons, R.J. (2006) Nonnegative matrix factorization for spectral data analysis. *Linear Algebra and Its Applications*, **416**(1), 29–47.

Plumbley, D., Abdallah, S., Bello, J. *et al.* (2002) Automatic music transcription and audio source separation. *Cybernetics and Systems*, **33**(6), 603–627.

Pu, X., Yi, Z., Zheng, Z. *et al.* (2005) Face recognition using fisher non-negative matrix factorization with sparseness constraints. Advances in Neural Networks–ISNN 2005, pp. 112–117.

Rutkowski, T.M., Zdunek, R., and Cichocki, A. (2007) Multichannel EEG brain activity pattern analysis in time–frequency domain with nonnegative matrix factorization support. *International Congress Series*, **1301**, 266–269.

Wang, Y., and Jia, Y. (2004) Fisher non-negative matrix factorization for learning local features. Proceedings of Asian Conference on Computer Vision, pp. 495–511.

Wang, G., Kossenkov, A.V., and Ochs, M.F. (2006) LS-NMF: a modified non-negative matrix factorization algorithm utilizing uncertainty estimates. *BMC Bioinformatics*, **7**(1), 175.

Wang, Z.N., Yu, X.C., and Zhang, L.B. (2008a) A remote sensing image fusion algorithm based on constrained nonnegative matrix factorization. CISP'08, Congress on Image and Signal Processing, Vol. 4, pp. 672–676.

Wang, Z.N., Yu, X.C., and Zhang, L.B. (2008b) Based on non-negative matrix factorization multispectral and panchromatic remote sensing image fusion. *Journal of Beijing Normal University (Natural Science Edition)*, **44**(4), 387–390, (in Chinese).

Wei, L. (2004) Blind signal separation based on non-negative matrix factorization algorithm. *The Electro-Optic and Control*, **11**(2), 38–53, (in Chinese).

Xue, Y., Tong, C.S., Chen, W.S. *et al.* (2006) A modified non-negative matrix factorization algorithm for face recognition. 18th International Conference on Pattern Recognition, 2006, ICPR 2006, Vol. 3, pp. 495–498.

Yan, J.J., Xia, C.M., and Zheng, J.R. (2007) Fusion method based on the NMF multi-spectral image and panchromatic image. *Computer Engineering*, **21**, 169–171, (in Chinese).

Zhang, Y., and Mu, Z.C. (2006) Ear recognition based on improved NMFSC. *Computer Applications*, **4**, 790–792, (in Chinese).

16

Sparse Component Analysis and Applications

Independent component analysis (ICA), which has shown good performance in the signal processing area, is most widely used for blind source separation (BSS) (Bell and Sejnowski, 1995; Cardoso, 1998; Hyvärinen, 1999a,b; Pearlmutter and Parra, 1996). However, it needs to satisfy certain constraints, for example, the source signals having a non-Gaussian distribution and being mutually independent. In actual conditions, many signals (like images) do not satisfy these constraints. With a new BSS approach desperately needed, sparse component analysis (SCA) has emerged just in time. SCA utilizes the sparseness of signals to extract sources and has higher precision of separation. If the source signals have a Gaussian distribution or are dependent, SCA can still extract the sparse sources effectively. SCA is also a promising approach for BSS when there are fewer sensors than sources present. Besides the strict theory of SCA, it has extensive application potential, and is one of the main approaches besides ICA and non-negative matrix factorization (NMF) for BSS.

An outline of this chapter is given below:

Section 16.1 states the concepts and fundamental model of SCA, discusses how to definite sparseness, and comprehensively summarizes the theory, development, and research achievements of SCA, introduces some fundamental algorithms with respect to iteration and clustering and discusses problems with the conventional method.

In Section 16.2, for clustering SCA, based on the fact that the columns of mixtures are clustered lengthwise, the orientations of the columns of the mixing matrix in a complete set have sparsity 1. We present a linear clustering sparse component analysis (LC-SCA) method and an improvement thereof (I-LC-SCA). We use I-LC-SCA for actual speech extraction and obtain satisfactory separation results.

In Section 16.3, based on the fact that columns of mixtures are clustered in a sub-plane expanded by the mixing matrix in a complete set with sparsity $m - 1$, we

Blind Source Separation: Theory and Applications, First Edition. Xianchuan Yu, Dan Hu and Jindong Xu.

present a PC-SCA method and use it for blind image separation. The experimental results show that PC-SCA can reconstruct the sources precisely and achieve better separation results than FastICA.

In Section 16.4 we discuss improvements to PC-SCA and present the complete plane clustering SCA (PCO-SCA), which is suitable for over-complete set BSS. It solves the underdetermined BSS problem of extracting $n(n = m + 1)$ source signals from m mixed signals with sparsity $m - 1$.

In Section 16.5, for unsparse signals (such as images), we present blind image separation based on a wavelet transformation and SCA method, which effectively solves the BSS problem for unsparse signals.

In Section 16.6 we introduce a new BSS algorithm based on feedback SCA, which is robust to mixing noise.

In Section 16.7 SCA is incorporated into the remote sensing field under the foundation of a traditional unsupervised classified method, and a remote sensing image analysis method based on SCA is presented. Experimental results show that SCA achieves higher classification accuracy than principal component analysis (PCA).

Section 16.8 presents a summary of the chapter.

16.1 Overview

16.1.1 Concepts and Fundamental Model of SCA

A basic problem in the signal processing field is matrix factorization, a simple example of which is linear matrix factorization (Georgiev *et al.*, 2005a):

$$\boldsymbol{X} = \boldsymbol{A}\boldsymbol{S}, \quad \boldsymbol{X} \in \mathbf{R}^{m\times T}, \boldsymbol{A} \in \mathbf{R}^{m\times n}, \boldsymbol{S} \in \mathbf{R}^{n\times T} \tag{16.1}$$

The unknown matrix $\boldsymbol{A}$ (dictionary) and $\boldsymbol{S}$ (source signals) have some specific properties, for instance:

1. the rows of $\boldsymbol{S}$ are statistically independent as far as possible – ICA;
2. the elements of $\boldsymbol{X}$, $\boldsymbol{A}$, and $\boldsymbol{S}$ are nonnegative – NMF;
3. $\boldsymbol{S}$ contains as many zeros as possible – SCA.

For SCA, $\boldsymbol{X} = [\boldsymbol{X}_1, \boldsymbol{X}_2, \cdots, \boldsymbol{X}_T]$ is the mixtures matrix (observed signals), $\boldsymbol{A}$ is the mixing matrix, and $\boldsymbol{S} = [\boldsymbol{S}_1, \boldsymbol{S}_2, \cdots, \boldsymbol{S}_T]$ is the source matrix (source signals). In addition, m denotes the number of mixtures, T the number of samples, and n the number of sources. The aim of SCA is to estimate the mixing matrix $\boldsymbol{A}$ and extract the source matrix $\boldsymbol{S}$ as accurately as possible, under the condition that sources satisfy the sparseness constraint.

How to depict the sparsity of a matrix in SCA is a key problem, with the sparsity of a vector forming the basis of the matrix. Now, there are several definitions for the sparsity of a vector (Chunqi, Jiyun, and Fung, 2005; Cichocki, 2003; Kopriva and Sersic, 2007):

Definition 16.1 Sparsity of a vector ($\sigma(v)$ **denotes the sparsity of signal vector** $v \in \mathbf{R}^N$

1. L_0 normal form

$$\sigma(v) = \#v \tag{16.2}$$

   where # is the number of non-zero elements.
2. L_1 normal form

$$\sigma(v) = \sum_{k=1}^{N} |V_k| \tag{16.3}$$

3. L_p normal form

$$\sigma(v) = ||v||_p = \sum_{k=1}^{N} \sqrt[p]{|V_k|^p} \tag{16.4}$$

4. Mixing normal form

$$\sigma(v) = \sqrt{N}\frac{||v||_2}{||v||_1} \tag{16.5}$$

5. Normal form presented by Du, Hu, and Yu (2005)

$$\sigma(v) = \frac{2}{\pi}\sum_{k=1}^{N} \tan^{-1}\left(\frac{|V_k|}{p}\right) \quad p > 0,$$

$$\text{or } \sigma(v) = \frac{2}{\pi}\sum_{k=1}^{N} \tan^{-1}\left(\frac{V_k^2}{p}\right) \quad p > 0,$$

$$\text{or } \sigma(v) = \sum_{k=1}^{N} \log\left(1 + \frac{V_k^2}{p}\right) \quad p > 0 \tag{16.6}$$

Owing to the mature research on the L_0 normal form, most SCA algorithms are currently based on this form. Thus, the following definitions of sparsity are all based on the L_0 normal.

Definition 16.2 Sparsity of a matrix For matrix S, if the sparsity of each column vector is at least m, then the sparsity of matrix S is m. That is, for matrix $S \in \mathbf{R}^{n\times T}$, $\sigma(S) = \min(\sigma(s_j))$, $\forall j = 1, 2, \ldots, T$.

Based on the above concept of sparsity, Georgiev *et al.* (2005a,b) presented two constraints for reconstructing sources about SCA:

1. Any $m \times m$ sub-matrix is non-singular.
2. The sparsity of the source matrix is at least $m - 1$.

Definition 16.3 $m \times m$ **sub-matrix** Given matrix $\boldsymbol{A} = (\boldsymbol{A}_1, \boldsymbol{A}_2, \cdots, \boldsymbol{A}_n) \in \mathbf{R}^{m \times n}$, the $m \times m$ sub-matrix is $(\boldsymbol{A}_{j_1}, \boldsymbol{A}_{j_2}, \cdots, \boldsymbol{A}_{j_m})$, where $j_1, j_2, \cdots, j_m \in \{1, 2, 3, 4, \cdots, n\}$ and $j_1 \neq j_2 \neq j_3 \neq \cdots \neq j_n$.

Definition 16.4 Non-singular For a non-null matrix $\boldsymbol{A}$, if there is a matrix $\boldsymbol{B}$ satisfying $\boldsymbol{AB} = \boldsymbol{BA} = \boldsymbol{I}$ ($\boldsymbol{I}$ is a unit matrix), then $\boldsymbol{A}$ is reversible and is called a non-singular matrix.

Conditions (1) and (2) are the premises for SCA to reconstruct the sources. Thus, the SCA algorithms presented in this chapter are all based on conditions (1) and (2). Obviously, we can improve some of the algorithms by developing constraints for specific cases.

16.1.2 Development of SCA

In 1996, Nature published a paper titled "Emergence of simple-cell receptive field properties by learning a sparse code for natural images" (Olshausen and Field, 1996), which first introduced the concept of a sparsity model. The paper pointed out that an image processed by human vision is based on a sparsity description. The components of the image itself depend on each other, and thus the sparsity model is more suitable for image coding than an independent model (Joseph and Kenneth, 2007). In 1997, Olshausen and Field (1997) stated that the human vision system (HVS) cannot use the concept of a basis image of the ICA model and that the over-complete description is more suitable for HVS. In 2000, Lewicki and Sejnowski (2000) presented an over-complete sparse component model, which first introduced the concepts of SCA. Since then, SCA has become a research hotspot.

Relatively quickly, as a result of the efforts of numerous experts and scholars, research on SCA yielded worthy achievements (Lewicki and Sejnowksi, 2000; Bofill and Zibulevsky, 2001; O'Grady, Pearlmutter, and Rickard, 2005; Li, Cichocki, and Amari, 2003a,b). These achievements can be categorized according to five aspects: signal over-complete description, sparsification method, measurement criteria for sparsity, BSS, and integration with other theory.

There are many methods for creating a signal over-complete description, such as the Method of Frame (MOF) (Daubechies, 1988), Matching Pursuit (MP) (Cichocki *et al.*, 2004; Bronstein *et al.*, 2003), Basis Pursuit (Chen *et al.*, 1998, 2001), Orthogonal Matching Pursuit (OMP) (Pati, Rezaiifar, and Krishnaprasad, 1993), and Best Orthogonal Basis (BOB) (Coifman and Wickerhauser, 1992). These methods were established from a linear combination of the over-complete basis. However, as they can only process low-dimensional signals, over-complete decomposition of a high-dimensional signal is still difficult.

At present, the method for sparsification includes a short time Fourier transform (STFT), wavelet transform, and discrete cosine transform (DCT). Sanei *et al.* (2005) proposed carrying out a Z transform on the mixtures, estimating the mixing matrix

A by K-means clustering, and finally reconstructing the sources by $S = A^{-1}X$. Zibulevsky and Pearlmutter (2001) suggested making the sources sparser using a wavelet packet transform, then estimating the mixing matrix by C-means clustering or the maximum posterior probability (MPP) method, and finally using shortest path decomposition to recover the sources. However, this method is only suitable for certain situations ($m = n$). Since the presented methods are usually only applicable in specific situations, we need to research a universal sparsification method.

The measurement criterion of sparsity is the foundation of the SCA algorithm. However, for the hardest definition, there are fewer papers than for the other aspects. Based on building the relationship between sparsity and the reconstruction, Li, Cichocki, and Amari (2004) discussed how to select an SCA algorithm to ensure good separation performance when sources are not sparse. Zibulevsky and Pearlmutter (2001) analyzed in detail the reasons for sparsity in favor of source signal reconstruction and proposed the K-means clustering SCA and MPP SCA. Georgiev *et al.* presented the concept of K-sparsity SCA and gave two conditions for the reconstruction of source signals (Georgiev *et al.*, 2004). Du *et al.* (2005) put forward the pervasive criterion of the SCA normalized function and recommended certain normalized function families from which it is easy to obtain an optimal solution. A pervasive criterion is the key for future development.

BSS is a fast growing research area of SCA and there have been many achievements (Li and Cichocki, 2003; Li *et al.*, 2006a,b). Smith and Lukasiak (2005) proposed array SCA, which involves dividing the mixed signals (after the Fourier transform) into k lengths, calculating the mean and variance of every signal (except the first signal) with respect to the first signal, and then utilizing a source cancelation approach and deflation technique to reconstruct the sources one by one until all the sources have been estimated. Shi, Tang and Tang (2005) presented an over-complete BSS algorithm based on a sparse mixed model, in which a gradient studying algorithm is used to estimate the mixing matrix and a linear programming algorithm is then applied to recover the sources. In the mixing matrix estimation step, the gradient studying algorithm needs to transform the rectangular coordinate system to a polar coordinate system. To limit the dimensions of the polar coordinates, this method can only process low-dimensional signals ($n \leq 3$) for BSS. Bradly and Mangasarian (2004) introduced a K-plane clustering algorithm (Washizawa and Cichocki, 2006; He and Cichocki, 2006) in detail and proved the effectiveness of the algorithm theoretically. He and Cichocki (2006) presented K-EVD (eigen value decomposition) SCA. Although this method requires only one source to be nonzero in most samples and the others to be zero, most signals cannot satisfy this sparsity condition in practice. Tan and Fu (2006) introduced the normal vector of plane clustering sparse component analysis (PC-SCA), while He and Cichocki (2006)developed the K-super-plane clustering algorithm, which can extract sources with a sparsity level of $m - 1$. Compared with conventional SCA, this method has a wider application area despite its complexity. In 2008, Washizawa and Cichocki (2008) presented K-orthodrome clustering SCA, while Woolfson, Bigan and Crow (2008) applied MHC to SCA, the idea of which is similar to the linear clustering method. In all, clustering SCA has greater separation accuracy than iteration ICA, but it is sensitive to noise, is not robust, and has a narrow application area.

In terms of the integration of SCA with other theories, SPCA (sparse principal component analysis) (Hui *et al.*, 2006) and SICA (sparse independent component analysis), amongst others, have appeared. Stadlthanner *et al.* (2005) set the diagonal elements of the mixing matrix to 1 using a normalization operation, regarded the other $m^2 - m$ elements as genes, and then utilized a genetic algorithm (with crossover and mutation) to find the optimized solution (mixing matrix $\boldsymbol{A}$). In 2007, d'Aspremont *et al.* proposed a new sparse principle component analysis method (SPCA), gave an optimal criterion for SPCA, and introduced an efficient iteration algorithm (d'Aspremont *et al.*, 2007). In the same year, Kim and Park presented a new sparse nonnegative matrix factorization method (SNMF), which can be used in cancer prediction and gene data analysis (Kim and Park, 2007). The integration methods mentioned above represent expansions of SCA, with this type of research destined to remain a hotspot for many years to come.

The development of SCA, summarized according to the five categories given above, is depicted in Figure 16.1.

General BSS Approaches Based on SCA

Currently, general BSS approaches based on SCA can be approximately divided into two classes:

1. Iteration optimization approaches utilize the MPP and genetic algorithms, among others, to iteratively find an optimal solution.
2. Clustering approaches utilize linear or plane clustering under different sparseness conditions to estimate the mixing matrix and then reconstruct the source signals.

16.1.3 Iteration Optimization Approach

Here as an example, we consider SCA based on MPP, which is one of the iteration optimization approaches.

For the SCA model $\boldsymbol{X} = \boldsymbol{AS}$, assume the source matrix is sparse in dictionary $\boldsymbol{\Phi}$ (Chen *et al.*, 1998; Mallat, 1998; Rubinstein *et al.*, 2010), then

$$s_i(t) = \sum_{k=1}^{K} C_{ik}\Phi_k(t) \tag{16.7}$$

Suppose sparse matrix C_{ik}, which is an independent stochastic variable, satisfies the following probability density function (Olshausen and Field, 1997):

$$p_i(C_{ik}) \infty \exp(-\beta_i h(C_{ik})) \tag{16.8}$$

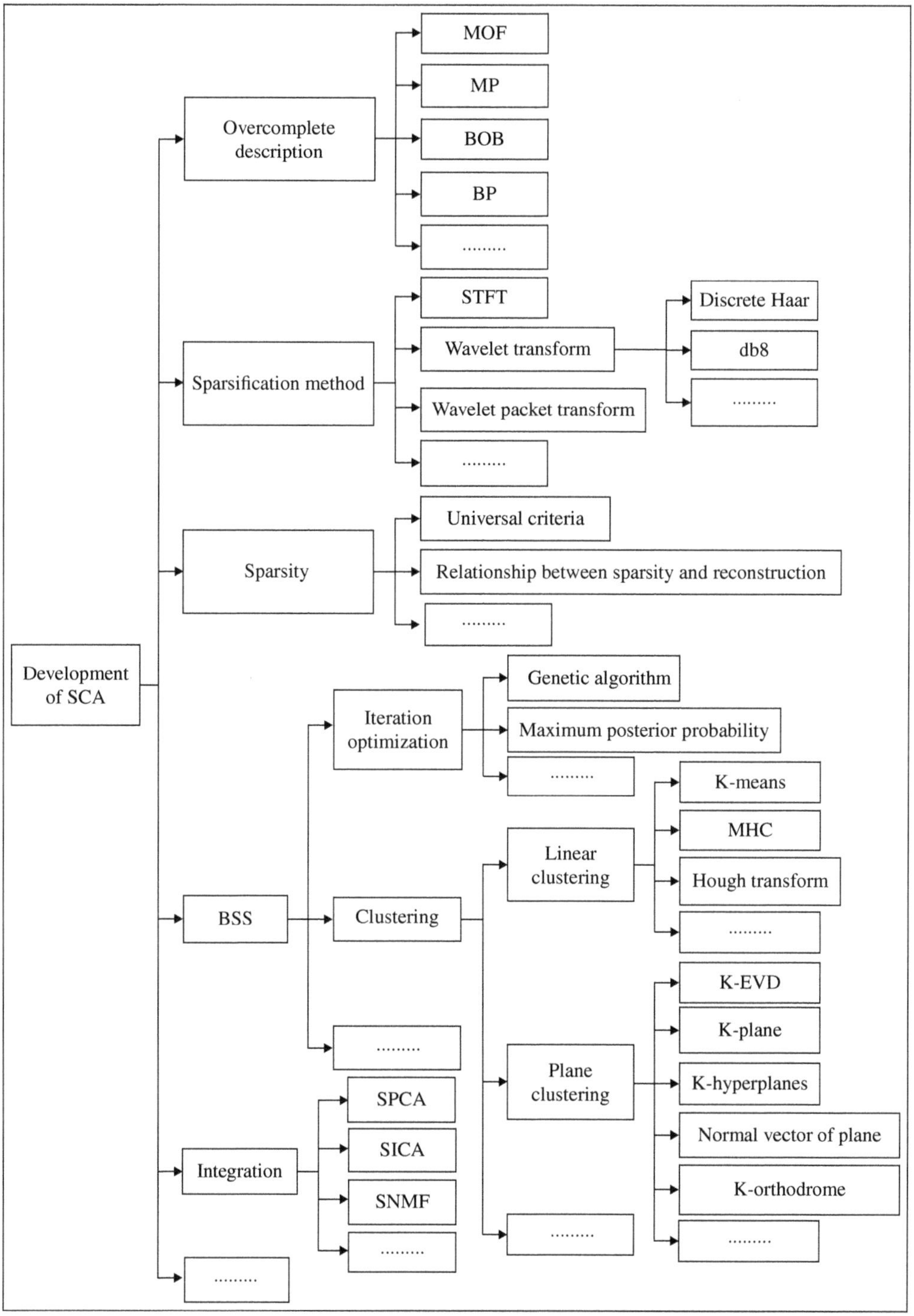

Figure 16.1 Development of SCA

where $h(c) = |c|^{1/\gamma}(\gamma \geq 1)$. The objective of SCA based on MPP is to maximize the probability function:

$$\max_{A,C} P(\boldsymbol{A}, \boldsymbol{C}|\boldsymbol{X}) \propto \max_{A,C} P(\boldsymbol{X}|\boldsymbol{A}, \boldsymbol{C})P(\boldsymbol{A})P(\boldsymbol{C}) \tag{16.9}$$

If the probability function of $P(\boldsymbol{A})$ is uniform, then Equation 16.9 can be abbreviated as:

$$\max_{A,C} P(\boldsymbol{X}|\boldsymbol{A}, \boldsymbol{C})P(\boldsymbol{C}) \tag{16.10}$$

For $P(\boldsymbol{X}|\boldsymbol{A}, \boldsymbol{C}) \propto \prod_{i,t} \exp(-\frac{(X_{it} - (\boldsymbol{AC\Phi})_{it})^2}{2\sigma^2})$ and Equation 16.8, we get the following optimization problem:

$$\min_{A,C} \frac{1}{2\sigma^2} \|\boldsymbol{AC\Phi} - \boldsymbol{X}\|_F^2 + \sum_{j,k} \beta_j h(C_{jk}) \tag{16.11}$$

We use optimization theory to solve Equation 16.11, and obtain a result that is the solution of SCA-based MPP (Donoho, 2001).

16.1.4 SCA-Based Clustering Approach

16.1.4.1 Conventional Estimation Algorithm for the Mixing Matrix

The hypothesis for accurately estimating the mixing matrix is that all bar one of the source signals are zero. The algorithm is as follows:

Algorithm

Input: sample set $\boldsymbol{X} = [\boldsymbol{X}_1, \boldsymbol{X}_2, \cdots, \boldsymbol{X}_T]$

Output: estimated mixing matrix $\boldsymbol{A} = [\boldsymbol{A}_1, \boldsymbol{A}_2, \cdots, \boldsymbol{A}_n]$

Steps:

1. Normalize columns $\boldsymbol{X}_i(i = 1, 2, \cdots, T)$ of mixing matrix $\boldsymbol{X}$, that is, $\boldsymbol{Y}_i = X_i/\|X_i\|_2$, and set $\boldsymbol{A}_1 = \boldsymbol{Y}_1, \boldsymbol{i} = 2, k = 1$.
2. If $k = n + 1$, terminate the loop; else calculate $c_{ij} = \boldsymbol{Y}_i^{\mathrm{T}} \times \boldsymbol{Y}_j (j = 1, 2, \cdots, i - 1)$. If $|c_{ij}| < 1, \forall j = 1, 2, \cdots, i - 1$, set $\boldsymbol{A}_k = \boldsymbol{Y}_i$, $k = k + 1$, $i = i + 1$ and repeat step 2; else set $i = i + 1$ and repeat step 2.

16.1.4.2 K-Hyperplane Clustering Method for Identification of the Mixing Matrix

This method is appropriate for a source matrix with sparsity $k = n - m + 1$. Under this condition, the columns of the mixtures lie in the sub-planes of the mixing matrix, while the orientations of the intersections of sub-planes are exactly the orientations of the columns of the mixing matrix. Hence, once all the hyperplanes have been estimated, mixing matrix $\boldsymbol{A}$ will be solved. The K-hyperplane clustering method is as follows:

Algorithm Normal vector estimation of K-hyperplane

Input: sample set $X = [X_1, X_2, \cdots, X_T]$
Output: normal vector of the K-hyperplane, $u_i (1 \le i \le k)$
Steps:

1. Initialize u_i and set $\|u_i\|_2 = 1$.
2. Start loop Calculate the distance $|u_i^T X_t|$ between each column $X_t (t = 1, 2, \cdots, T)$ and hyperplane H_i. Assign each column to the closest hyperplane.
3. Jump out of loop If the mean value of all distances is less than the threshold set in advance, then go to step 5; else go to step 4.
4. Update classes center
   a. For all $i (1 \le i \le k)$, assume that the hyperplane H_i class includes n elements: $X_1, X_2, \cdots, X_n$. Set $Y^{(i)} = (X_1, X_2, \cdots, X_n), C = Y^i \times (Y^i)^T$;
   b. Calculate the eigenvector corresponding to the minimum eigenvalue of C;
   c. Set $u_i = v / \|v\|_2$.
5. End loop, u_i is the normal vector of hyperplane H_i.

16.2 Linear Clustering SCA (LC-SCA)

Theorem 16.1 For a complete set whose sparsity is 1, the columns of mixtures are clustered along the columns directions of the mixing matrix.

Proof: For source matrix S, assume it is not zero at (i, j). That is

$$\begin{cases} S(i,j) \neq 0 & \\ S(k,j) = 0 & k \neq i \end{cases} \qquad 1 \le j \le T \tag{16.12}$$

Then $X(:,j) = S(i,j) \times A(:,j)$, $X(:,j)$, and $A(:,j)$ are collinear.

We know, when $S(i,j) \neq 0$, columns of X and the columns of mixing matrix A are collinear. The linear cluster center of mixtures determines the direction of mixing matrix A columns, and the number of columns is the number of classes.

16.2.1 Estimation of Mixing Matrix

For a complete set whose sparsity is 1, the columns of mixtures are clustered along the columns direction of the mixing matrix (Theorem 16.1). So, we can estimate the mixing matrix by a linear clustering method (Cao and Yu, 2010).

16.2.2 Identification of Source

After the mixing matrix is estimated, we can reconstruct the source images, that is, $S = A^{-1}X$, utilizing the given mixed images and the estimated mixing matrix A.

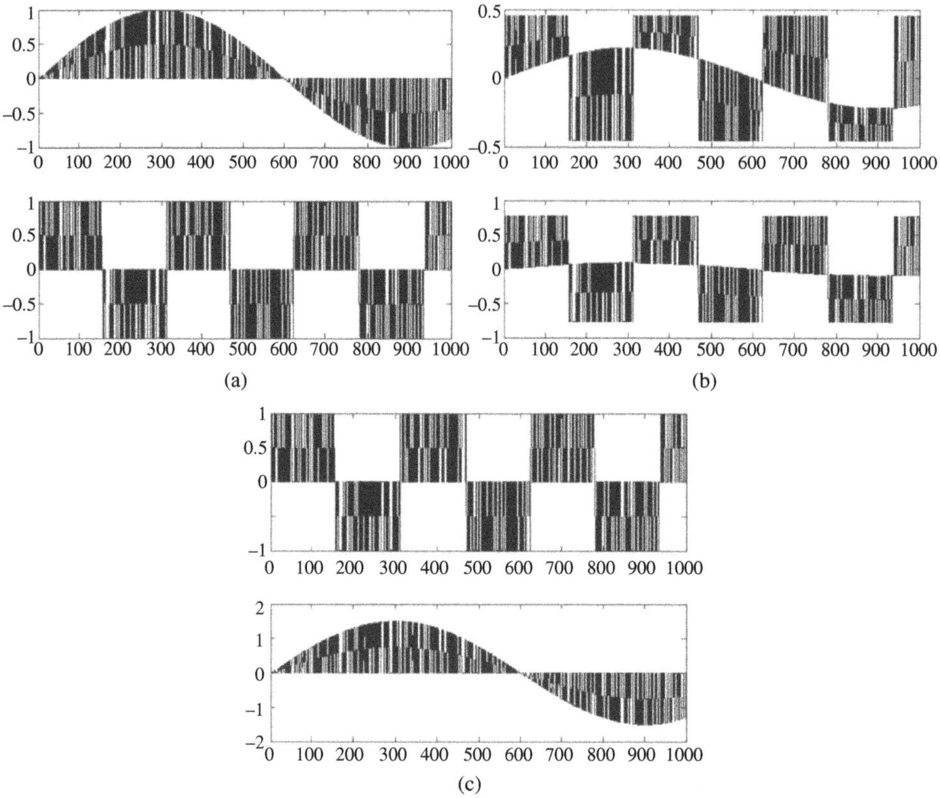

Figure 16.2 The BSS simulation of two channels of mixed signals. (a) Original signals, (b) mixed (observed signals), (c) separated results by LC-SCA

16.2.3 BSS Based on LC-SCA

16.2.3.1 BSS for Two Channels of Mixed Signals

In this experiment, we use MATLAB® software to produce a sinusoidal signal and a square signal. The sparsification signals are shown in Figure 16.2a, and the randomly mixed signals in Figure 16.2a, and the separated results by LC-SCA are shown in Figure 16.2c.

We cluster the column vectors of two mixed signals and the result is shown in Figure 16.3. Clearly, the column vectors of mixtures are clustered in two linear directions, which is equal to the number of columns of the mixing matrix $\boldsymbol{A}$. That is in correspondence with Theorem 16.1.

The estimated mixing matrix by LC-SCA is $\boldsymbol{A}_{\text{LC-SCA}} = \begin{pmatrix} 0.45259 & 0.14647 \\ 0.7662 & 0.060863 \end{pmatrix}$, and the normalized result is $\varphi(\boldsymbol{A}_{\text{LC-SCA}}) = \begin{pmatrix} 0.5086 & 0.9234 \\ 0.8610 & 0.3837 \end{pmatrix}$.

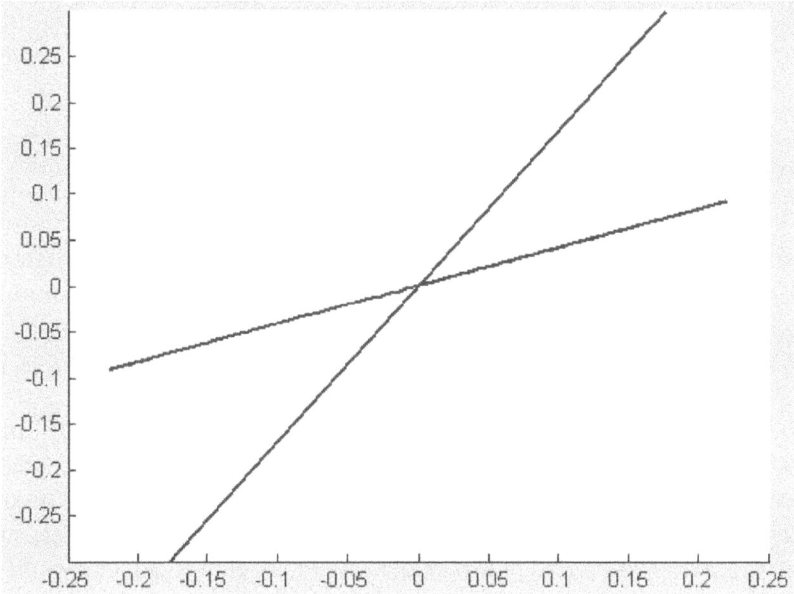

Figure 16.3 Column vectors of mixtures are clustered in linear directions

The original mixing matrix $\boldsymbol{A}$ and the normalized result $\varphi(\boldsymbol{A})$ are given as follows.

$$\boldsymbol{A} = \begin{pmatrix} 0.22017 & 0.45259 \\ 0.091489 & 0.7662 \end{pmatrix},$$

$$\varphi(\boldsymbol{A}) = \begin{pmatrix} 0.9234 & 0.5086 \\ 0.3837 & 0.8610 \end{pmatrix}.$$

We can see that $\varphi(\boldsymbol{A})$ and $\varphi(\boldsymbol{A}_{\text{LC-SCA}})$ are the same. The mixing matrix has been estimated accurately.

We use normalized correlation coefficients (NCCs) between separated results and sources to analyze the performance, and NCC is shown as Equation 16.13.

$$\text{NCC}(X, Y) = \frac{\sum_{i,j} |X(i,j) - \overline{X}| \times |Y(i,j) - \overline{Y}|}{\sqrt{\sum_{i,j} |X(i,j) - \overline{X}|^2} \sqrt{\sum_{i,j} |Y(i,j) - \overline{Y}|^2}}$$

$$\text{NCC} = \frac{\sum_{i} |v(i) - \overline{v}| \times |v\prime(i) - \overline{v\prime}|}{\sqrt{\sum_{i} |v(i) - \overline{v}|^2} \sqrt{\sum_{i} |v\prime(i) - \overline{v\prime}|^2}} \tag{16.13}$$

where $v(i)$ and $v\prime(i)$ are the amplitudes of the two signals, and $\overline{v}$ and $\overline{v\prime}$ are the corresponding mean values. The NCC value is greater if the degree of correlation between

Table 16.1 NCC index between separated results and source signals

Source signals	Separated result 1	Separated result 2
Signal 1	−0.0029	**1**
Signal 2	**1**	−0.0029

the two images is higher, which means that they are more similar. When the value of NCC is 1, the two signals are considered to have the same "form" (although the total amplitude may differ).

The NCC index between the separated results and source signals is given in Table 16.1.

From Table 16.1 we can see that the NCC index is 1 for separated result 1 and signal 2 (and similarly, for separated result 2 and signal 1). That is, LC-SCA can extract the source from the mixtures precisely up to the permutation and scaling of sources.

From the above results (Figure 16.2, $\varphi(\boldsymbol{A})$ and $\varphi(\widetilde{\boldsymbol{A}})$, and Table 16.1), the mixed signals are separated with good performance and the source signals are extracted precisely.

16.2.3.2 BSS for Three Channels of Mixed Signals with One Channel of Gaussian Noise

Based on Section 16.2.3.1, we produce a Gaussian noise source as a channel to participate in the mixing system. The source signals are shown in Figure 16.4a and the mixed signals in Figure 16.4b. Note that the third source signal has a Gaussian distribution, which does not satisfy the constraint of FastICA. The separated results by FastICA are shown in Figure 16.4d. We can clearly see that FastICA cannot extract the source signals from the mixtures. However, LC-SCA can extract the source signals from the mixtures precisely. The separated results by LC-SCA are shown in Figure 16.4c.

In addition, the mixing matrix $\boldsymbol{A}$ produced by MATLAB®, and the estimated mixing matrices $\boldsymbol{A}_{\text{LC-SCA}}$ and $\boldsymbol{A}_{\text{FastICA}}$ by LC-SCA and FastICA, respectively, are given as follows:

$$\boldsymbol{A} = \begin{pmatrix} 0.6178 & 0.4363 & 0.5961 \\ 0.1161 & 0.4328 & 0.3361 \\ 0.6082 & 0.6332 & 0.2109 \end{pmatrix},$$

$$\boldsymbol{A}_{\text{LC-SCA}} = \begin{pmatrix} 0.3101 & 0.4078 & 0.4363 \\ 0.1749 & 0.0767 & 0.4328 \\ 0.1097 & 0.4015 & 0.6332 \end{pmatrix},$$

$$\boldsymbol{A}_{\text{FastICA}} = \begin{pmatrix} -0.1020 & 0.2626 & -0.2833 \\ -0.1852 & 0.1727 & -0.0958 \\ -0.1213 & 0.3879 & -0.1865 \end{pmatrix}.$$

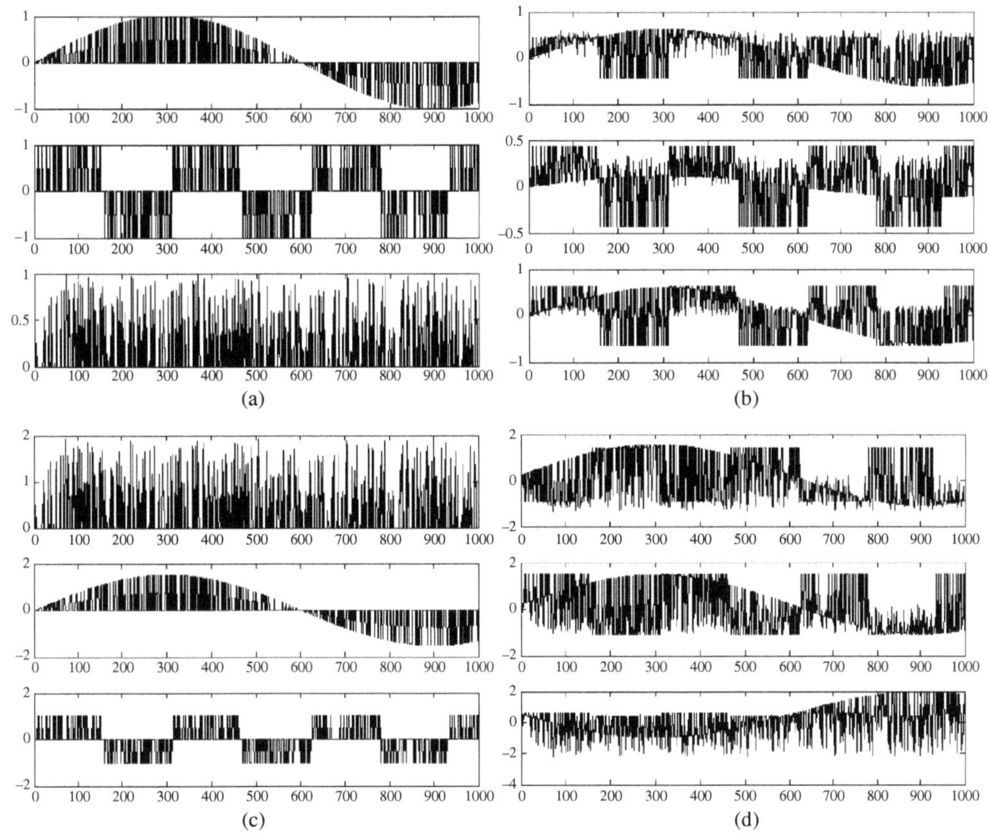

Figure 16.4 Simulation involving mixing a channel of white Gaussian noise and two channels of signals. (a) Original signals, (b) mixed (observed signals), (c) separated results by LC-SCA, (d) separated results by FastICA

For ease of comparison, $\boldsymbol{A}$, $\boldsymbol{A}_{\text{LC-SCA}}$, and $\boldsymbol{A}_{\text{FastICA}}$ are normalized as follows.

$$\varphi(\boldsymbol{A}) = \begin{pmatrix} 0.7063 & 0.4945 & 0.8324 \\ 0.1328 & 0.4905 & 0.4694 \\ 0.6953 & 0.7176 & 0.2945 \end{pmatrix},$$

$$\varphi(\boldsymbol{A}_{\text{LC-SCA}}) = \begin{pmatrix} 0.8324 & 0.7063 & 0.4945 \\ 0.4694 & 0.1328 & 0.4905 \\ 0.2945 & 0.6953 & 0.7176 \end{pmatrix},$$

$$\varphi(\boldsymbol{A}_{\text{FastICA}}) = \begin{pmatrix} -0.4183 & 0.5260 & -0.8038 \\ -0.7598 & 0.3460 & -0.2717 \\ -0.4977 & 0.7769 & -0.5293 \end{pmatrix}.$$

Table 16.2 NCC indexes between sources and separated results using LC-SCA

NCC index	Result 1	Result 2	Result 3
Source 1	−0.0376	**1.0000**	0.0011
Source 2	0.0104	0.0011	**1.0000**
Source 3	**1.0000**	−0.0376	0.0104

Table 16.3 NCC indexes between sources and separated results using FastICA

NCC index	Result 1	Result 2	Result 3
Source 1	0.5787	0.5541	−0.5984
Source 2	−0.6643	0.7461	0.0467
Source 3	−0.5006	−0.3833	−0.7762

We can see that $\varphi(\boldsymbol{A}_{\text{LC-SCA}})$ and $\varphi(\boldsymbol{A})$ have the same form up to the permutation, that is, LC-SCA can estimate the mixing matrix effectively. However, the estimated matrix $\varphi(\boldsymbol{A}_{\text{FastICA}})$ by FastICA differs greatly from the source mixing matrix $\varphi(\boldsymbol{A})$.

Tables 16.2 and 16.3 give the NCC indexes between the sources and separated signals recovered by LC-SCA and FastICA, respectively. The separated results by LC-SCA have higher NCC indexes than those by FastICA.

From the above, we can see that the ICA method cannot recover the sources when there are Gaussian signals included in the mixing system. On the other hand, the LC-SCA method can separate the mixtures very well; three sources are recovered precisely up to the permutation and scaling.

16.2.3.3 BSS for Dependent Signals

Using MATLAB®, we produced two source signals (Figure 16.5a) that are mutually dependent; that is, $s_2(t) = s_1(t + 300)$. The randomly mixed signals are shown in Figure 16.5b. The source signals were extracted by LC-SCA and FastICA, with the results shown in Figure 16.5c,d, respectively. It is clear that the dependent source signals are recovered precisely by LC-SCA, but not by FastICA.

The original mixing matrix $\boldsymbol{A}$ and its normalized result $\varphi(\boldsymbol{A})$ are given as follows:

$$\boldsymbol{A} = \begin{pmatrix} 0.2243 & 0.1050 \\ 0.8011 & 0.8786 \end{pmatrix} \quad \varphi(\boldsymbol{A}) = \begin{pmatrix} 0.2696 & 0.1187 \\ 0.9630 & 0.9929 \end{pmatrix}.$$

The estimated mixing matrix $\boldsymbol{A}_{\text{LC-SCA}}$ by LC-SCA and its normalized result $\varphi(\boldsymbol{A}_{\text{LC-SCA}})$ are given as:

$$\boldsymbol{A}_{\text{LC-SCA}} = \begin{pmatrix} 0.0690 & 0.1488 \\ 0.5777 & 0.5313 \end{pmatrix} \quad \varphi(\boldsymbol{A}_{\text{LC-SCA}}) = \begin{pmatrix} 0.1187 & 0.2696 \\ 0.9929 & 0.9630 \end{pmatrix},$$

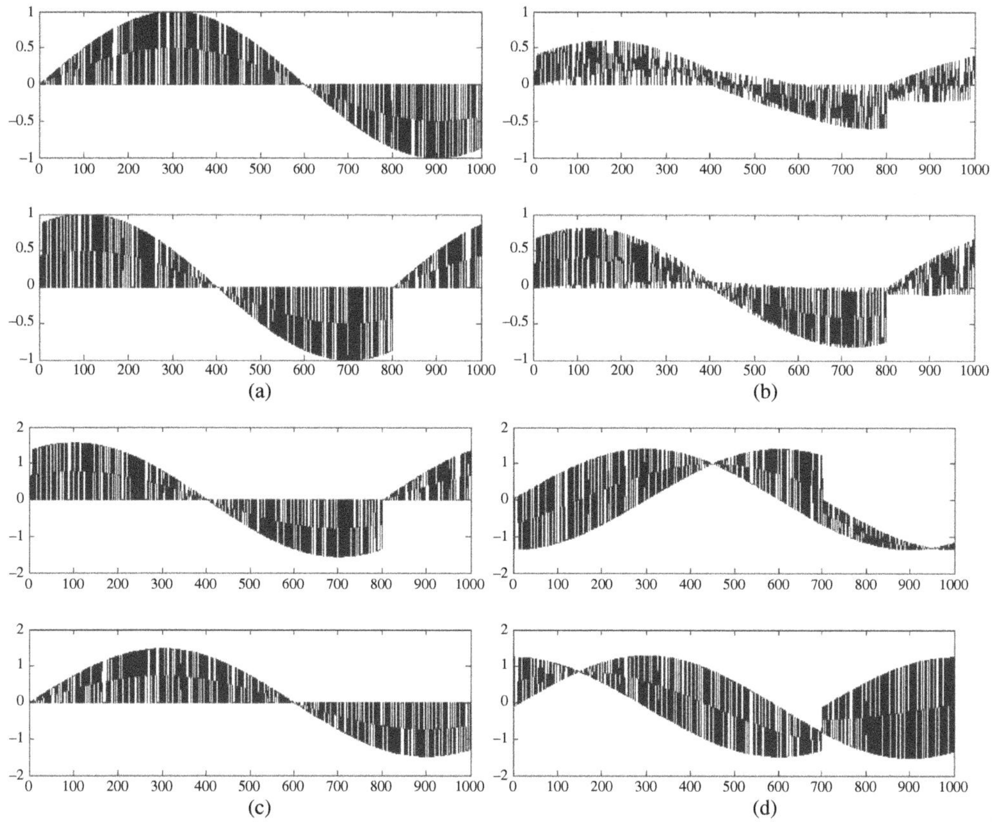

Figure 16.5 BSS experiments for dependent signals. (a) Original signals, (b) mixed (observed signals), (c) separated results by LC-SCA, (d) separated results by FastICA

while the estimated mixing matrix $\boldsymbol{A}_{\text{FastICA}}$ by FastICA and its normalized result $\varphi(\boldsymbol{A}_{\text{FastICA}})$ are given as:

$$\boldsymbol{A}_{\text{FastICA}} = \begin{pmatrix} 0.0416 & 0.1192 \\ -0.0376 & 0.6077 \end{pmatrix} \; \varphi(\boldsymbol{A}_{\text{FastICA}}) = \begin{pmatrix} 0.7414 & 0.1925 \\ -0.6711 & 0.9813 \end{pmatrix}.$$

From the values of $\varphi(\boldsymbol{A})$, $\varphi(\boldsymbol{A}_{\text{LC-SCA}})$, and $\varphi(\boldsymbol{A}_{\text{FastICA}})$, we can see that LC-SCA estimates the mixing matrix precisely, whereas FastICA cannot.

Table 16.4 gives the NCC indexes between the sources and separated results by LC-SCA and FastICA.

LC-SCA can extract the dependent signals precisely, while the extracted results using FastICA have a low correlation with the source signals; in other words, FastICA cannot extract the dependent signals exactly.

Table 16.4 Comparison of LC-SCA and FastICA in terms of NCC index

NCC	LC-SCA result 1	LC-SCA result 2	FastICA result 1	FastICA result 2
Source 1	−0.0063	**1.0000**	0.7035	**0.7107**
Source 2	**1.0000**	−0.0063	**−0.7151**	0.6990

16.2.4 Concluding Remarks about BSS-Based LC-SCA

In a complete set with sparsity 1, the columns of mixtures are clustered along the orientations of the columns of the mixing matrix. The NCC between the sources and corresponding extracted signals by LC-SCA is 1.0000. LC-SCA can separate the mixtures precisely up to the permutation and scaling of sources. If the sources are dependent or Gaussian, the SCA method also works very well, whereas the FastICA method fails.

16.2.5 Improved Linear Clustering SCA

The LC-SCA method requires the sparsity of the source to be zero for any sample, something that most signals cannot satisfy, including speech signals and electroencephalograms (EEGs). These signals only satisfy a sparsity of 1 some of the time. To solve this problem, we present an improved LC-SCA algorithm (I-LC-SCA), which is suitable for sources that partly satisfy the condition of sparsity 1.

I-LC-SCA takes full advantage of sources with sparsity 1 at some time. Samples that do not satisfy the condition do not join the clustering process. After the mixing matrix has been estimated, the method uses the entire mixing matrix to recover the sources for those samples that do not satisfy the sparsity condition.

16.2.5.1 Estimation of the Mixing Matrix by I-LC-SCA

Algorithm Algorithm for estimating the mixing matrix by I-LC-SCA

Input: sample set $\boldsymbol{X} = [X_1, X_2, \cdots, X_T]$
Output: estimated mixing matrix $\boldsymbol{A} = [\boldsymbol{A}_1, \boldsymbol{A}_2, \cdots, \boldsymbol{A}_n]$
Steps:

1. Unify the direction For each column $\boldsymbol{X}_j (j = 1, 2, \cdots, T)$ of mixed signals, if $X_{1j} < 0$, set $\boldsymbol{X}_j = -\boldsymbol{X}_j$ and record the index j of this column.
2. Delete the zero columns For each column $\boldsymbol{X}_j (j = 1, 2, \cdots, T)$ of the mixed image signals, if $\|\boldsymbol{X}(:, j)\|_1 = 0$, delete the jth column of $\boldsymbol{X}$ to obtain a new mixed matrix $\boldsymbol{X}'$, and record index j of this column.

3. Linear clustering For each pair of arbitrary columns X'_i and X'_j, if $\cos(X'_i, X'_j) = \frac{X'_i \cdot X'_j}{\|X'_i\| \times \|X'_j\|} = 1$, then columns X'_i and X'_j are collinear. If, and only if, column X'_i and X'_j are collinear, then $X'_i \in \theta(k), X'_j \in \theta(k)$. In this way, all the column vectors of $X\prime$ can be clustered in the linear directions and we get $\theta(k)$ $k = 1, 2, \cdots, t(t \geq m)$.
4. Estimate the mixing matrix $\boldsymbol{A}$For each class $\theta(k)$, if the sample number is less than the pre-set threshold α, we delete the class. Otherwise, we calculate the clustering center. Assuming that class $\theta(k)$ includes C_k elements $X'_{C_1}, X'_{C_2}, \cdots, X'_{C_k}$, the clustering center vector $\boldsymbol{M}_k = \frac{X'_{C_1} + X'_{C_2} + \cdots + X'_{C_k}}{C_k}$. Then the clustering center matrix $\boldsymbol{M}$ can be considered as the mixing matrix A, that is, $\boldsymbol{A} = \boldsymbol{M}$ up to the permutation and scaling of the sources.

16.2.5.2 I-LC-SCA Source Recovery

After the mixing matrix has been estimated, it is used to recover sources by I-LC-SCA. The detailed algorithm is as follows:

Algorithm Source recovery by I-LC-SCA

Input: sample set $\boldsymbol{X} = [\boldsymbol{X}_1, \boldsymbol{X}_2, \cdots, \boldsymbol{X}_T]$ and estimated mixing matrix $\boldsymbol{A}$.
Output: source set $\boldsymbol{S} = [\boldsymbol{S}_1, \boldsymbol{S}_2, \cdots, \boldsymbol{S}_T]$
Steps:

1. For each column $\overline{\boldsymbol{X}}(:, j)(j = 1, 2 \cdots T)$ of mixed signals, find vector $\boldsymbol{A}(:, i)$ that is collinear with $\overline{\boldsymbol{X}}(:, j)(j = 1, 2 \cdots T)$ of the mixed signals matrix, that is, $\overline{\boldsymbol{X}}(:, j) = \lambda \boldsymbol{A}(:, i)$. Then $\begin{cases} S(i,j) = \lambda \\ S(k,j) = 0 \quad k = 1, 2 \cdots n, k \neq i \end{cases}$. Otherwise, $\boldsymbol{S}(:, j) = \boldsymbol{A}^{-1} \times \overline{\boldsymbol{X}}(:, j)$.
2. Insert a zero column: using the recording tab of the zero column that was removed in the mixing matrix estimation step, recover the column with zeros.
3. Using its recording tab, restore the direction of the column whose direction was changed in the mixing matrix estimation step.

16.2.5.3 BSS Based on I-LC-SCA

BSS for Two Channels of Mixed Signals
In this experiment, we used MATLAB® software to produce two source signals and 1000 samples. Figure 16.6c shows the separation results when only 20 samples satisfy the sparsity conditions (i.e., sparsity is 1/50). Figure 16.7c shows the separation results when only 10 samples satisfy the sparsity conditions (i.e., sparsity is 1/100). From the two figures, it is clear that I-LC-SCA can still extract the source signals when the sparsity is very low.

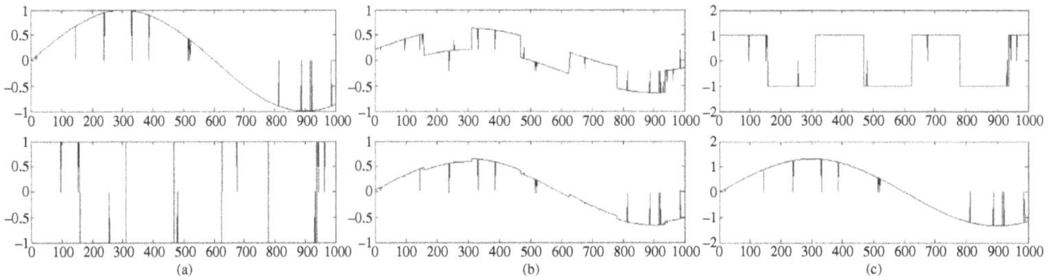

Figure 16.6 Experimental results when sparsity is 1/50. (a) Source signals, (b) mixed signals, (c) separated results by I-LC-SCA

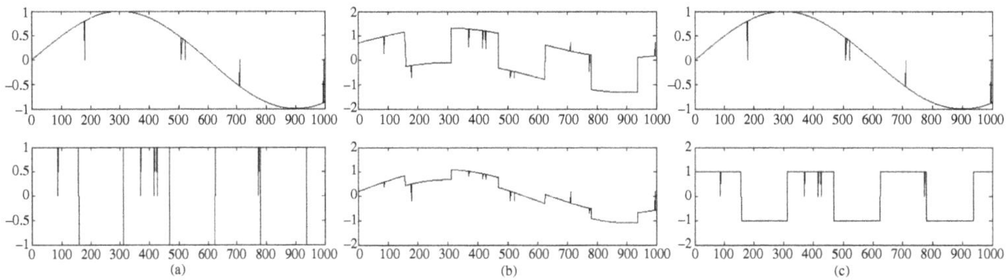

Figure 16.7 Experimental results when sparsity is 1/100. (a) Source signals, (b) mixed signals, (c) separated results by I-LC-SCA

I-LC-SCA for Speech Signal Separation

We chose source signals from the TIMIT standard speech bank, which comprises 2342 speech signal recordings. We randomly selected two speech recordings obtaining the signals "*This was easy for us*" (s_1) and "*She is thinner than I am*" (s_2). The separation results by I-LC-SCA are shown in Figure 16.8c, and those by FastICA in Figure 16.8d.

From Figure 16.8 we can see that both I-LC-SCA and FastICA can separate the speech signals effectively. For I-LC-SCA, the separated result 1 corresponds to the original signal 2, while separated result 2 corresponds to the original signal 1. This is the same for FastICA, although separated result 2 by FastICA is the inverse of the original signal 1.

The NCC indexes of the experiment are given in Table 16.5.

Table 16.5 NCC indexes of the I-LC-SCA speech signal separation experiment

NCC	I-LC-SCA result 1	I-LC-SCA result 2	FastICA result 1	FastICA result 2
s_1	−0.0025	**0.9864**	−0.0077	**−1.0000**
s_2	**0.9735**	−0.0080	**0.9999**	0.0139

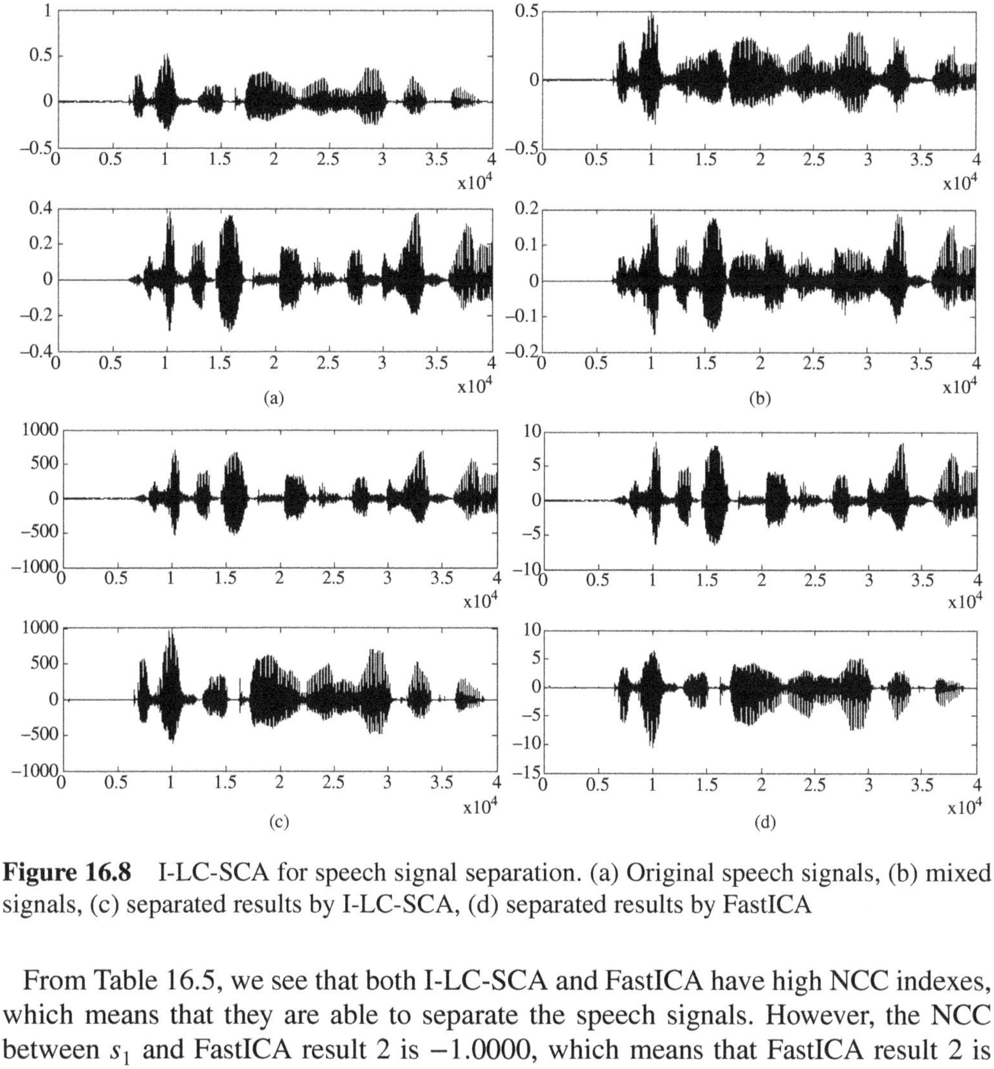

Figure 16.8 I-LC-SCA for speech signal separation. (a) Original speech signals, (b) mixed signals, (c) separated results by I-LC-SCA, (d) separated results by FastICA

From Table 16.5, we see that both I-LC-SCA and FastICA have high NCC indexes, which means that they are able to separate the speech signals. However, the NCC between s_1 and FastICA result 2 is -1.0000, which means that FastICA result 2 is the inverse of s_1. Thus, FastICA cannot ascertain the phase congruency between the separated and source signals.

16.2.5.4 Concluding Remarks about I-LC-SCA

Signals that partly satisfy the sparsity conditions, such as speech signals, cannot be separated directly by the conventional LC-SCA method. We can use segmental signals satisfying the sparsity for clustering to estimate the mixing matrix accurately. When identifying the source signals, the calculation method is the same as before for sparse samples, utilizing every column of the mixing matrix to estimate those sources that are not sparse samples. We proposed the I-LC-SCA method to perform practical speech signal extraction and obtained satisfactory separation results.

16.3 Plane Clustering SCA (PC-SCA)

The conventional SCA is based on a single-dominant-component assumption whereby, at each time, there is only one dominant component with the others zero or negligible. In this case, the scatter plot of observed signals shows clear geometrical line orientations. However, if there are two or more dominant components at each time instant, there will be no line orientations in the scatter plot of observed signals and the conventional SCA algorithms will fail to estimate the mixing matrix. For multipledominant-component SCA, hyperplane clustering must be used. Babaie-Zadeh *et al.* (2004) were the first to address this problem. Georgiev, Theis, and Cichocki (2004) pointed out that as long as the number of nonzero components in source signals is smaller than the number m of observed signals, SCA is solvable and they proposed an algorithm for small-scale problems without the presence of noise. Later, they proposed an alternative algorithm, which is more suitable for noisy data and larger scale problems.

Theorem 16.2 In a complete set with sparsity $m-1$, the columns of mixed signals are clustered along the orientation of the sub-plane of the mixing matrix and the number of columns of the mixing matrix determines the number of classes.

Proof: Here we merely prove the condition of three-dimensional source signals (i.e., $n=3$). If $n>3$, we need only expand the dimensions.

Assume source signals $S(j_1,i)=0$ at the ith $(1\le i\le N)$ sample, then

$$\begin{cases} X(1,i)=S(j_2,i)\times A(1,j_2)+S(j_3,i)\times A(1,j_3) \\ X(2,i)=S(j_2,i)\times A(2,j_2)+S(j_3,i)\times A(2,j_3) \\ X(3,i)=S(j_2,i)\times A(3,j_2)+S(j_3,i)\times A(3,j_3) \end{cases} \text{i.e., } \boldsymbol{X}_i=S(j_2,i)\times \boldsymbol{A}_{j_2}+S(j_3,i)\times \boldsymbol{A}_{j_3}.$$

Thus, it can be seen that $\boldsymbol{X}_i$ is in the plane based on $\boldsymbol{A}_{j_2}$ and $\boldsymbol{A}_{j_3}$. $\boldsymbol{X}_i$, $\boldsymbol{A}_{j_2}$, and $\boldsymbol{A}_{j_3}$ are coplanar. Since i could denote arbitrary samples, when the j_1th $(1\le j_1\le 3)$ source signal is equal to zero, the columns of the corresponding mixed signals are coplanar. In a similar way, when the j_2th $(1\le j_2\le 3)$ source signal is equal to zero, the columns of the corresponding mixed signals are in the same plane based on $\boldsymbol{A}_{j_1}$ and $\boldsymbol{A}_{j_3}$; when the j_3th $(1\le j_3\le 3)$ source signal is equal to zero, the columns of the corresponding mixed signals are in the same plane based on $\boldsymbol{A}_{j_1}$ and $\boldsymbol{A}_{j_2}$. The number of classes is $C_n^{n-1}=n$ and this is exactly equal to the number of columns of the mixing matrix.

16.3.1 Estimation of the Mixing Matrix by PC-SCA

According to Theorem 16.2, the columns of mixed signals are clustered along the plane and the number of classes is just the number of columns of the mixing matrix. The orientation of the intersection of all the planes is the orientation of the columns of the mixing matrix. Therefore, we can calculate the mixing matrix using the following plane clustering algorithm.

Algorithm Estimation of mixing matrix by PC-SCA

Input: sample set $\boldsymbol{X} = [\boldsymbol{X}_1, \boldsymbol{X}_2, \cdots, \boldsymbol{X}_T]$
Output: estimated mixing matrix $\boldsymbol{A} = [\boldsymbol{A}_1, \boldsymbol{A}_2, \cdots, \boldsymbol{A}_n]$
Steps:

1. Unify the direction. For each column $\boldsymbol{X}_j (j = 1, 2, \cdots, T)$ of mixed signals, if $X_{1j} < 0$ then set $\boldsymbol{X}_j = -\boldsymbol{X}_j$ and record the index j of this column.
2. Delete the zero columns. For each column $\boldsymbol{X}_j (j = 1, 2, \cdots, T)$ of the mixed image signals, if $\|\boldsymbol{X}(:, j)\|_1 = 0$, delete the jth column of $\boldsymbol{X}$ to obtain a new mixed matrix $\boldsymbol{X}\prime$, and record the index j of this column.
3. Plane clustering. Assume $\overline{\boldsymbol{X}}_i, \overline{\boldsymbol{X}}_j, \overline{\boldsymbol{X}}_k$ are coplanar, that is,
$$\boldsymbol{X}_i = k_1 \overline{\boldsymbol{X}}_j + k_2 \overline{\boldsymbol{X}}_k. \tag{16.14}$$
The above m equations include $m - 1$ unknowns. If the answer has been found, the rank of $\boldsymbol{M} = (\overline{\boldsymbol{X}}_j, \overline{\boldsymbol{X}}_k, -\overline{\boldsymbol{X}}_i)$ should be less than m. So, we just need to traverse all the columns of the mixed signals and find those columns satisfying the inequality rank$(\boldsymbol{M}) < m$.
4. Calculate the norm vectors of all the planes. Assuming $\overline{\boldsymbol{X}}_{i_1}, \overline{\boldsymbol{X}}_{i_2}, \overline{\boldsymbol{X}}_{i_3} \cdots \overline{\boldsymbol{X}}_{i_k}$ are coplanar and setting $\boldsymbol{P} = (\boldsymbol{X}_{i_1}, \boldsymbol{X}_{i_2}, \boldsymbol{X}_{i_3} \cdots \boldsymbol{X}_{i_k})$, the norm vector of $\boldsymbol{P}$ is $\alpha = \text{eig}(\boldsymbol{P} \times \boldsymbol{P}^{\text{T}})$, where eig$(\cdot)$ denotes calculating the eigenvector of the minimum eigenvalue.
5. Calculate the equation for each plane. After the norm vector α has been obtained, the equation of the plane with m dimensions can be denoted as $\alpha^{\text{T}} \times x = 0$, where $x = (x_1, x_2, \cdots, x_m)^{\text{T}}$.
6. Estimate mixing matrix $\boldsymbol{A}$. Calculate the intersections of all the planes. The number of intersections is the number of columns of the mixing matrix while the orientations of the intersections are the orientations of the columns of the mixing matrix.

16.3.2 Source Signal Recovery

After the mixing matrix has been estimated, it is used to recover sources by PC-SCA. The detailed algorithm is given below.

Algorithm Source signal recovery by PC-SCA

Input: sample set $\boldsymbol{X} = [\boldsymbol{X}_1, \boldsymbol{X}_2, \cdots, \boldsymbol{X}_T]$ and estimated mixing matrix $\boldsymbol{A}$.
Output: source $\boldsymbol{S} = [\boldsymbol{S}_1, \boldsymbol{S}_2, \cdots, \boldsymbol{S}_T]$
Steps:

1. Write down all the $C_n^1 = n$ sub-matrices $\boldsymbol{L}_i = (\boldsymbol{A}_1, \boldsymbol{A}_2, \cdots, \boldsymbol{A}_{i-1}, \boldsymbol{A}_{i+1}, \cdots \boldsymbol{A}_n)$ of $\boldsymbol{A}$.

2. For each column $X_j(1 \leq j \leq n)$ of the mixed signals, if X_j and L are coplanar, calculate S_j by the following formula. Assuming $Q_j = L^T \times (L \times L^T)^{-1} \times X_j$, where

$$Q_j = (Q_{1j}, Q_{2j}, \cdots Q_{n-1,j})^T, \begin{cases} S_{kj} = Q_{kj} & (k < i) \\ S_{kj} = 0 & (k = i) \\ S_{(k+1)j} = Q_{kj} & (k > i) \end{cases}.$$

3. Insert the zero column: using its recording tab, recover the zero column that was removed in the mixing matrix estimation step.
4. Using the recording tab of the column, restore the direction of the column whose direction was changed in the mixing matrix estimation step.

In summary, the PC-SCA algorithm requires a rigorous sparsity condition, namely, that the sparsity of an arbitrary source signal sample is at least 1. Similar to LC-SCA, we can improve the PC-SCA algorithm by clustering the segmental signals satisfying the sparsity restriction. For further details on the improved PC-SCA, refer to Section 16.5.

16.3.3 BSS Based on PC-SCA

In this experiment, we used natural images downloaded from the Internet to test the effectiveness of the algorithm. Figure 16.9a shows the three source images (s_1, s_2, s_3 downloaded from http://images.google.cn) with size 100×100 pixels, while the mixed results (x_1, x_2, x_3) are shown in Figure 16.9b. We used both PCA-SCA and FastICA to separate the mixed images and the separated results are shown in Figure 16.9c,d, respectively.

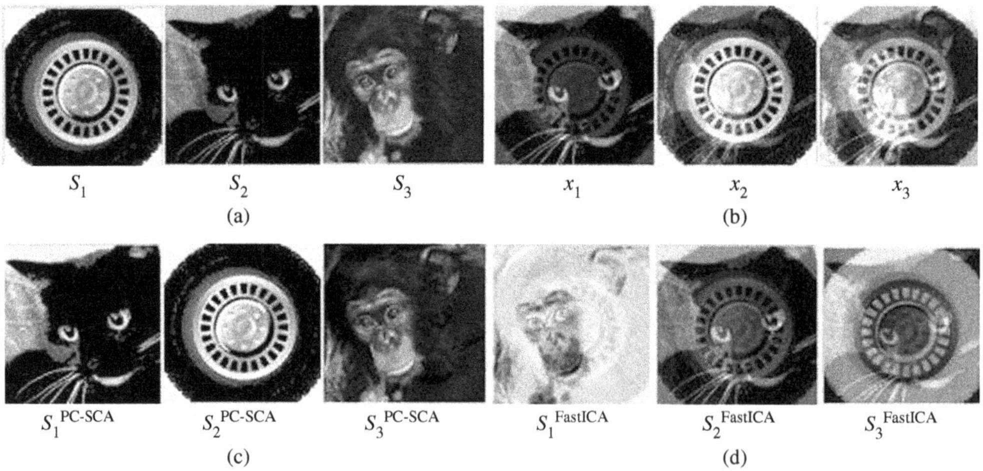

Figure 16.9 Blind image separation experiment using PC-SCA and FastICA. (a) Source images, (b) mixed images, (c) separated results by PC-SCA, (d) separated results by FastICA

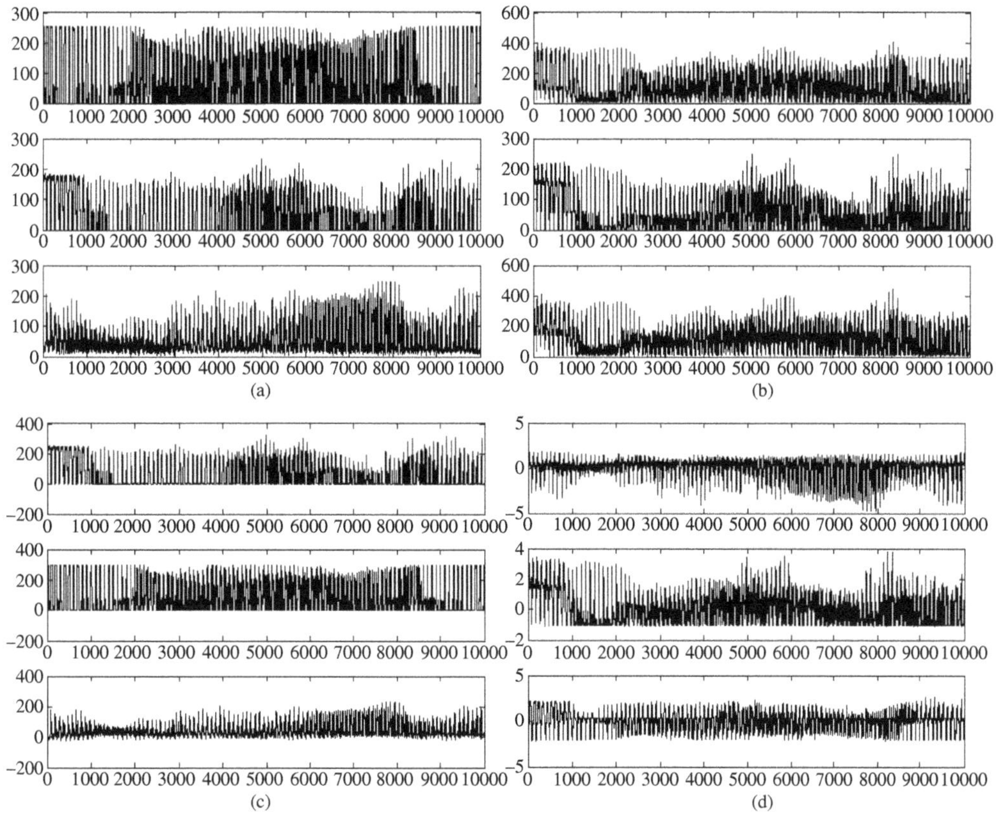

Figure 16.10 Blind image separation using PC-SCA (denoted by 1-D samples). (a) Source signals, (b) mixed signals, (c) separated results by PC-SCA, (d) separated results by FastICA

From Figure 16.9, we can see that PC-SCA extracts the source images very well, whereas FastICA cannot separate the signals and s_1^{FastICA} and s_3^{FastICA} are the inverse of the source images. This is confirmed in Figure 16.10d.

The NCC indexes between the source images and separated results are shown in Table 16.6. Images s_1^{FastICA} and s_3^{FastICA} separated by FastICA have a negative correlation with the source images (s_3 and s_1), while the separated results by PC-SCA all have positive correlations with the source images. Moreover, the NCC indexes for PC-SCA (1.0000) are all greater than those for FastICA, which means that PC-SCA separates signals more precisely than FastICA.

16.3.4 Concluding Remarks about PC-SCA

In a complete set with sparsity $n - 1$, the columns of mixed signals are clustered along the orientation of the sub-plane of the mixing matrix, and the number of columns of the mixing matrix determines the number of classes. According to this characteristic,

Table 16.6 NCC indexes for the image separation experiment

NCC	$s_1^{\text{PC-SCA}}$	$s_2^{\text{PC-SCA}}$	$s_3^{\text{PC-SCA}}$	s_1^{FastICA}	s_2^{FastICA}	s_3^{FastICA}
s_1	−0.0236	**1.0000**	0.1631	0.1239	0.5658	**−0.8152**
s_2	**1.0000**	−0.0236	−0.2130	0.0546	**0.8025**	0.5942
s_3	−0.0034	0.2178	**0.9762**	**−0.9397**	0.2400	−0.2435

PC-SCA is used in blind image separation. The experimental results show that PC-SCA can reconstruct the source signals precisely. Compared with FastICA, PC-SCA yields better separation results.

16.4 Over-Complete SCA Based on Plane Clustering (PCO-SCA)

Theorem 16.3 In an over-complete set with sparsity $m-1$, the columns of mixed signals are clustered along the orientation of the sub-plane of the mixing matrix. (This is sometimes referred to as the underdetermined case in the literature.) (Dong *et al.*, 2013)

Proof: Here we merely prove the condition of three source signals and four mixtures (i.e., $m=3, n=4$). For other cases, it is necessary only to expand the dimensions. Assume that at $i-th(1 \le i \le N)$ sample, $S(j_1,i)$ and $S(j_2,i) \ne 0$ and $S(j_3,i)$ and $S(j_4,i)=0$, then

$$\begin{cases} X(1,i) = S(j_1,i)\times A(1,j_1) + S(j_2,i)\times A(1,j_2) \\ X(2,i) = S(j_1,i)\times A(2,j_1) + S(j_2,i)\times A(2,j_2) \\ X(3,i) = S(j_1,i)\times A(3,j_1) + S(j_2,i)\times A(3,j_2) \end{cases} \text{i.e., } \boldsymbol{X}_i = S(j_1,i)\times \boldsymbol{A}_{j_1} + S(j_3,i)\times \boldsymbol{A}_{j_2}.$$

Thus, it can be seen that $\boldsymbol{X}_i$ is in the plane based on $\boldsymbol{A}_{j_1}$ and $\boldsymbol{A}_{j_2}$, and hence, $\boldsymbol{X}_i, \boldsymbol{A}_{j_2}$, and $\boldsymbol{A}_{j_3}$ are coplanar. Since i could denote an arbitrary sample, when $S(j_3,i)$ and $S(j_4,i)=0$, the columns of the corresponding mixed signals are all coplanar. In a similar way, when $S(j_1,i)$ and $S(j_2,i)=0$, the columns of the corresponding mixed signals are in the same plane based on $\boldsymbol{A}_{j_3}$ and $\boldsymbol{A}_{j_4}$. Thus, $C_n^2 = C_4^2 = 6$ sub-matrices can be formed and the columns of mixtures are clustered along the orientations of these six sub-matrices.

16.4.1 Estimation of Mixing Matrix by PCO-SCA

From Theorem 16.3, we get that there are $C^2_{n(n-1)/2}$ intersections for the C_n^2 planes. Since any column of the estimated mixing matrix can appear $C^1_{n-1} = n-1$ times in the $C^2_{n(n-1)/2}$ intersections, the one that appears $n-1$ times is the column of the mixing matrix. Hence, we can estimate the mixing matrix by following the plane clustering algorithm.

Algorithm Estimation of mixing matrix by PCO-SCA

Input: sample set $\boldsymbol{X} = [\boldsymbol{X}_1, \boldsymbol{X}_2, \cdots, \boldsymbol{X}_T]$
Output: estimated mixing matrix $\boldsymbol{A} = [\boldsymbol{A}_1, \boldsymbol{A}_2, \cdots, \boldsymbol{A}_n]$
Steps:

1. Unify the direction For each column $\boldsymbol{X}_j(j = 1, 2, \cdots, T)$ of the mixed signals, if $X_{1j} < 0$ then set $\boldsymbol{X}_j = -\boldsymbol{X}_j$ and record the index j of this column.
2. Delete the zero columns For each column $\boldsymbol{X}_j(j = 1, 2, \cdots, T)$ of the mixed image signals, if $\|\boldsymbol{X}(:, j)\|_1 = 0$, delete the jth column of $\boldsymbol{X}$ to yield a new mixed matrix $\boldsymbol{X}\prime$, and record the index j of this column.
3. Plane clustering Assume $\overline{\boldsymbol{X}}_i, \overline{\boldsymbol{X}}_j, \overline{\boldsymbol{X}}_k$ are coplanar, that is,
$$\boldsymbol{X}_i = k_1\overline{\boldsymbol{X}}_j + k_2\overline{\boldsymbol{X}}_k \tag{16.15}$$
The above m equations include $m - 1$ unknowns. If the answer has been found, the rank of $\boldsymbol{M} = (\overline{\boldsymbol{X}}_j, \overline{\boldsymbol{X}}_k, -\overline{\boldsymbol{X}}_i)$ should be less than m. Thus, we just need to traverse all the columns of the mixed signals to find the columns that satisfy the inequality $\text{rank}(\boldsymbol{M}) < m$.
4. Calculate the norm vectors of all the planes Assume $\overline{\boldsymbol{X}}_{i_1}, \overline{\boldsymbol{X}}_{i_2}, \overline{\boldsymbol{X}}_{i_3} \cdots \overline{\boldsymbol{X}}_{i_k}$ are coplanar and set $\boldsymbol{P} = (\boldsymbol{X}_{i_1}, \boldsymbol{X}_{i_2}, \boldsymbol{X}_{i_3} \cdots \boldsymbol{X}_{i_k})$, then the norm vector of $\boldsymbol{P}$ is $\alpha = \text{eig}(\boldsymbol{P} \times \boldsymbol{P}^{\text{T}})$, where $\text{eig}(\cdot)$ denotes calculating the eigenvector of the minimum eigenvalue.
5. Calculate the equation of each plane Once the norm vector α has been obtained, the equation of the plane with m dimensions can be denoted as $\alpha^{\text{T}} \times x = 0$, where $x = (x_1, x_2, \cdots, x_m)^{\text{T}}$.
6. Estimate mixing matrix $\boldsymbol{A}$ Calculate the intersections of all the planes. The orientation of the intersections with frequency of occurrence equal to or greater than $n - 1$, is the orientation of the columns of mixing matrix $\boldsymbol{A}$.

16.4.2 Source Signal Recovery

After the mixing matrix has been estimated, it can be used to recover the sources by PCO-SCA. The detailed algorithm is as follows.

Algorithm Source signal recovery by PCO-SCA

Input: sample set $\boldsymbol{X} = [\boldsymbol{X}_1, \boldsymbol{X}_2, \cdots, \boldsymbol{X}_T]$ and estimated mixing matrix $\boldsymbol{A}$.
Output: source set $\boldsymbol{S} = [\boldsymbol{S}_1, \boldsymbol{S}_2, \cdots, \boldsymbol{S}_T]$
Steps:

1. Write down all the $C_n^1 = n$ sub-matrices $\boldsymbol{L}_i = (\boldsymbol{A}_1, \boldsymbol{A}_2, \cdots, \boldsymbol{A}_{i-1}, \boldsymbol{A}_{i+1}, \cdots \boldsymbol{A}_n)$ of $\boldsymbol{A}$.

2. For each column $\boldsymbol{X}_j$ $(1 \le j \le n)$ of mixed signals, if $\boldsymbol{X}_j$ and $\boldsymbol{L}$ are coplanar and supposing $\boldsymbol{L}$ comprises columns j_1 and $j_2 (j_1 < j_2)$ of the mixing matrix, we can calculate $\boldsymbol{S}_j$ by the following formula. Assuming $\boldsymbol{Q}_j = \boldsymbol{L}^{\mathrm{T}} \times (\boldsymbol{L} \times \boldsymbol{L}^{\mathrm{T}})^{-1} \times \boldsymbol{X}_j$, where

$$\boldsymbol{Q}_j = (\boldsymbol{Q}_{1j}, \boldsymbol{Q}_{2j}, \cdots \boldsymbol{Q}_{n-1,j})^{\mathrm{T}}, \text{ then } \begin{cases} S_{kj} = Q_{kj} & (k < j_1) \\ S_{kj} = 0 & (k = j_1) \\ S_{(k+1)j} = Q_{kj} & (j_1 < k < j_2) \\ S_{(k+2)j} = 0 & (k = j_2) \\ S_{(k+2)j} = Q_{kj} & (k > j_2) \end{cases}.$$

3. Insert a column with zeros: using the recording tab of the column with zeros that were removed in the mixing matrix estimation step, recover the column with zeros.
4. Restore the direction of the column whose direction was changed in the mixing matrix estimation step using its recording table.

16.4.3 BSS Based on PCO-SCA

To test the effectiveness of the PCO-SCA algorithm, we generated four source signals (Figure 16.11a) using MATLAB® software and employed a random 3×4 matrix as the mixing matrix to mix the source signals, resulting in three mixtures as shown in Figure 16.11b. We used PCO-SCA and FastICA to separate the mixtures and the results are shown in Figure 16.11c,d, respectively.

From Figure 16.11, we can see that PCO-SCA can separate the sources in underdetermined conditions (where $m = 3$ and $n = 4$). The normalized estimated mixing matrix $\varphi(\boldsymbol{A}_{\text{PCO-SCA}})$ is given as follows:

$$\varphi(\boldsymbol{A}_{\text{PCO-SCA}}) = \begin{pmatrix} 0.0975 & 0.8216 & 0.6975 & 0.7665 \\ 0.9688 & 0.0239 & 0.6656 & 0.4175 \\ 0.2278 & 0.5696 & 0.2654 & 0.4881 \end{pmatrix}.$$

The normalized original mixing matrix $\varphi(\boldsymbol{A})$ produced by MATLAB® is as follows:

$$\varphi(\boldsymbol{A}) = \begin{pmatrix} 0.097519 & 0.8216 & 0.69753 & 0.76648 \\ 0.9688 & 0.023905 & 0.66558 & 0.41747 \\ 0.22785 & 0.56957 & 0.26543 & 0.48808 \end{pmatrix}.$$

Thanks to $\varphi(\boldsymbol{A})$ and $\varphi(\boldsymbol{A}_{\text{PCO-SCA}})$, BSS based on PCO-SCA can estimate the mixing matrix precisely.

We also carried out NCC analysis for PCO-SCA with the results given in Table 16.7. From this table, we can see that the NCC index is 1.0000, which means that there is a complete positive correlation.

Similarly, we carried out NCC analysis for FastICA with the results shown in Table 16.8. From this table, we can see that the NCC indexes are smaller than those in Table 16.7 and that they are sometimes negative. Figure 16.11d and Table 16.8

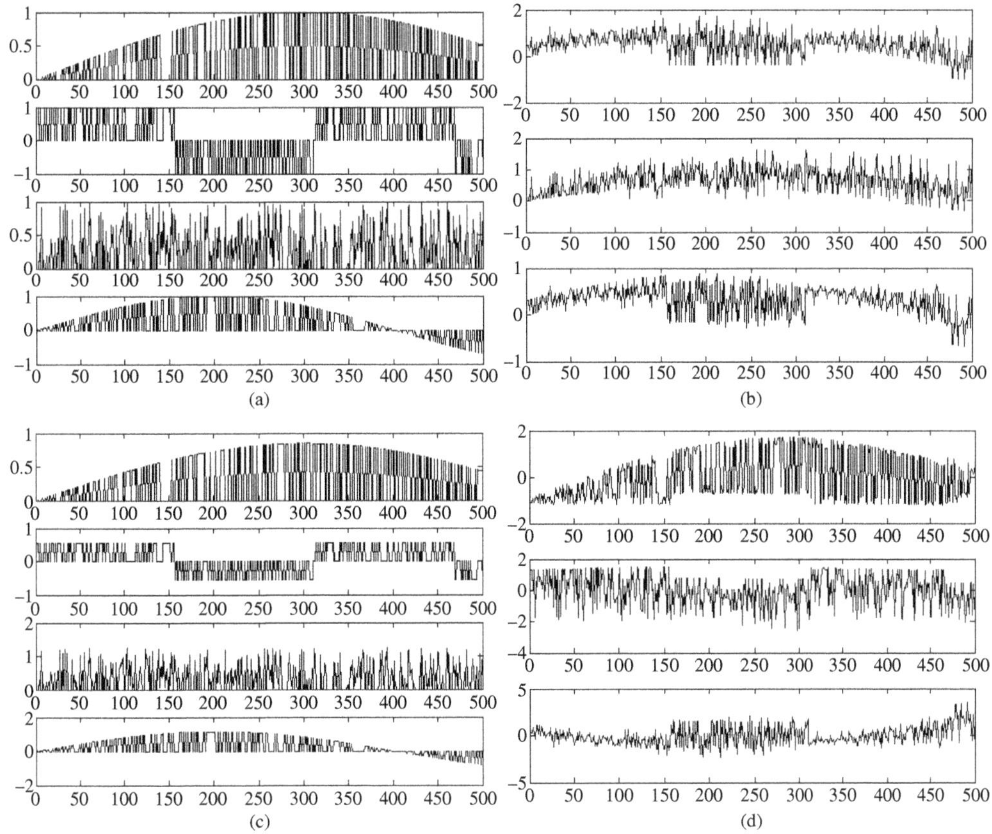

Figure 16.11 Experimental results of BSS based on PCO-SCA. (a) Source signals, (b) mixed signals, (c) separated results by PCO-SCA, (d) separated results by FastICA

Table 16.7 NCC indexes between sources and separated results by PCO-SCA

NCC	$s_1^{\text{PCO-SCA}}$	$s_2^{\text{PCO-SCA}}$	$s_3^{\text{PCO-SCA}}$	$s_4^{\text{PCO-SCA}}$
s_1	**1.0000**	−0.1435	−0.1681	−0.1593
s_2	−0.1435	**1.0000**	−0.0520	−0.0802
s_3	−0.1681	−0.0520	**1.0000**	−0.1282
s_4	−0.1593	−0.0802	−0.1282	**1.0000**

Table 16.8 NCC indexes between sources and separated results by FastICA

NCC	s_1^{FastICA}	s_2^{FastICA}	s_3^{FastICA}
s_1	**0.9805**	0.0429	−0.0111
s_2	−0.2951	0.5427	**−0.4943**
s_3	−0.1951	**−0.8628**	−0.3628
s_4	−0.0249	0.1501	**−0.6152**

also show that FastICA only produced three results, which is less than the number of sources. Thus, FastICA cannot separate the signals in undetermined conditions.

16.4.4 Concluding Remarks about PCO-SCA Algorithm

In an over-complete set with sparsity $m - 1$, the columns of mixed signals are clustered along the orientation of the sub-plane of the mixing matrix. According to this characteristic, we can use the sparseness of the signals to offset the defect of insufficient mixtures and realize BSS for an over-complete set. We used PCO-SCA to extract four source signals from three mixtures and the separated results have a full positive correlation with the source signals, which means that PCA-SCA can reconstruct the sources precisely.

16.5 Blind Image Separation Based on Wavelets and SCA (WL-SCA)

The BSS algorithms discussed above can only operate on sparse sources. However, since some signals, such as image signals, are not sufficiently sparse, the orientation of the scatter diagram will not be very clear and it cannot be carried out directly by SCA (Fadili *et al.*, 2010). So, how do we solve blind image separation using SCA? First, we need to make the source signals sparser (Yu *et al.*, 2010).

16.5.1 Idea behind the WL-SCA Algorithm

With respect to the sparseness property of wavelet high frequency coefficients of images, we can solve the sparsification problem using a wavelet transform as the processing stage. The WL-SCA algorithm is as follows.

Algorithm Blind image separation based on WL-SCA

Input: m mixed images
Output: m source images
Steps:

1. Transform the mixed images from the spatial domain to the frequency domain using a Haar wavelet.
2. Select the high frequency part of each mixed image, and obtain the mixed signal matrix $\boldsymbol{X}$.
3. Utilize the LC-SCA or PC-SCA algorithm given above to estimate mixing matrix $\boldsymbol{A}$.

Given the mixed images, we can recover the source images by LC-SCA or PC-SCA.

16.5.2 Simulation of Blind Image Separation Based on WL-SCA

We used standard gray scale images (Lena s_1 and Cameraman s_2, 256×256 pixels) as shown in Figure 16.12a, as the source images in our experiment. Mixed images were randomly linearly combined by MATLAB®, as shown in Figure 16.12b. The separation results using WL-SCA (left: $s_1^{\text{WL-}SCA}$, right: $s_2^{\text{WL-}SCA}$) and FastICA (left: s_1^{FastICA}, right: s_2^{FastICA}) are shown in Figure 16.12c,d, respectively.

First, the mixed images were transformed by a Haar wavelet, the results of which are shown in Figure 16.13. We selected the diagonal component for processing the mixed signals at hand. Using the improved BSS based on LC-SCA, we obtained $\boldsymbol{A}_{\text{WL-SCA}}$ and the normalized result $\varphi(\boldsymbol{A}_{\text{WL-SCA}})$ as given below:

$$\boldsymbol{A}_{\text{WL-SCA}} = \begin{pmatrix} 5.5374 & 226.3094 \\ 22.6099 & 298.6839 \end{pmatrix} \quad \varphi(\boldsymbol{A}_{\text{WL-SCA}}) = \begin{pmatrix} 0.2379 & 0.6039 \\ 0.9713 & 0.7970 \end{pmatrix}.$$

Figure 16.12 Blind image separation experiments based on WL-SCA and FastICA. (a) Source images, (b) mixed images, (c) separated results by WL-SCA, (d) separated results by FastICA

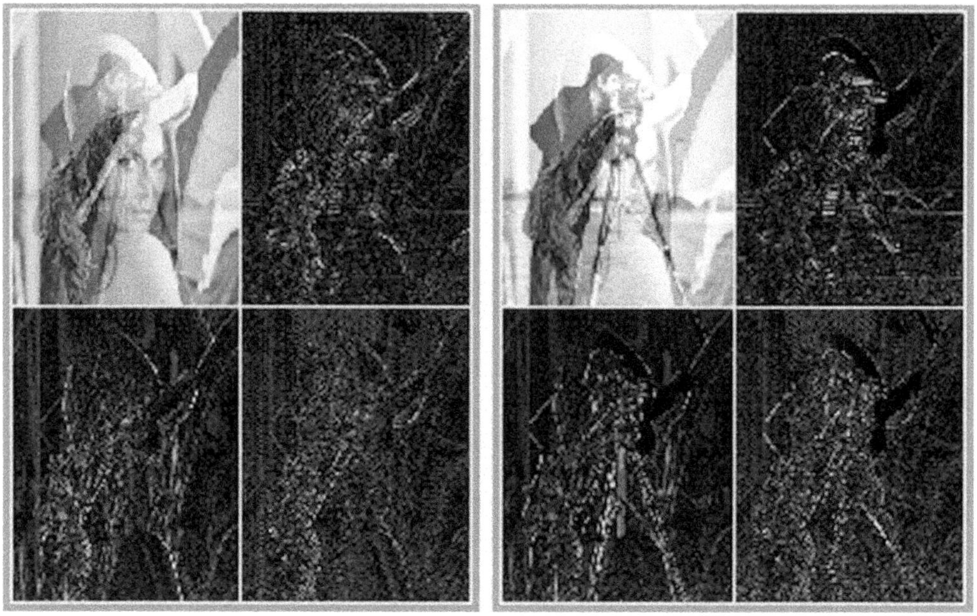

Figure 16.13 Results of the transformed mixed image using a Haar wavelet

Using FastICA to estimate the mixing matrix, we obtained $\boldsymbol{A}_{\text{FastICA}}$ and the normalized result $\varphi(\boldsymbol{A}_{\text{FastICA}})$ as follows:

$$\boldsymbol{A}_{\text{FastCA}} = \begin{pmatrix} -5.2247 & -52.3884 \\ -16.0975 & -42.5349 \end{pmatrix} \varphi(\boldsymbol{A}_{\text{FastICA}}) = \begin{pmatrix} -0.3087 & -0.7763 \\ -0.9512 & -0.6303 \end{pmatrix}.$$

The actual mixing matrix $\boldsymbol{A}$ and the normalized result $\varphi(\boldsymbol{A})$ are as follows:

$$\boldsymbol{A} = \begin{pmatrix} 0.6357 & 0.1978 \\ 0.8390 & 0.8075 \end{pmatrix} \varphi(\boldsymbol{A}) = \begin{pmatrix} 0.6039 & 0.2379 \\ 0.7970 & 0.9713 \end{pmatrix}.$$

Comparing $\varphi(\boldsymbol{A})$ with $\varphi(\boldsymbol{A}_{\text{FastICA}})$ and $\varphi(\boldsymbol{A}_{\text{WL-SCA}})$, we see that WL-SCA can estimate the mixing matrix precisely, whereas the estimated mixing matrix by FastICA differs somewhat from the original mixing matrix.

After mixing matrix $\boldsymbol{A}$ has been estimated, we can use the mixed images $\boldsymbol{X}$ to identify the original images using the LC-SCA or PC-SCA algorithm. Table 16.9 gives the NCC (Equation 16.13) indexes between the sources and separated signals by WL-SCA and FastICA.

From Table 16.9, we can see that $s_1^{\text{WL-SCA}}$ corresponds to s_2 and $s_2^{\text{WL-SCA}}$ corresponds to s_1. WL-SCA can extract the original images precisely, as confirmed by Figure 16.12c. For FastICA, the NCC between $\boldsymbol{s}_1^{\text{FastICA}}$ and $\boldsymbol{s}_2$ is 0.9998, which means that $\boldsymbol{s}_2$ is mostly extracted. The NCC between $\boldsymbol{s}_2^{\text{FastICA}}$ and $\boldsymbol{s}_1$, however, is −0.9874, which means that these have a negative correlation; $\boldsymbol{s}_2^{\text{FastICA}}$ is the inverse of $\boldsymbol{s}_1$, as confirmed by Figure 16.12d.

Table 16.9 Comparison of NCC indexes between sources and separated signals by WL-SCA and FastICA

NCC	$s_1^{\text{WL-SCA}}$	$s_2^{\text{WL-SCA}}$	s_1^{FastICA}	s_2^{FastICA}
s_1	0.1802	**1.0000**	0.1585	**−0.9874**
s_2	**1.0000**	0.1802	**0.9998**	−0.0220

16.5.3 Concluding Remarks about WL-SCA

For those images that do not satisfy the sparseness constraint of SCA, we can make the images sparser using a wavelet transform. The mixing matrix is estimated by high frequency coefficients of the wavelet in the frequency domain, and then the original images are identified by LC-SCA or PC-SCA in the spatial domain. The experimental results show that WL-SCA can extract the original images (Lena and Cameraman) precisely. Compared with FastICA, WL-SCA yields better separation results.

16.6 New BSS Algorithm Based on Feedback SCA

16.6.1 A Shortcoming of the BSS Model

The classical model for BSS is not suitable for actual conditions. Considering a classical "Cocktail Party" situation, there are not only speaker voices, but also a variety of other noises (such as steam whistles, soft music, footsteps, and the sounds of clinking glasses, amongst others). When we use microphones to record the sounds of a party, the received signals usually contain a mixture of speaker voices and other noises. As such, the model for BSS should be as shown in Figure 16.14.

However, most researchers consider the added noise $\boldsymbol{N}$ in the system but do not take the mixing noise into account, which really is an element of the mixing system and is usually not sparse. Thus, it is more valuable to carry out research on mixing noise conditions.

We take WGN as the mixing noise owing to the fact that the superposition of a large number of noises is close to a Gaussian distribution, and WGN has the largest entropy with the same energy for all kinds of noise that could produce maximum interference. If a method works well under WGN conditions, it will work even better under other

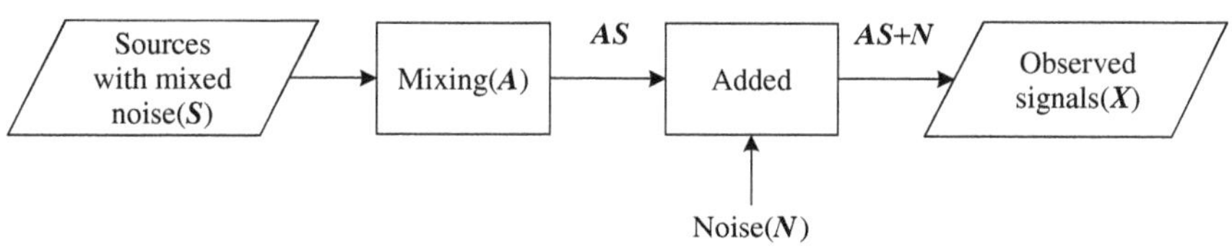

Figure 16.14 Model for SCA

Figure 16.15 Standard gray images (5 columns on the left) and WGN images (rightmost column)

circumstances. As a result, we focus on the conditions of one channel with WGN to devise an effective BSS method.

16.6.2 Analysis of SCA Algorithm

We carried out an experiment with mixed GWN conditions to test the anti-noise performance of the WL-SCA algorithm. Various standard gray images (256 × 256 pixels) and WGN images (256 × 256 pixels) produced by MATLAB®'s "wgn" order with the "dBW" parameter were adopted as the original sources (see Figure 16.15).

The mixed images (shown in Figure 16.16a with a mixing noise intensity of 30 dBW and those in Figure 16.16b with a mixing noise intensity of 100 dBW) were generated by linear superposition of the source images according to formula (1). The mixing matrix $\boldsymbol{A}$ was produced by the "rand (3)" function in MATLAB®. From Figure 16.16, we can see that with a noise intensity of 100 dBW, the mixing signals are completely polluted by the noise and the original signals cannot be identified by the naked eye.

We used the WL-SCA algorithm to separate the mixed images with different noise intensities (Figure 16.16). The separated results are shown in Figure 16.17, from which we can see that the algorithm cannot separate all the original images.

However, after running some experiments, we found that there is always at least one separated image that looks perfect. To check the results objectively, we calculated the NCC (Equation 16.13) between the separated results and the sources. The NCC value is larger if the degree of correlation between the two images is higher, which means that they are more similar. If the value of NCC is 1, the two images are considered to have the same "form" (although the total amplitude may differ).

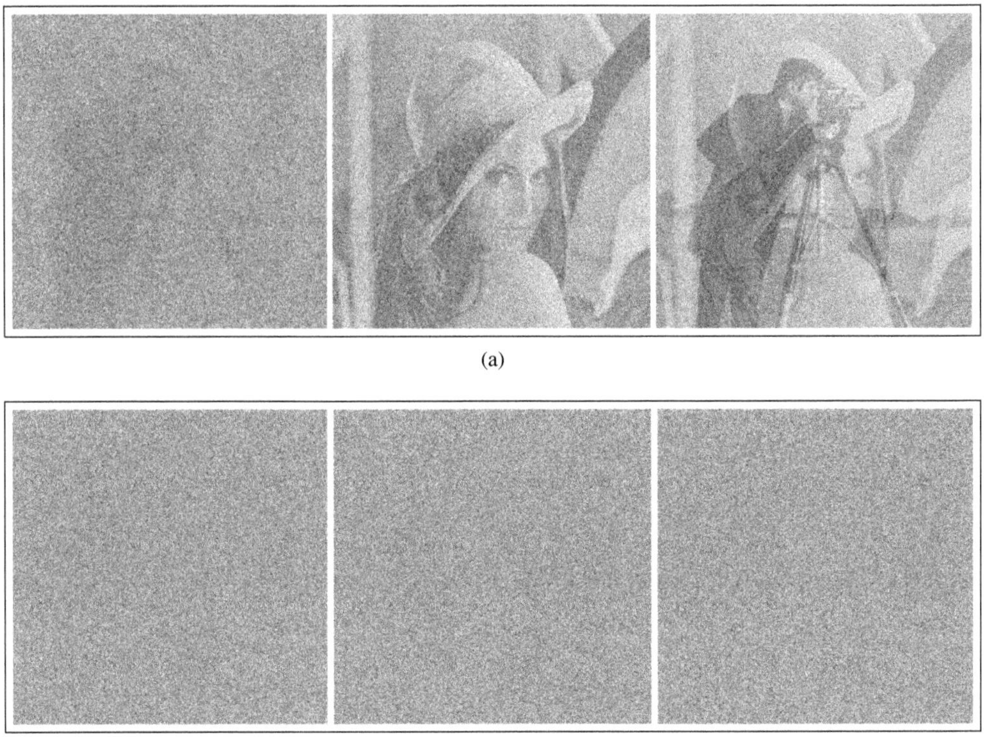

(a)

(b)

Figure 16.16 Linear mixed results using $\boldsymbol{X} = \boldsymbol{AS}$ with mixing noise intensity of (a) 30 dBW and (b) 100 dBW

Table 16.10 gives the NCC values between the separated and source images with a mixing noise intensity of 100 dBW. From this table, we can see that the NCC value is 1.0000 between the second separated image with good depiction and the original Lena image, which means they have the same "form" (excluding amplitude).

We also carried out various experiments with three signal channels (standard gray images) and one channel of noise. In this case, the SCA algorithm also succeeded in extracting at least one signal channel.

16.6.3 BSS Algorithm Based on Feedback SCA (FSCA)

See (Yu et al., 2013)

16.6.3.1 Basic Idea behind FSCA

From the experimental results in Section 16.6.2, we find that if we remove the signal that is considered to be perfectly extracted (the pure component), we can reduce

(a)

(b)

Figure 16.17 Separated results using SCA with mixing noise intensity of (a) 30 dBW and (b) 100 dBW

Table 16.10 NCC values between the original and separated images using the SCA algorithm

Sources	Separated image 1	Separated image 2	Separated image 3
Lena	0.0867	**1.0000**	0.6152
Cameraman	0.9949	0.1866	0.8890
Noise (100 dBW)	0.0018	0.0039	0.0000

the number of signals participating in the mixing system by one. If we repeat the procedure continuously, all of the signals can be extracted effectively step by step.

16.6.3.2 How to Identify the Pure Component

Since the sources are unknown, how to identify the pure component extracted by the SCA algorithm is a key issue. The pure component has a lower correlation with the mixtures than the impure components. Therefore, considering the robustness,

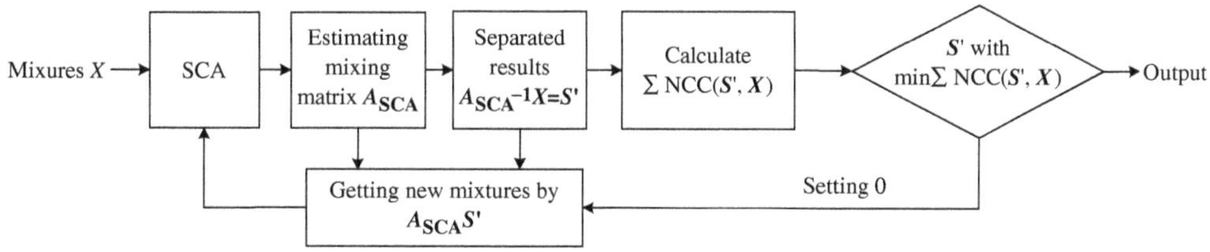

Figure 16.18 Flowchart of FSCA

we use the sum of the NCC values between one extracted component S' and all mixtures X ($\sum_m \text{NCC}(S', X)$) to identify the purely separated components. The one with min $\sum_m \text{NCC}(S', X)$ is considered to be the purest source.

16.6.3.3 Feedback SCA Algorithm

Based on the above, a BSS algorithm called feedback sparse component analysis (FSCA) is proposed. The flowchart is depicted in Figure 16.18.

16.6.4 Experimental Results and Analysis

16.6.4.1 Anti-WGN Testing

To evaluate the performance of the proposed FSCA method, we carried out a blind image separation experiment of artificially mixed images with three sources (Figure 16.15 shows the WGN image with intensity 40 dBW and two other standard images). The mixed images are shown in Figure 16.16a. The extraction results by FastICA and FSCA are shown in Figure 16.19a,b, respectively.

Kurtosis is adopted to measure the sparsity of the sources. For univariate data $v_1, v_2, \cdots v_N$, the formula for kurtosis is given as:

$$\text{Kurtosis} = \frac{\sum_{i=1}^{N} (v_i - \overline{v})^4}{(N-1)\sigma^4} \tag{16.16}$$

where $\overline{v}$ is the mean, σ is the standard deviation, and N is the number of data points.

Table 16.11 gives a comparison between FastICA and FSCA based on the NCC values, with the kurtosis values for the diagonal coefficients of a one level wavelet (Haar) transform of the source shown in the first column. From Table 16.11, we can see that the FSCA algorithm has better performance than the classical FastICA method in the case of mixed WGN images.

We also carried out another separation experiment mixing three standard images (Bridges, Columbia, and Airplane) and one WGN image (40 dBW). The mixed images

(a)

(b)

Figure 16.19 Separated results using (a) the FastICA algorithm and (b) the FSCA algorithm

Table 16.11 Comparison between FastICA and FSCA based on NCC

Sources (*Kurtosis*)	FastICA Separated image 1	FastICA Separated image 2	FastICA Separated image 3	FSCA Separated image 1	FSCA Separated image 2	FSCA Separated image 3
Lena (*17.5310*)	**0.9873**	0.1588	0.0042	**1.0000**	0.1865	0.0182
Cameraman (*25.5540*)	0.3402	**0.9403**	0.0085	0.1865	**1.0000**	0.0173
Noise (40 dBW) (*2.5057*)	0.0281	0.0149	**0.0034**	0.0182	0.0173	**1.0000**

(a)

(b)

Figure 16.20 (a) Mixed images using three standard images and one WGN image and (b) separated results by FSCA

Table 16.12 NCC between separated and source images

Sources (*Kurtosis*)	Separated image 1	Separated image 2	Separated image 3	Separated image 4
Boats (*16.1951*)	**1.0000**	0.0718	0.0584	0.0023
Airplane (*15.0809*)	0.0718	**1.0000**	0.1022	0.0050
Columbia (*20.0960*)	0.0584	0.1022	**1.0000**	0.0063

and separated results are shown in Figure 16.20a,b, respectively, while the statistical results of NCC and kurtosis are presented in Table 16.12. It is obvious that the FSCA algorithm can extract all of the sources effectively.

16.6.4.2 Anti-speckle Noise Testing

We carried out a simple test on the robustness and expansibility of FSCA as a byproduct. Under speckle noise conditions, the separated results show that the algorithm also performs well. The speckle noise image (shown in Figure 16.21) is similar to pulse noise or spiced salt noise. The mixed images are shown in Figure 16.22a, and the separated results in Figure 16.22b.

Figure 16.21 Simulated speckle noise image created using Photoshop software

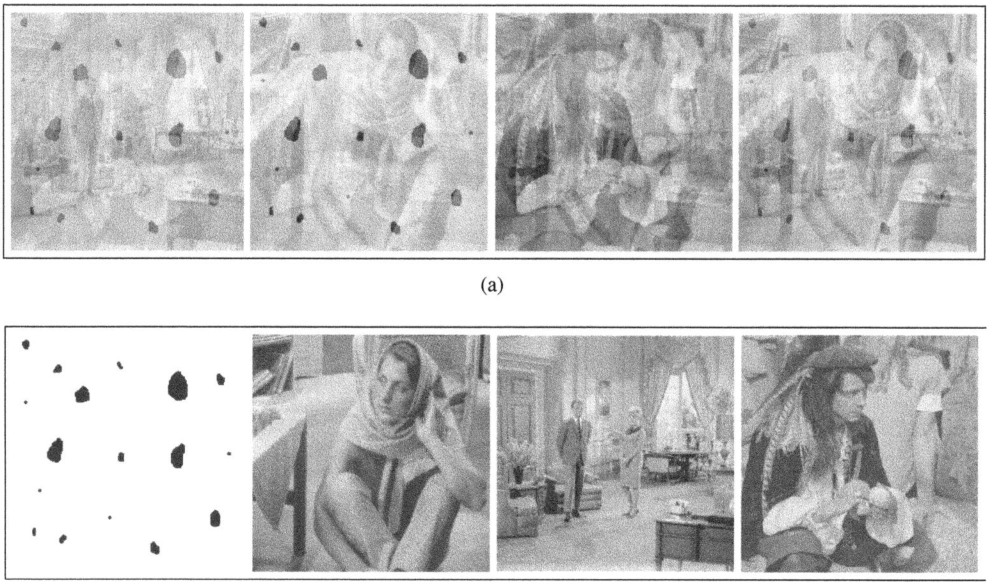

(a)

(b)

Figure 16.22 (a) Mixed images with speckle noise and (b) separated results by FSCA

16.6.5 Concluding Remarks about FSCA

In this section, we proposed a novel blind image separation method based on FSCA. The FSCA algorithm extends the SCA method by incorporating a feedback mechanism and achieves better performance in the case of mixing noise or sources with low levels of sparseness. Compared with the traditional FastICA and SCA, the FSCA method shows better performance (higher separation precision and better visual effects).

16.7 Remote Sensing Image Classification Based on SCA

With natural resources highly exploited, land utilization and cover are changing rapidly. How to obtain land information effectively has become a key question. Remote sensing image classification, as a method for obtaining land information, remains a hotspot of research (Liu *et al.*, 2008; Chen, Xu, and Cai, 2008; Huang *et al.*, 2007a, 2007b). Nevertheless, remote sensing image classification remains a difficult problem (owing to the complexity of the natural environment, there are some undetermined properties in obtaining land cover information from remote sensing images) (Yu, Cao and Yang, 2009b; Zhang, Chen, and Weng, 2008; Huang *et al.*, 2007a, 2007b). General methods for remote sensing image classification are mainly divided into supervised and unsupervised approaches (Zhong, Li, and Li, 2007; Chunqi *et al.*, 2005; Shi *et al.*, 2005). The unsupervised approaches are more automatic and do not require good knowledge of the classified regio. However, the problem of "same things with a different spectrum and the same spectrum with different things" limits the classification accuracy of these approaches. Supervised approaches can utilize specialized prior knowledge to obtain higher classification accuracy. However, because these methods do not consider global features of the training data, it is easy to fall into local optima.

Before unsupervised classification, preprocessing the data using PCA can yield better results. However, since PCA is based on second order statistics and does not consider the higher order statistics which provide the details, the classification results are not good enough when processing overlapping spectra of different land objects. SCA is not restricted by the effect of source overlapping and mutual independence. It takes full advantage of the sparseness and reduces redundant higher order information (Tan *et al.*, 2006; Du *et al.*, 2005; Li *et al.*, 2006c). In this section, we use SCA as a preprocessing step before classification. Compared with PCA, it achieves higher classification accuracy (Yu *et al.*, 2010).

16.7.1 Remote Sensing Image Classification Algorithm Based on SCA

In the process of remote sensing imaging, we can assume that there are several source signals that are independent and separable in the spectrum. Regarding the interference of weather, lighting, and landforms, amongst others, we could perform blind source analysis to reduce the interference and redundancy among the data, and then recover the essence to process the data correctly. In SCA, the source component to be estimated is a linear combination of the observed mixtures:

$$Z_i = \sum_{k=1}^{m} c_k X_i = \sum_{j=1}^{n} A_i S_{ij} \tag{16.17}$$

where X_i is the ith row of mixture $\boldsymbol{X}$, A_i is the ith row of mixing matrix $\boldsymbol{A}$, $\boldsymbol{S}$ is the source signal matrix, and $c_1, c_2, \cdots, c_m$ are the parameters that must be identified.

Before carrying out SCA on remote sensing images, we first introduce Theorem 16.4:

Theorem 16.4 For $\Omega = \{n|x_1(n) = 0, x_2(n) \neq 0\}$, $x_1(n)$, and $x_2(n)$ are non-overlapping. Then, $J(t) = \delta(x_1 + t \times x_2)$ attains its maximum at $t = 0$ with the following conditions satisfied:

1. $x_1(n) \times x_2(n) = 0 \quad \forall n,$
2. $E(x_1(n) \times x_2(n)) = 0 \quad \forall n,$
3. $|e_\rho| < \frac{\|S_1\|_2}{\sqrt{K}} \delta(S_1) \sum_{k \in \Omega} |S_2(k)|.$ where $e_\rho = \sum_{k=1}^{K} S_1(k)S_2(k) - \delta(S_1) \sum S_1'(k)S_2(k),$ $S_1'(k) = \frac{\|S_1\|_2}{\sqrt{K}} \mathrm{sgn}(S_1(k)).$

If there are only two inputs and two outputs of the system, Equation 16.17 can be written as

$$z(n) = y_1(n) + cy_2(n) \tag{16.18}$$

Because there is only one variable c in Equation 16.18, we only need to solve $\frac{d(\delta(z,c))}{d(c)} = 0$ and determine c.

For a system with more than two source components, we can use the following equation to identify the sources:

$$z(n) = X_1(n) + \sum_{k=2}^{m} c_k X_k(n) \tag{16.19}$$

Then Equation 16.17 can be written as:

$$\delta(z, c) = \sqrt{N} \frac{\|z\|_2}{\|z\|_1} \tag{16.20}$$

According to Theorem 16.4, when the sparsity function $\delta(z, c)$ reaches a local maximum, we can determine $c_1, c_2, \cdots, c_k$ and Equation 16.19 to reconstruct the sources.

16.7.2 Experiment for Remote Sensing Image Classification Based on SCA

We used the LANDSAT TM remote sensing image of Zhaoqing, Guangdong (taken in 2002) with seven bands (see Figure 16.23) as the experimental data (Cao *et al.*, 2009).

Figures 16.24 and 16.25 are the results of preprocessing by PCA and SCA, respectively. Comparing Figures 16.24 and 16.25, we can see that the preprocessed result by SCA is better than that by PCA.

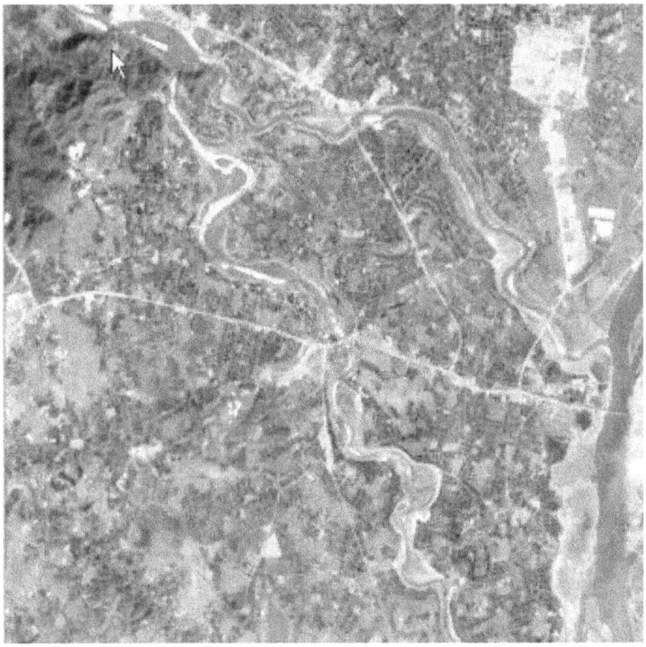

Figure 16.23 TM image of Zhaoqing (taken in 2002)

Figure 16.24 Preprocessed image by PCA

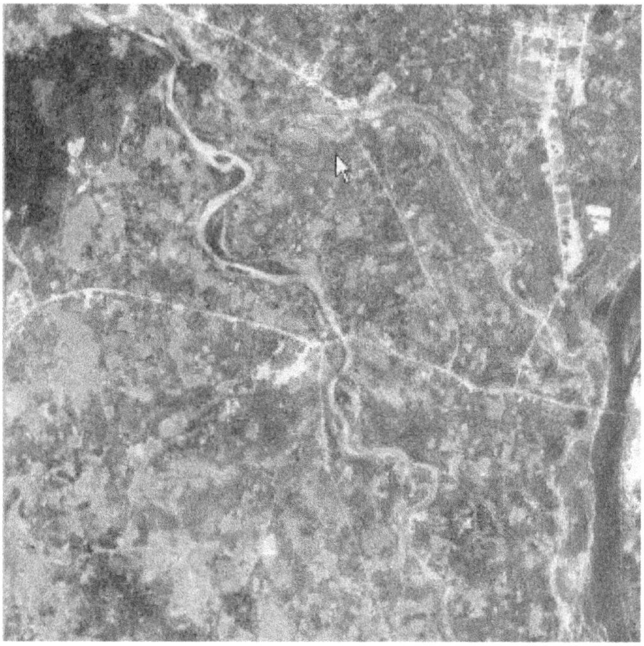

Figure 16.25 Preprocessed image by SCA

After preprocessing, we classified the preprocessed results using the same unsupervised classification method; the classified results are shown in Figures 16.26 and 16.27. We can see that the classified result based on SCA is more accurate than that by PCA; for example, for PCA, there are many vegetation classes that are incorrectly classified as water classes, whereas this does not occur using SCA.

Figure 16.28 compares the classification accuracy of PCA and SCA. It is clear that the classification accuracy of SCA is higher than that of PCA. Thus, the SCA method is better than PCA and can effectively enhance divisibility of different ground features.

Tables 16.13 and 16.14 give the classification accuracy after preprocessing using SCA and PCA, respectively. These tables also show that SCA is better than PCA with an average classification accuracy of 91.4% compared with 76.63% for PCA.

The main reason for this is that in remote sensing images, the spectra of different ground features may overlap and may be correlated. Since the correlation among features usually hides in higher-order statistics and because PCA can only eliminate second-order correlations, the classification results are unsatisfactory. SCA does not require independence of the signals and, therefore, it achieves better classification results.

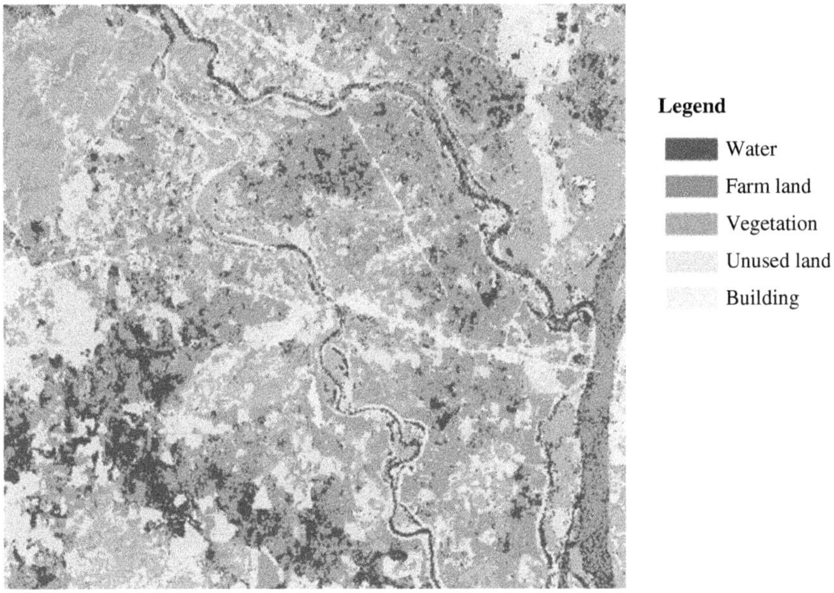

Figure 16.26 Pseudo-color image of the classified result based on PCA

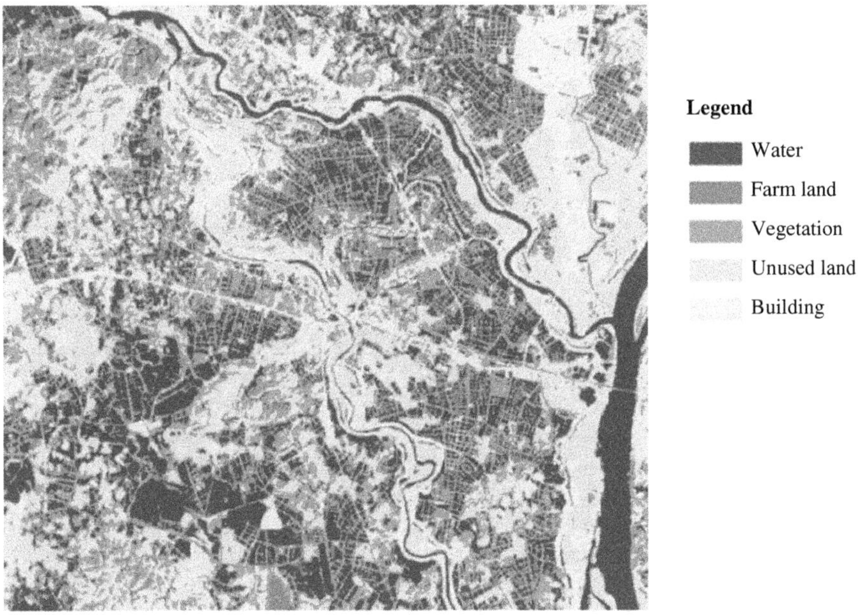

Figure 16.27 Pseudo-color image of the classified result based on SCA

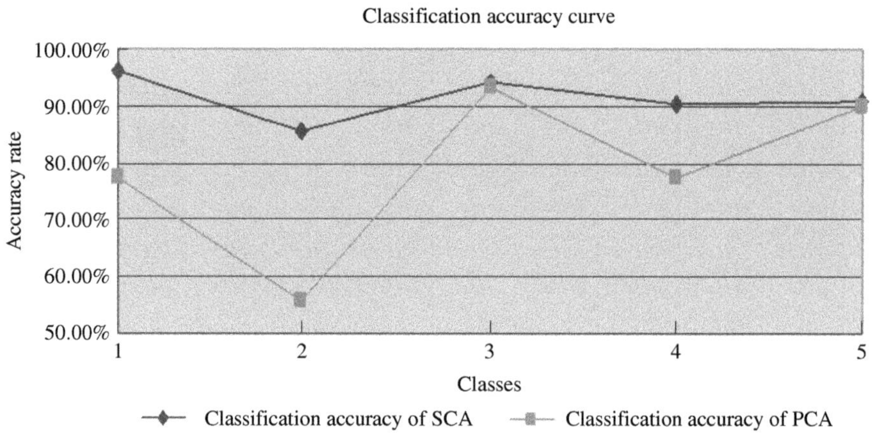

Figure 16.28 Classification accuracy of PCA and SCA

Table 16.13 Classification accuracy after SCA preprocessing

Classes	Sampling points	Classified points	Correctly classified points	Producer accuracy (%)	User accuracy (%)
Class 1	59	53	51	86.44	96.23
Class 2	52	56	48	92.31	85.71
Class 3	51	51	48	94.12	94.12
Class 4	70	74	67	95.71	90.54
Class 5	24	22	20	83.33	90.91

Total average classification accuracy = 91.4%

Table 16.14 Classification accuracy after PCA preprocessing

Classes	Sampling points	Classified points	Correctly classified points	Producer accuracy (%)	User accuracy (%)
Class 1	59	40	31	52.54	77.5
Class 2	52	90	50	96.15	55.56
Class 3	51	44	41	80.39	93.18
Class 4	70	62	48	68.57	77.41
Class 5	24	20	18	75	90.00

Total average classification accuracy = 76.63%

16.7.3 *Concluding Remarks about Remote Sensing Image Classification Based on SCA*

In this section, we presented a remote sensing image classification method based on SCA. Using a conventional unsupervised classification method, SCA can reduce interference such as that caused by landforms and lighting, and the correlation of different spectra to enhance distinguishability of remote sensing images. The experimental results show that the classification accuracy with SCA preprocessing is higher than that with PCA.

16.8 Chapter Summary

In this chapter, we discussed the concepts, fundamental model, and development of SCA. We systematically reviewed some well-known SCA methods and presented several new methods (including LC-SCA, I-LC-SCA, PC-SCA, PCO-SCA, WL-SCA, and FSCA). Some practical examples of speech signals and remote sensing images were presented toward the end of the chapter.

Based on the fact that in a complete set with sparsity 1, the columns of mixtures are clustered along the orientations of the columns of the mixing matrix, LC-SCA uses linear clustering of the columns of mixtures to estimate mixing matrix $\boldsymbol{A}$, and then extracts the source signals. Then, based on the fact that in a complete set with sparsity 1 the columns of mixtures are clustered along the orientations of the sub-planes of the mixing matrix, and the number of columns determines the number of classes, PC-SCA clusters the mixtures along the orientations of the planes to estimate the mixing matrix and extract the source signals. PCO-SCA is an improved version of PC-SCA that solves the undetermined problem. WL-SCA utilizes a wavelet to transform mixed images from the spatial domain to the frequency domain and obtain sparse wavelet coefficients. It then uses LC-SCA to separate the mixed images. The algorithms above have been ordered according to the degree to which they satisfy the sparsity constraints to form an entire algorithm series for BSS: that is, from complete satisfaction to partial satisfaction and, finally, to completely not satisfying the sparsity constraints. Considering the aspects of Gaussian noise, dependent signals, speech signals, natural images, and remote sensing images, SCA can separate signals more precisely than FastICA. Therefore, SCA is a better prospect for use in feature extraction and fusion of remote sensing images, information recognition of mineral resources, and disaster monitoring, among others.

References

Babaie-Zadeh, M., Mansour, A., Jutten, C., and Marvasti, F. (2004) A geometric approach for separating several speech signals. Proceedings of ICA 2004, Granada, Spain, pp. 798–806.

Bell, A.J., and Sejnowski, T.J. (1995) An information-maximization approach to blind separation and blind deconvolution. *Neural Computation*, **7**(6): 1129–1159.

Bofill, P., and Zibulevsky, M. (2001) Underdetermined blind source separation using sparse representation. *Signal Processing*, **81**(11): 2353–2362.

Bradly, P.S., and Mangasarian, O.L. (2004) K-plan clustering. Proceedings of 2004 IEEE International Symposium on Circuits and Systems, pp. 12–30.

Bronstein, A.M., Bronstein, M., Zibulevsky, M. *et al.* (2003) Blind separation of reflections using sparse ICA. Proceedings of the International Conference ICA 2003, Nara, Japan, pp. 227–232.

Cao, T.T. and Yu, X.C. (2010) Sparse component analysis based on linear clustering and application to blind source separation. *Journal of Beijing Normal University (Natural Science)*, **46**(1): 44–48 (in Chinese).

Cao, T.T., Yu, X.C., Yang, C.P. *et al.* (2009) Remote sensing image classification based on sparse component analysis. MIPPR 2009: Multispectral Image Acquisition and Processing, Proceedings of SPIE, pp. 1–10.

Cardoso, J.F. (1998) Blind signal separation: statistical principles. *Proceedings of the IEEE*, **86**(10), 2009–2025.

Chen, S.S., Donoho, D.L., and Saunders, M.A. (2001) Atomic decomposition by basis pursuit. *SIAM Journal on Scientific Computing*, **43**(1): 129–159.

Chen, S.S., Donoho, D.L. and Saunders, M.A. (2001) Atomic Decomposition by Basis Pursuit. *SIAM Review*, **43**(1): 129–159.

Chen, Y., Xu, R.S., Cai, R. *et al.* (2008) Application progress of remote sensing technology in earthquake research. *Progress in Geophysics*, **23**(4): 1273–1281 (in Chinese).

Chunqi, C., Jiyun, R., and Fung, P.C.W. (2005) Novel sparse component analysis approach to free radical EPR spectra decomposition. *Journal of Magnetic Resonance*, **175**(2): 242–255.

Cichocki, A. (2003) Blind source separation algorithms with matrix constraints. *IEICE Transactions on Fundamentals of Electronics, Communications and Computer Science*, **E86-A**(3): 522–531.

Cichocki, A., Li, Y., Georgiev, P. *et al.* (2004) Beyond ICA: Robust sparse signal representations. Proceedings of 2004 IEEE International Symposium on Circuits and Systems, pp. 684–687.

Coifman, R.R., and Wickerhauser, M.V. (1992) Entropy-based algorithms for best-basis selection. *IEEE Transactions on Information Theory*, **38**(2), 713–718.

d'Aspremont, A., Ghaoui, L., and Jordan, M. (2007) A direct formulation for sparse PCA using semidefinite programming. *SIAM Review*, **49**(3): 434–448.

Daubechies, I. (1988) Time-frequency localization operators: a geometric phase space approach. *IEEE Transactions on Information Theory*, **34**(4): 605–612.

Dong, T.B., Lei, Y.K., and Yang, J.S. (2013) An algorithm for underdetermined mixing matrix estimation. *Neurocomputing*, **104**(15): 26–34.

Donoho, D.L. (2001) Sparse components of images and optimal atomic decompositions. *Constructive Approximation*, 353–382.

Du, X.Y., Hu, W.D. and Yu, W.X. (2005) A Criterion for the construction of a regularization function in sparse component analysis. *Circuits Systems Signal Processing*, **24**(4): 315–325.

Fadili, J.M., Starck, J.L., Bobin, J. *et al.* (2010) Image decomposition and separation using sparse representations: an overview. *Proceedings of the IEEE*, **98**(6), 983–994.

Georgiev, P. and Cichocki, A. (2004) Sparse component analysis of overcomplete mixtures by improved basis pursuit method. Proceedings of 2004 IEEE International Symposium on Circuits and Systems (ISCAS 2004), Vancouver, Canada, Vol. V, pp. 37–40.

Georgiev, P., Theis, F., Cichocki, A. (2004) Blind source separation and sparse component analysis of overcomplete mixtures. ICASSP 2004, IEEE Signal Processing Society, pp. 493–496.

Georgiev, P., Theis, F., Cichocki, A. *et al.* (2005a) Optimization algorithms for sparse representations and applications. *Multiscale Optimization Methods*, **3**(2), 56–62.

Georgiev, P., Theis, F., Cichocki, A. *et al.* (2005b) Sparse component analysis: a new tools for data mining. *Data Mining in Biomedicine*, **8**(16): 79–85.

He, Z.S. and Cichocki, A. (2006) K-subspace clustering and its application in sparse component analysis. The 14th European Symposium on Artificial Neural Networks, pp. 98–102.

Huang, X., Zhang, L.P., and Li, P.X. (2007a) Classification of high spatial resolution remotely sensed imagery based upon fusion of multiscale features and SVM. *Journal of Remote Sensing*, **11**(1): 48–54, (in Chinese)

Huang, X., Zhang, L.P., and Li, P.X. (2007b) Classification of high spatial resolution remotely sensed imagery based on the fusion of spectral and shape features. *Journal of Remote Sensing*, **11**(2): 193–200, (in Chinese).

Hui, Z., Hastie, T., and Tibshirani, R. (2006) Sparse principal component analysis. *Journal of Computational and Graphical Statistics*, **15**(2): 16–20.

Hyvärinen, A. (1999a) Fast and robust fixed-point algorithms for independent component analysis. *IEEE Transactions on Neural Networks*, **10**(3): 626–634.

Hyvärinen, A. (1999b) Survey on independent component analysis. *Neural Computing Surveys*, **2**, 94–128.

Joseph, F.M., and Kenneth, K.D. (2007) Learning sparse overcomplete codes for images. *Journal of VLSI Signal Processing*, **46**(1): 1–13.

Kim, H., and Park, H. (2007) Sparse non-negative matrix factorizations via alternating non-negativity-constrained least squares for microarray data analysis. *Gene Expression*, **23**(12), 1495–1502.

Kopriva, I., and Sersic, D. (2007) Wavelet packets approach to blind separation of statistically dependent sources. *Neurocomputing*, **4**(2): 56–59.

Lewicki, M. and Sejnowksi, T. (2000) Learning overcomplete representations. *Neuralcomputing*, **12**(2): 337–365.

Li, Y.Q., and Cichocki, A. (2003) Sparse representation of images using alternating linear programming. Proceedings of the Seventh International Symposium on Signal Processing and Its Applications, pp. 57–60.

Li, Y.Q., Amari, S., Cichocki, A. *et al.* (2006a) Underdetermined blind source separation based on sparse representation. *IEEE Transactions on Signal Processing*, **54**(2): 423–437.

Li, Y.Q., Cichocki, A., Amari, S. *et al.* (2006b) Blind estimation of channel parameters and source components for EEG signals: a sparse factorization approach. *IEEE Transactions on Neural Networks*, **17**(2): 419–431.

Li, Y.Q., Cichocki, A., Amari, S. *et al.* (2006c) Probability estimation for recoverability analysis of blind source separation based on sparse representation. *IEEE Transactions on Information Theory*, **52**(7): 3139–3152.

Li, Y.Q., Cichocki, A., Shunichi, A. *et al.* (2006d) Analysis of source sparsity and recoverability for SCA based blind source separation. Proceedings of the 6th International Conference on Independent Component Analysis and Blind Source Separation, pp. 831–837.

Li, Y.Q., Cichocki, A., and Amari, S. (2003a) Sparse component analysis for blind source separation with less sensors than sources. Proceedings of 4th International Symposium on Independent Component Analysis and Blind Signal Separation, pp. 89–94.

Li, Y.Q., Cichocki, A. and Amari, S. (2003b) Sparse representation and its applications in blind source separation. Seventeenth Annual Conference on Neural Information Processing Systems (NIPS 2003), pp. 98–102.

Li, Y.Q., Cichocki, A. and Amari, S. (2004) Analysis of sparse representation and blind source separation. *Neural Computation*, **16**(6): 1193–1234.

Liu, P.X., Li, X., He, J.Q. *et al.* (2008) Classification of remote sensing images based on Ant Colony optimization. *Journal of Remote Sensing*, **12**(2): 253–262, (in Chinese).

Mallat, S. (1998) *A Wavelet Tour of Signal Processing*, Academic Press, pp. 42–45.

O'Grady, P.D., Pearlmutter, B.A., and Rickard, S.T. (2005) Survey of sparse and non-sparse methods in source separation. *International Journal of Imaging Systems and Technology*, **15**(1): 18–33.

Olshausen, B.A. and Field, D.J. (1996) Emergence of simple-cell receptive field properties by learning a sparse code for natural images, *Nature*, **381**(13): 607–609.

Olshausen, B.A. and Field, D.J. (1997) Sparse coding with an overcomplete basis set: a strategy employed by V1. *Vision Research*, **37**(23): 3311–3325.

Pati, Y.C., Rezaiifar, R.P., and Krishnaprasad, S. (1993) Orthogonal matching pursuit: recursive function approximation with applications to wavelet decomposition. Proceedings of 27th Asilomar Conference on Signals, Systems and Computers, pp. 42–15.

Pearlmutter, B.A. and Parra, L.C. (1996) A context-sensitive generalization of ICA. ICONIP'96, pp. 151–157.

Rubinstein, R., Zibulevsky, M. and Elad, M. (2010) Double sparsity learning sparse dictionaries for sparse signal approximation. *IEEE Transactions on Signal Processing*, **58**(3): 1553–1564.

Sanei, S., Constantinides, S., Took, C.C. *et al.* (2005) Incorporating frequency warping into sparse component analysis. Proceedings of the 13th European Signal Processing Conference, pp. 15–18.

Shi, Z.W., Tang, H.W., and Tang, Y.Y. (2005) Blind source separation of more sources than mixtures using sparse mixture models. *Pattern Recognition Letters*, **26**(16): 2491–2499.

Smith, D., and Lukasiak, J. (2005) A sequential approach to sparse component analysis. Proceedings of the 7th Workshop on Multimedia Signal Processing, pp. 1–4.

Stadlthanner, K., Theis, F.J., Puntonet, C.G. *et al.* (2005) Hybridizing sparse component analysis with genetic algorithms for blind source separation, in *ISBMDA 2005*, Springer-Verlag, Berlin, Heidelberg, pp. 137–148.

Tan, B.H., and Fu, Y.L. (2006) Mixture matrix identification of underdetermined blind source separation based on plane clustering algorithm, in *ICIC 2006*, Springer-Verlag, Berlin, Heidelberg, pp. 60–65.

Washizawa, Y., and Cichocki, A. (2006) On line K-plane clustering learning algorithm for sparse component analysis. 2006 IEEE International Conference on Acoustics, Speech, and Signal Processing, pp. 41–44.

Washizawa, Y., and Cichocki, A. (2008) Sparse blind identification and separation by using adaptive K-Orthodrome clustering. *Neurocomputing*, **71**(10): 2321–2329.

Woolfson, M.S., Bigan, C., and Crow, J.A. (2008) Method to separate sparse components from signal mixtures. *Digital Signal Processing*, **18**(6): 985–1012.

Yu, X.C., Cao, T.T., Hu, D. *et al.* (2010) Blind image separation based on wavelet transformation and sparse component analysis. *Journal of Beijing University of Posts and Telecommunications*, **33**(2): 58–63 (in Chinese).

Yu, X.C., Chu, X.F., and Cao, H.Z. (2009a) Decomposition of mixed pixels of SAR remote sensing images base on Partheno-Genetic Algorithm. *Chinese Journal of Geophysics*, **52**(11): 2886–2892, (in Chinese).

Yu, X.C., Cao, T.T., and Yang, C.P. (2009b) Remote sensing image classification based on sparse component analysis. *Progress in Geophysics*, **24**(6): 2274–2279 (in Chinese).

Yu, X.C., Xu, J.D., Hu, D., and Xing, H.H. (2013) A new blind image source separation algorithm based on feedback sparse component analysis. *Signal Processing*, **93**(1): 288–296.

Zhang, L., Chen, X.H., and Weng, Y. (2008) Clonal selection algorithm for classification of remote sensing imagery. *Acta Scientiarum Naturalium Universitatis Sunyatseni*, **47**(3): 128–132, (in Chinese).

Zhong, Y.F., Li, L.P., and Li, P.X. (2007) Classification of multi-spectral remote sensing image based on multiple-valued immune network. *Chinese Journal of Computers*, **30**(12): 2181–2198, (in Chinese).

Zibulevsky, M., and Pearlmutter, B.A. (2001) Blind source separation by sparse decomposition in a signal dictionary. *Neural Computation*, **13**(4): 863–872.

Index

Blind Source Separation: Theory and Applications, First Edition. Xianchuan Yu, Dan Hu and Jindong Xu.

www.ingramcontent.com/pod-product-compliance
Ingram Content Group UK Ltd.
Pitfield, Milton Keynes, MK11 3LW, UK
UKHW050943290926
15470UKWH00005B/1610

9 781118 679845